Physics at KAON

D. Frekers · D.R. Gill · J. Speth (Eds.)

Physics at KAON

Hadron Spectroscopy, Strangeness, Rare Decays

Proceedings of the International Meeting, Bad Honnef, 7–9 June 1989

With 146 Figures

Springer-Verlag Berlin Heidelberg New York
London Paris Tokyo Hong Kong

Dr. Dieter Frekers
Dr. David R. Gill
TRIUMF, 4004 Wesbrook Mall, Vancouver BC, V6T 2A3, Canada

Professor Dr. Josef Speth
Institut für Kernphysik, Forschungszentrum Jülich,
Postfach 19 13, D-5170 Jülich, Fed. Rep. of Germany

This book originally appeared as a supplement to Vol. 46, 1990, of the journal
Zeitschrift für Physik C – Particles and Fields, 46
(ISSN 0170-9739) © Springer-Verlag Berlin Heidelberg 1990

Programme Committee
A. Citron, D. Frekers, O. Häusser, K. Kilian, P. Kitching, K. Kleinknecht,
T. Mayer-Kuckuk, B. Povh, V. Soergel, U. Wienands.

ISBN 3-540-52241-7 Springer-Verlag Berlin Heidelberg New York
ISBN 0-387-52241-7 Springer-Verlag New York Berlin Heidelberg

Printing: Druckhaus Beltz, 6944 Hemsbach/Bergstr.
Binding: J. Schäffer GmbH & Co. KG., 6718 Grünstadt
2156/3150-543210 – Printed on acid-free paper

Preface

"Physics at KAON", an international meeting jointly organized by the KFA Jülich and TRIUMF, was held in the Physikzentrum Bad Honnef from June 7 through June 9, 1989. This was one of a series of meetings – the first one in Europe – in which plans for the medium energy physics laboratory KAON were presented and some aspects of the physics at this new facility were discussed.

The meeting focussed mainly on the topics of hadron spectroscopy, K-meson scattering, strangeness in nuclei, and rare decays. Also presented were some of the research programs at SATURNE and COSY which may well lead to KAON physics in the future. These proceedings include articles which summarize our current experimental and theoretical knowledge in the various areas, as well as papers which describe lines of research feasible with KAON.

The large number of participants – limited, in fact, by the capacity of the Physikzentrum – clearly demonstrates the great interest of the European physics community in the research avenues which will be opened by the high-intensity hadron facilities.

March 1990 D. Frekers, D.R. Gill, J. Speth

Contents

Z. Phys. C – Particles and Fields 46, S1 (1990)

Zeitschrift für Physik C Particles and Fields

Opening remarks

E. Vogt

TRIUMF, 4004 Wesbrook Mall, Vancouver, B.C., Canada V6T 2A3

It is a privilege to join Professor Joachim Treusch, of Jülich, in opening this International Meeting on Physics at KAON. Although KAON is to be located in Vancouver, Canada, it is intended to serve the world.

The purpose of this meeting is, primarily, to address the physics of KAON. We have a very full schedule of talks dedicated to this purpose. The whole field of subatomic physics has been stimulated by the standard model of quarks and leptons and unified forces. The new questions which have arisen from this leap forward in knowledge have, in turn, led to very large and ambitious new accelerator projects now planned or under construction. Most of these new projects are colliders operating at the highest energies. KAON is an essential component of this new world network of accelerator projects, complementing the colliders by offering not the highest energy but rather the highest intensity of various secondary beams: kaons, antinucleons, other hadrons (hyperons, pions, etc.) and neutrinos. At the intensity frontier one approaches many of the same physics issues which confront the colliders. We expect, at this conference, a rich feast of the new physics.

It is not only the menu of physics which is important for this meeting but also the special circumstances which led to this gathering taking place. KAON not only opens new opportunities for the medium-energy community of the world but also it is the very engagement of the interest of this community which is essential to the birth of KAON. We are then, all of us, collectively and individually, prospective benefactors of KAON's science and midwives in its creation. What makes this so is the special nature under which KAON will be constructed and operated. I will have occasion at the end of this meeting (Summary talk) to expand on the special circumstances pertaining to KAON's internationalization and the essential role of foreign partners for Canada's KAON.

On the way toward the funding of KAON a number of science workshops have been held recently in North America and one in Japan. This is the first in Europe. The interest in Europe is intense, as manifested in the many Europeans who aggregated for the European Hadron Facility (EHF) Proposal. Hadron physics is very strong in Europe. Some of that strength will now focus on Canada's KAON and hopefully be reflected in the future emergence of complementary facilities here in Europe. This is then the right place and the right time to discuss and review the current hot topics of the field. We have much to look forward to in the very interesting three-day program which lies ahead.

Z. Phys. C – Particles and Fields 46, S3–8 (1990)

Zeitschrift für Physik C Particles and Fields

The TRIUMF kaon factory accelerators

M. K. Craddock *

TRIUMF, 4004 Wesbrook Mall, Vancouver B. C., Canada V6T 2A3

Abstract

To accelerate a 100 μA proton beam from the TRIUMF H^- cyclotron to 30 GeV a five-ring accelerator complex is proposed. Each accelerator is followed by a storage ring for time-matching – the cw cyclotron by the Accumulator, the 3 GeV 50 Hz Booster by the Collector, and the 30 GeV 10 Hz Driver by the Extender – the latter providing the slow-extracted beam for coincidence experiments. Under the current $11 million pre-construction study prototypes are being built of various components of the Booster ring – a fast-cycling dipole magnet, a dual-frequency magnet power supply, ceramic beam pipes, rf cavities (both parallel and perpendicular bias versions) and an extraction kicker. In addition the lattice designs for all five rings and the shielding and remote handling requirements are being reviewed. These activities will allow construction to start in 1990.

1. Introduction

The TRIUMF Kaon-Antiproton-Otherhadron-Neutrino Factory accelerators have been described in full in the original proposal [1] and outlined in various papers [2,3]. The basic aim is to accelerate a 100 μA beam of protons to 30 GeV, roughly 100 times more than available at present.

The project is currently the subject of an $11M pre-construction Engineering Design and Impact Study funded jointly by the governments of Canada and British Columbia. This began in October 1988 and is planned to take 15 months. It will enable prototypes of the major components to be built, the cost estimates to be updated, the international contributions to be better defined and various impact studies to be carried out. The various projects are listed below, together with the names of the group leaders and other engineers and physicists involved.

Project Leader	A. Astbury
Accelerator Design	M.K. Craddock; R. Baartman, S. Koscielniak, G.H. Mackenzie, J.R. Richardson, R.V. Servranckx and U. Wienands
Systems Integration	E.W. Blackmore; G. Clark, M. Zanolli (CERN)
RF Systems	R. Poirier; R. Burge, T. Enegren
Magnets	A.J. Otter; C. Haddock, P. Schwandt (IUCF)
Magnet Power Supplies	K. Reiniger;
Beam Pipe & Vacuum	C.J. Oram
Kickers	G. Wait; M. Barnes
Controls	D. Dohan; W.K. Dawson, B. Frammery (CERN), D. Schultz (LANL)
Shielding & Safety	I.M. Thorson; D. Axen (U.B.C.)
Cyclotron Beam Extraction	M. Zach; G. Dutto, R.E. Laxdal, J. Pearson
Experimental Areas	J. Beveridge; J. Doornbos, G. Stinson, L. Criegee (DESY)
Targets	T.A. Hodges
Science Workshops	P. Kitching (Univ. of Alberta)
International Consultations	P. Dyne (ISTC); E.W. Vogt
Project Management	G. Ritchie; G. Ridout (UMA Spantec Ltd); V.K. Verma
Building Design	Phillips, Barratt, Kaiser Ltd.; Chernoff Thomson Ltd.
Tunnel Design	Stewart EBA Ltd.; Sandwell Swan Wooster Ltd.

*On leave from Physics Department, University of British Columbia.

Services & Power	D.W. Thompson Ltd.; Hipp Engineering Ltd.
Industry Development	D. Williams; A. Stretch (Monenco Ltd.); J. Carey
Economic Assessment	Company to be appointed
Legal Studies	Company to be appointed
Environmental Studies	Company to be appointed

The TRIUMF H^- cyclotron, which routinely delivers 150 μA beams at 500 MeV, provides a ready-made and reliable injector. It would be followed by two fast-cycling synchrotrons, interleaved with 3 storage rings, as follows:

- A Accumulator: accumulates cw 450 MeV beam from the cyclotron over 20 ms periods
- B Booster: 50 Hz synchrotron; accelerates beam to 3 GeV; circumference 214 m
- C Collector: collects 5 Booster pulses and manipulates longitudinal emittance
- D Driver: main 10 Hz synchrotron; accelerates beam to 30 GeV; circumference 1072 m
- E Extender: 30 GeV stretcher ring for slow extraction for coincidence experiments

The energy-time plot in figure 1 shows how this arrangement allows the B and D rings to run continuous acceleration cycles without flat bottoms or flat tops. The use of a Booster permits a smaller normalized emittance and hence reduces the aperture and cost of the Driver magnets for a given space charge tune shift. The use of a Booster also simplifies the rf design by separating the requirements for large frequency swing and high voltage (33% and 720 kV respectively for the Booster, and 3% and 2550 kV for the Driver). These high rf voltages are associated with the high cycling rates; the use of an asymmetric magnet cycle with a rise 3 times longer than the fall in the Driver (figure 1) reduces the voltage required by one-third, and the number of cavities in proportion. In the Booster the saving is less because more voltage is needed for bucket creation.

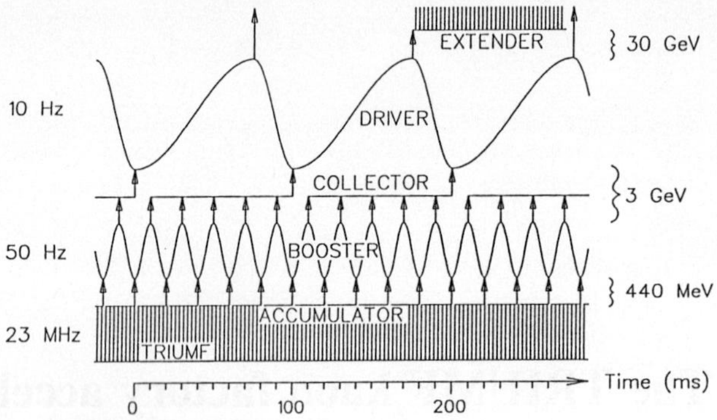

Figure 1: Energy-time sequence for the five rings.

Figure 2 gives a schematic layout of the lattices of the five rings, showing the arrangement of magnets, rf cavities and beam transfer lines. Although illustrated side by side for convenience, in practice the Accumulator would be mounted above the Booster in the small tunnel and the Collector above the Driver in the main tunnel. The optimum location for the Extender is under study, as discussed further below. Identical lattices and tunes are used for the rings in each tunnel. This is a natural choice providing structural simplicity, similar magnet apertures and straightforward matching for beam transfer.

Separated-function magnet lattices are used with the dispersion modulated so as to drive its mean value towards zero, enabling transition to be kept above top energy in all rings. This avoids transition-crossing problems, such as emittance mismatch and change of rf phase under high beam loading. Racetrack lattices

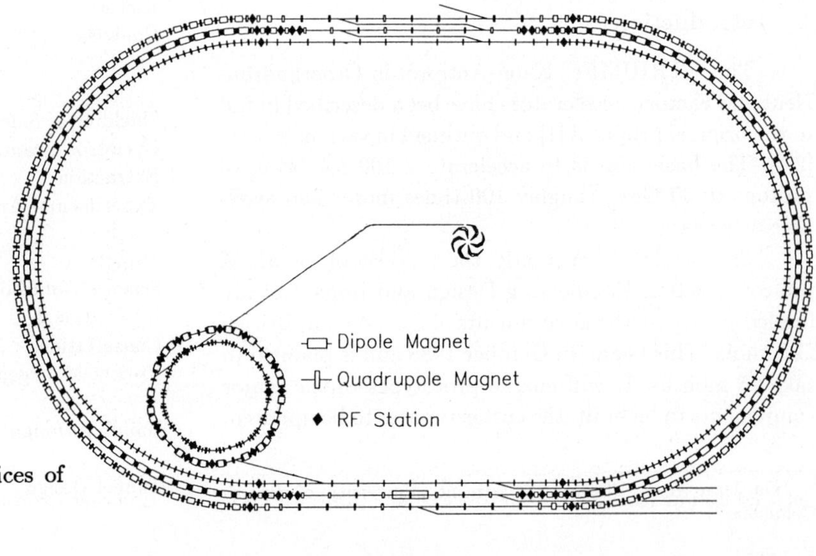

Figure 2: Schematic layout of the lattices of the five rings.

have now been adopted for the C,D and E rings but superperiodicity-6 lattices are proposed for the A and B rings.

Injection into the Accumulator is achieved by stripping the H^- beam from the cyclotron, enabling many turns to be injected into the same area of phase space. The small emittance beam from the injector is in fact "painted" over the much larger three-dimensional acceptance of the Accumulator to limit the space charge tune shift. Painting also enables the optimum density profile to be obtained and the number of passages through the stripping foil to be limited.

2. Beam Dynamics

In order to cut beam loss at slow extraction well below the usual 1%, racetrack lattices have now been adopted for the C,D and E rings (Servranckx *et al.*[4]). These provide long straights with high β (100 m) at the septa and room for an additional pre-septum and for collimators downstream. Tracking simulations, which include power supply noise effects, suggest that the beam loss can be kept below 0.2%. The 180° arcs contain 24 cells, and are second-order achromats, normally tuned to $5 \times 2\pi$. The tune for the whole ring may be varied by ± 1 in each plane independently. A half-integer resonance may be used for extraction, to simplify the collimation process. Such a racetrack lattice is also convenient for the Driver synchrotron, allowing either for the insertion of Siberian snakes, or for tuning for low depolarization without snakes, using high-periodicity arcs and spin-transparent straight sections (Wienands [5]). Investigation of the properties of the lattice in detail show that its dynamic aperture is as large as for the old circular design. Two methods of separating the Extender extraction straight from those of the C and D rings, to provide room for shielding, are under study. The simplest, but most expensive, would keep the same lattice but use a larger circumference to give ~ 3 m horizontal separation all round. A cheaper but less optically friendly alternative would use a single bypass, starting 45° into the arc. Racetrack lattices have also been investigated for the Booster and Accumulator rings [4], but have so far failed to provide a better overall solution than the superperiodicity-6 lattice of the reference design.

Studies continue to determine the optimum strategy for painting the beam at injection. Developments in stripping foil construction suggest that two-sided "corner" foils may be usable, reducing the number of foil interceptions during accumulation (Mackenzie [6]). Better models are now available for scattering within thin foils (Butler *et al.*[7]).

The stability of unusual beam density distributions, such as those formed during painting or debunching, is under study (Baartman [8]). For distributions $\rho(p, \phi)$ hollow in longitudinal phase space, simulations [9] reveal an intensity threshold for instability. The stability criterion can be expressed in terms of the slope $d\rho/dp$. The merits of lower-frequency rf systems for beam stability have also been studied [10]. The effects of space charge and of feedback control loops can be included in our longitudinal tracking codes, and have also been studied analytically.

3. Magnet Development

A preliminary design for a Booster dipole magnet has been prepared (Otter *et al.*[11]) and a prototype is under construction. The magnet will be 3 m long with a pole gap of 10.7 cm and will cycle at 50 Hz between 0.27 T and 1.05 T with a field uniformity $< \pm 2 \times 10^{-4}$ over ± 5 cm. The prototype will be built from 26-gauge laminations of M17 (non-grain oriented) steel with 10-turn coils, each containing 12 square hollow copper conductors in a vertical array. Studies continue on the various other magnets needed in the accelerators and beam lines.

4. Magnet Power Supplies

As explained above, dual-frequency magnet excitation is being considered for the synchrotrons, with a rise time three times longer than the fall. To test the performance of such a system a high-power test stand has been set up (figure 3). Four magnets from the decommissioned NINA synchrotron are used, one as the load and three in series as the resonant 81 mH choke. A 1000 μF capacitor bank may be switched in parallel with a 125 μF bank to change the resonant frequency from 100 Hz to 33 Hz. Dual-frequency operation was achieved recently and replacement of the present converted dc supply by a custom-built ac supply will enable tests to be continued to full current. A two-magnet cell will also be tested (Reiniger [12]). A new power distribution scheme for the one reference and 24 Booster dipoles has been worked out, based on 5 cells with 5 magnets each.

Figure 3: High-power test stand for dual-frequency magnet excitation studies.

5. Kickers

The kicker with the most challenging specifications is probably that for extraction from the Booster ring – about 30 kV across an 8 cm gap over a length of 2 m with a rise time ≤80 ns and operating at 50 Hz. Starting from scratch and with long delivery times on some items it seemed impractical to attempt to build a true prototype within 12 or 15 months. Instead we are gaining experience in delay-line kicker technology by putting together a somewhat similar system with the help of some critical components obtained on loan. A pulse forming network has been brought into operation to provide 40 kV pulses, 200 to 700 ns long with ~30 ns rise and fall times. The present repetition rate of 20 Hz will be raised to 50 Hz with the arrival of a new power supply. Materials are on order for the construction of a prototype kicker magnet.

A 1 MHz chopper will also be built for installation in the injection line from the cyclotron to create the 110 ns beam gap needed for kicker rise and fall. The chopper must provide 40 kV over 1 m with rise and fall < 35 ns. Our aim is to have a prototype chopper operating by the end of 1989.

6. Radio Frequency Systems

The reference design for the Booster cavities is based on those used in the Fermilab booster. A full-scale prototype cavity is almost complete and should be ready for tests with an air tuner soon (Poirier *et al.*[13]). Under our collaboration with LAMPF their booster cavity, which employs perpendicularly-biased microwave ferrite, is now also at TRIUMF, being prepared for testing under ac bias conditions – a crucial test of its viability. Under dc bias it has produced relatively high voltages (140 kV), potentially reducing the number of cavities required and hence the impedance presented to the beam and the likelihood of inducing coupled-bunch instabilities. Enegren *et al.*[14] have studied the higher-order modes for both cavities and report on several damping schemes. To reduce the stray magnetic field seen by the beam in the LAMPF cavity, both shielding and redesign of the bias coils are being investigated (Haddock *et al.*[15]). The revised design is shown in figure 4 and should be ready for testing in the autumn.

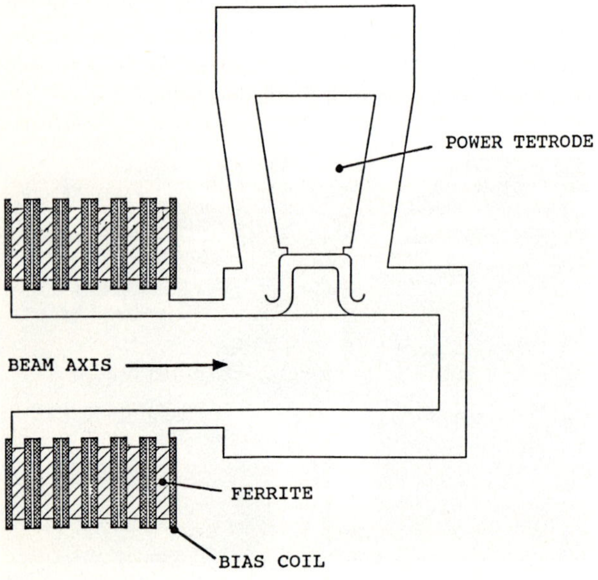

Figure 4: Los Alamos rf cavity with new coil arrangement for biasing the ferrite radially instead of axially.

Control of the rf systems under high beam loading is a crucial topic. The effect of fast feedback has been modelled analytically by Koscielniak[16]. Burge and Enegren [17] have described the operation of a generic regulator they have built for phase and amplitude control. This will be used in the low-level control system which TRIUMF is building for the LAMPF main ring cavity, along with the solid-state driver amplifier.

7. Beam Pipe & Vacuum

The vacuum and impedance requirements for all five rings are being carefully reviewed (Oram [18]). The high circulating beam current makes beam-induced multipactoring and ion desorption from the walls the most critical processes. A hydrocarbon-free system is required, with all metal elements pre-baked to 300°C, and pumps spaced no more than 5 m apart, automatically producing vacua better than 10^{-8} Torr. An additional concern in the Extender ring, where the beam may be debunched, is the possibility of electron-proton oscillations; electrostatic collector plates will be needed to suppress these.

Ceramic chambers must be used within the fast-cycling magnets but must contain a conducting shield. Two shielding schemes are being considered and 4 m long prototypes incorporating each are being constructed for the Booster dipoles. SAIC (San Diego) is building a chamber with longitudinal silver stripes painted on the inside walls, while RAL (U.K.) is building one incorporating a separate wire cage, as used in ISIS.

8. Computer Control System

A comprehensive review of both hardware and software options was carried out by Dawson *et al.*[19] in 1987, and recommended a segmented Ethernet communications backbone linking commercial workstations used as operator interfaces with microprocessor-based equipment controllers. A test platform is being assembled based on a VAX3200 workstation with a bridged Ethernet connexion to 2 VME crates. It will be tried out on selected cyclotron systems requiring beam control, such as the injection line or the new H$^-$ extraction system. Structural analysis techniques are being used to specify a logical model of the entire control system organized in a hierarchical structure. The possibility of incorporating expert systems techniques is also under study.

9. H$^-$ Extraction from the Cyclotron

To extract H$^-$ ions (instead of stripping them to protons as in normal operation) a conventional extraction system is being developed. With 18 kV on an rf deflector, which excites the ν_r=3/2 resonance, and 50 kV on the electrostatic deflector, 90% of the beam (66 μA macropulses at 1% duty factor) has been transmitted through the latter (Laxdal *et al.*[20]). The other 10% is stripped by a narrow foil shadowing the septum and protecting it from irradiation (figure 5); the resulting protons may be dumped or steered into an experimental beam line. In recent tests the average beam current was successfully raised to 10 μA. Design of the 4-segment magnetic channel which will steer the H$^-$ beam out of the cyclotron is now under way and one segment will be built and tested this year. Detailed design of the front end of the external beam line is also under way.

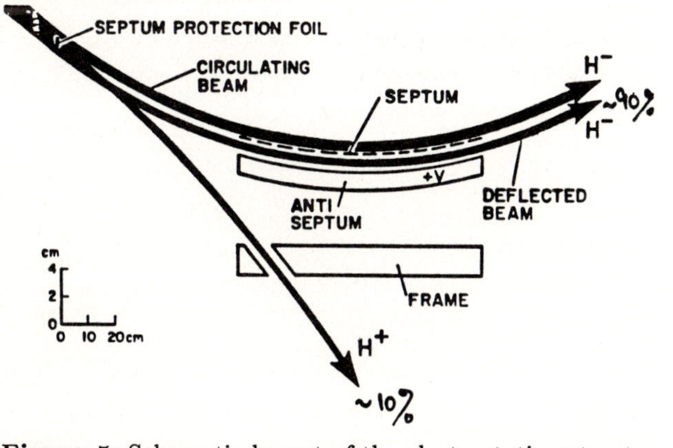

Figure 5: Schematic layout of the electrostatic extraction septum for H$^-$ ions, showing the stripping foil which protects the septum from irradiation.

10. Conclusion

A wide variety of activities is under way to finalize the design of the KAON Factory accelerators. Prototypes are being built of many components of the Booster synchrotron and a comprehensive review of the whole design is to be completed by the end of 1989, ready for a start to construction in 1990.

11. Acknowledgements

It is a pleasure to acknowledge the efforts of all those who have worked to improve the KAON Factory design recently (a partial list is given above). The advice and help of H. Baumann, D. Fiander, J. Griffin, M. Harold, R. Hohbach, E. Jones, C.W. Planner, G.H. Rees, H. Sasaki, H.D. Schönauer, T. Suzuki, E.J.N. Wilson and of our colleagues at the other hadron facilities has been especially appreciated. The author is also particularly grateful to Jana Thomson for the accuracy of the typing.

References

[1] KAON Factory Proposal, TRIUMF, September, 1985.

[2] M.K. Craddock, R. Baartman *et al.*, IEEE Trans. Nucl. Sci. NS-32, 1707 (1985).

[3] M.K. Craddock, Proc. Int. Workshop on Hadron Facility Technology, Santa Fe, February 1987, ed. H.A. Thiessen, Los Alamos Report LA-11130-C, pp 8-31 (1987).

[4] R.V. Servranckx, U. Wienands, *et al.*, "Racetrack Lattices for the TRIUMF KAON Factory", Proc. 1989 Particle Accelerator Conference, IEEE (in press).

[5] U. Wienands, "Polarized Beams at the KAON Factory", Proc. AHF Accelerator Design Workshop, Los Alamos, February 1989 (in press).

[6] G.H. Mackenzie, "TRIUMF Stripper Lifetime and Desirable Tests for the KAON Factory", *ibid.*

[7] M. Butler, S. Koscielniak and G.H. Mackenzie, "A Tabulation of Coulomb Scattering Cross Sections *etc.*", Proc. 1989 Particle Accelerator Conference, IEEE (in press).

[8] R. Baartman "Coasting Beam Instability Theory Applied to Bunched Beams", Proc. AHF Accelerator Design Workshop, Los Alamos, February 1989 (in press).

[9] R. Baartman, F. Jones, S. Koscielniak and G.H. Mackenzie, "Stability of Beams Hollow in Longitudinal Phase Space", Proc. 1989 Particle Accelerator Conference, IEEE (in press).

[10] R. Baartman, "The Case for Low-frequency RF Systems," Proc. AHF Accelerator Design Workshop, Los Alamos, February 1989 (in press).

[11] A.J. Otter, C. Haddock and P. Reeve, "Prototype Magnet Designs and Loss Measurements for the Dual Frequency Booster Synchrotron *etc.*" Proc. 1989 Particle Accelerator Conference, IEEE (in press).

[12] K. Reiniger, "The Generation of a Reference Design for Booster Magnet Excitation for TRIUMF KAON Factory", *ibid.*

[13] R.L. Poirier, T.A. Enegren and C. Haddock, "Perpendicular-Biased Ferrite Tuned RF Cavity for the TRIUMF KAON Factory Booster Ring", *ibid.*

[14] T.A. Enegren, R. Poirier *et al.*, "Higher-order Mode Damping in KAON Factory RF Cavities", *ibid.*

[15] C. Haddock, R.L. Poirier and T. Enegren, "Biasing of Ferrite Tuners," Proc. AHF Accelerator Design Workshop, Los Alamos, February 1989.

[16] S. Koscielniak, "System Stability for Beam-loaded Cavity with Fast Feedback," *ibid.*

[17] R.S. Burge, T.A. Enegren and J. Miszczak, "Amplitude and Phase Regulation of the RF Separator", Proc. 1989 Particle Accelerator Conference, IEEE (in press).

[18] C.J. Oram, "Vacuum and Beam Pipe Considerations", Proc. AHF Accelerator Design Workshop, Los Alamos, February 1989 (in press).

[19] W.K. Dawson *et al.*, "A Conceptual Design for the TRIUMF KAON Factory Control System", TRI-87-1 (1987).

[20] R.E. Laxdal *et al.*, "Progress towards H^- Extraction at TRIUMF", Proc. 1st European Particle Accelerator Conference, Rome 1988 (in press).

Z. Phys. C – Particles and Fields 46, S9–20 (1990)

Zeitschrift für Physik C Particles and Fields

Experimental facilities

P. Kitching

TRIUMF, 4004 Wesbrook Mall, Vancouver B.C., Canada V6T 2A3

Abstract

The beamlines and experimental area outlined in our original proposal are being updated and refined as part of the Kaon Factory Project Definition Study. A major part of the input for this process comes from a series of workshops being held for this purpose. The present status of the designs for beamlines will be described, together with a description of a possible experimental hall layout.

Let me begin by warning you that my talk will be considerably more tentative than those of the two previous speakers, because one of our major motivations in holding workshops such as this is to get the views of interested potential users as to the beamlines and facilities we should be planning to build in order to exploit the machine most effectively. I expect, therefore, that what we eventually build will be substantially modified by the series of workshops we are holding this year as part of the Project Definition Study. Having said all that, let me try to describe to you the present state of our thinking with regard to the experimental beamlines and areas. I will review the properties of the primary beam and the range of physics requirements which the facility might be expected to satisfy before outlining the beamlines and layout described in our initial Kaon Factory proposal. I will then tell you about some of the changes we have made since then and give a particular example of optimizing a particular beamline. I will conclude with some remarks about how we feel the beams should be laid out and our thinking concerning the experimental hall.

The primary proton beam will be 100 μA of unpolarized beam at 30 GeV, pulsed if extracted from the D-ring and CW if extracted from the E-ring. In addition, we hope to have a variable energy (up to 30 GeV) CW polarized proton beam whose intensity will be at least 1 μA. The time structure of the beams is shown in table 1, and we would normally expect to utilize the fast extraction from the D-ring for neutrino experiments and the slow extraction from the E-ring for all other experiments.

Table 1: Time Structure of 30 GeV/c Beam

Fast Extraction from D – Ring	Macrostructure 3.6 μsec/100 msec Microstructure 16 nsec (63 MHz)
Slow Extraction from E – Ring	Macrostructure 85% Duty Cycle (10 Hz) Microstructure - Two Modes Available 1) Bunched 1 nsec every 16 nsec 2) Debunched $\geq$ 50% micro-duty cycle

The main physics opportunities offered by the Kaon Factory are listed in table 2, together with the particle beams needed to exploit them. As you can see, we will need the full range from stopping kaons up to 20 GeV. Since a given channel cannot cover more than about a factor of two in momentum effectively, this means we must provide many secondary channels. The beamlines contained in our original proposal are shown in table 3 together with those in the LAMPF AHF proposal and the European EHF proposal. Also shown are existing beamlines at the Brookhaven AGS and at KEK. All follow the pattern that the maximum momentum of each is about twice that of the previous channel. The particle fluxes which would be available from the scheme in our proposal is shown in fig. 1 for $K^{\pm}, \pi^{\pm}$ and $\overline{p}$. (For comparison, the maximum flux available from LESB1 at

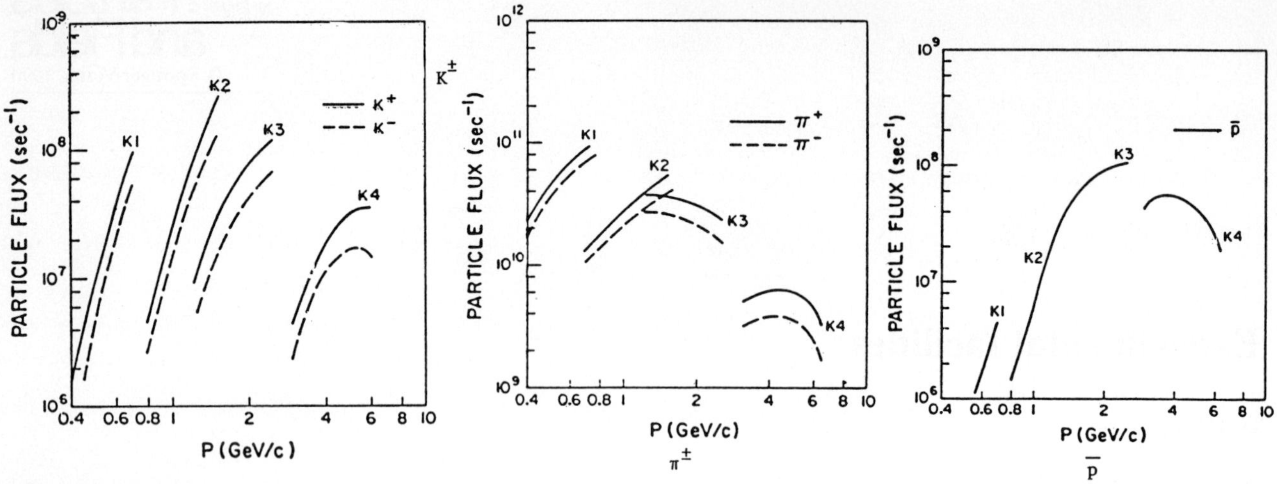

Figure 1: KAON proposal

Table 2: Physics Opportunities and Requirements

Rare Decays	K^0, stopping K^+
CP Violation	few GeV/c K^0
Neutrino Physics	1-2 GeV/c neutrinos beam stop neutrinos?
Meson Spectroscopy	8-15 GeV/c $K^\pm$ up to 20 GeV/c $\pi^\pm$
Baryon Spectroscopy	0.5 - 2.5 GeV/c $\pi^\pm$ 1-6 GeV/c K $^\pm$
Kaon-Nucleon Scattering	0.3-2.5 GeV/c $K^\pm$
Kaon-Nuclear Reactions	0.3-1.0 GeV/c $K^\pm$
Hypernuclei	0.3-2.5 GeV/c $K^\pm$ 1.0-1.5 GeV/c $\pi^\pm$
Spin Physics	3-30 GeV/c $\vec{p}$
Antiproton Physics	?-10 GeV/c $\bar{p}$
Low Energy Muon Physics	Low energy $\mu^\pm$

BNL is about 5×10^5 K^+/sec.) The design aim in all these secondary channels has been to reduce the pion contamination, keeping the π/K ratio to 1:1 or better. A typical example is shown in fig. 2, which shows the design for the 1.5 GeV/c line.

Anticipated K_L^0 beam rates are shown in fig. 3, together with the results of a study of a tertiary beam produced by a secondary pion beam. The tertiary beam is not really competitive, since, although it has less neutron contamination, it is two orders of magnitude lower in flux.

Table 3: Kaon Beamlines

	Beamline GeV/c (max)	$\Delta p/p$ %	$\Delta\Omega$ [msr]	Length [m]
Proposed				
TRIUMF	K0.7	5	6	18
	K1.5	3.8	1.6	30
	K2.5	4	0.5	54
	K6	3	0.16	115
LAMPF	K0.8	5	5	18
AHF	K1.5	3	2	25
	K2.5	5	1.0	35
	K6	5	0.07	75
	K35	1	0.05	130
EHF	S0.8	5	5	18
	S1.5	3	2	25
	S3	5	1	35
	S6	5	0.2	60
	S20	5	0.1	100
Existing				
Brookhaven	LESB2(0.8)	6	13	15
	LESB1(1.1)	4	2.6	15.3
	New(2.0)	6	1.3	31
	MESB(6.0)	6	0.3	81
KEK	K3(1.0)	6	7.2	14.5
	K2(2.0)	6	1.0	27.9

The experimental hall in our proposal, as shown in fig. 4, was intended mainly to provide a reference design for costing purposes and is almost certainly not optimized for physics. A more detailed view of one of the

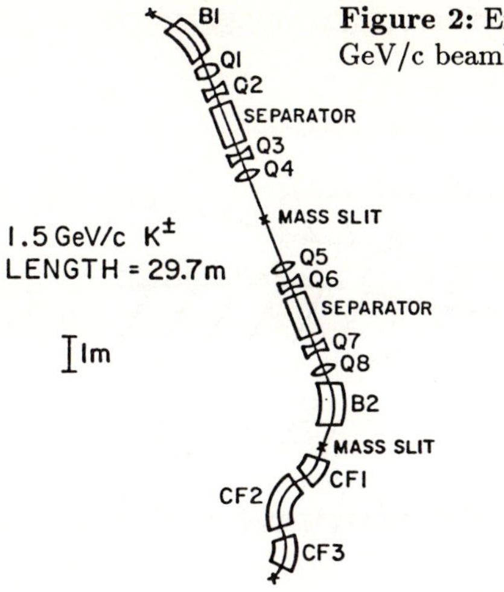

Figure 2: Example for 1.5 GeV/c beamline

SOLID ANGLE 1.6 msr

P/P 3%

DESIGN AIM K/π 1:1 or better

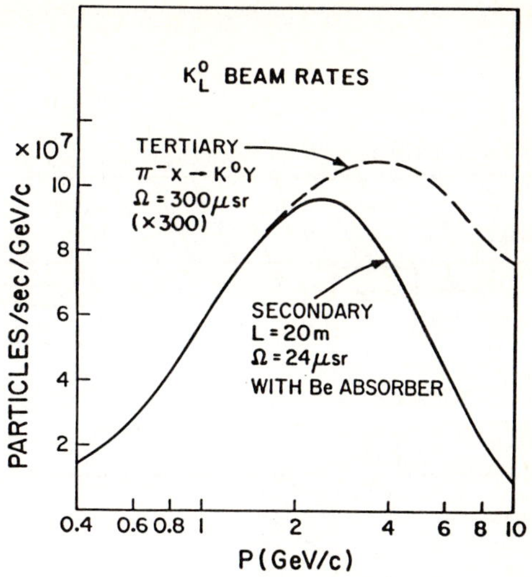

Figure 3: K^0_L beam rates

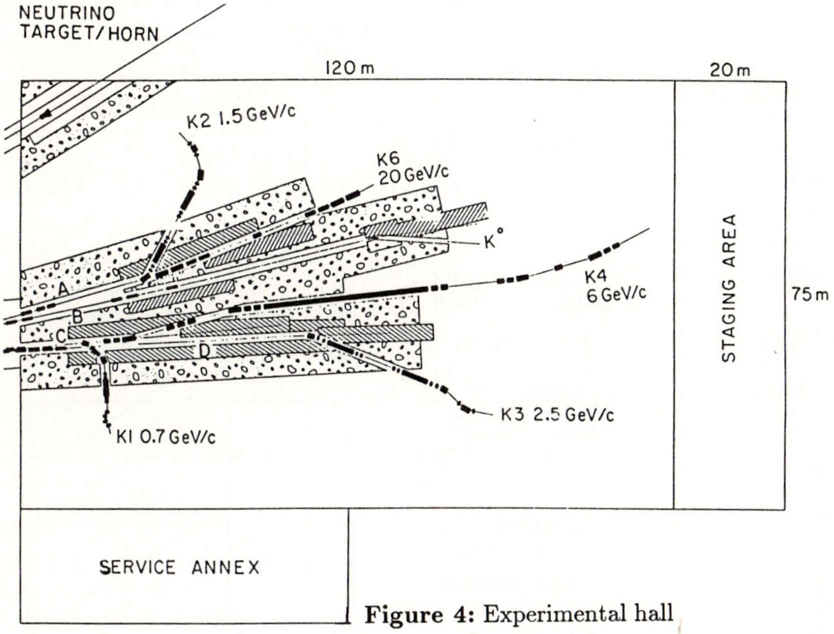

Figure 4: Experimental hall

target stations is shown in fig. 5, with two forward take-off channels, one low momentum and one high momentum, and the beam dump placed as close as possible to the target. A cross section of the target area is shown in fig. 6 to illustrate how one might deal with the technical problem associated with handling 3 MW of beam power. Components in high radiation areas must be adequately shielded and must be capable of being remotely installed and removed for repair. We chose to adapt a "canyon" approach with vertical installation and access.

The neutrino facility was placed with the "hot" components in the experimental area for crane access, and is schematically shown in fig. 7. The neutrino fluxes to be expected from such a facility are shown in fig. 8, with the maximum occurring around 1 GeV/c.

As far as experimental facilities are concerned, the Project Definition Study, which began last year, aims primarily to establish the scale and costs, i.e. how large a building will we need, where is the best place to locate it on the site, and what are the presently perceived beam lines contained in it. We must strive to maintain maximum flexibility; in particular we need to keep in mind the possibility of an upgrade to 100 GeV/c at some later date.

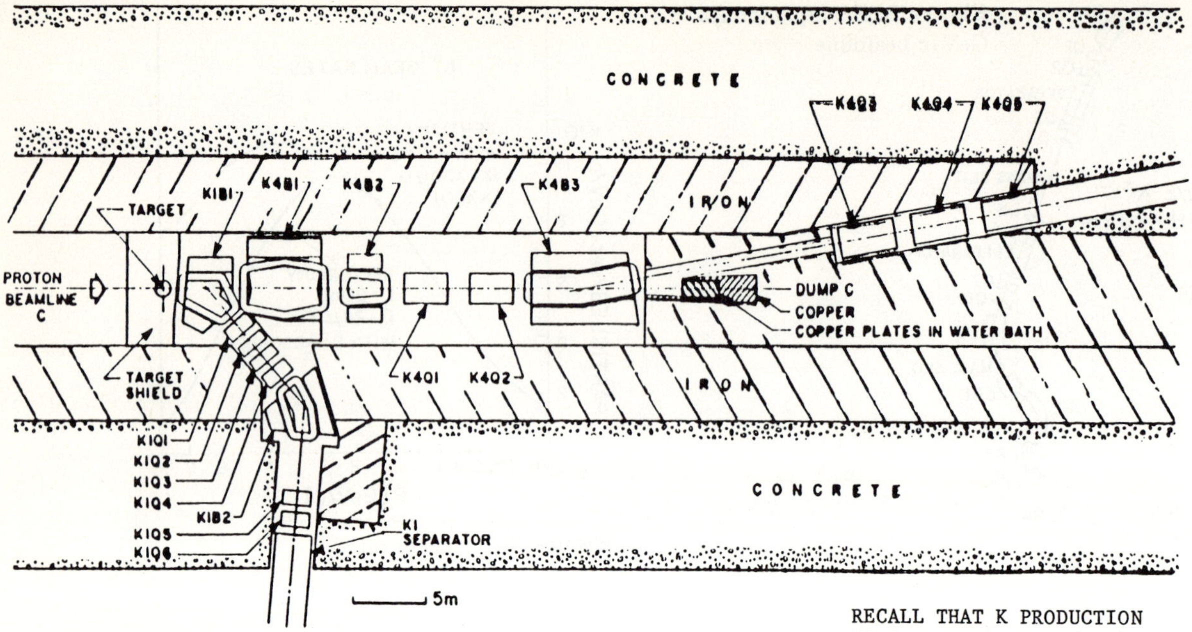

2 Forward Takeoff Channels
1 Low Momentum & 1 High Momentum
Operation completely decoupled
Beam Dump as close as possible

RECALL THAT K PRODUCTION
IS VERY FORWARD PEAKED.

	θ_{max}
1 GeV/c	15°
4 GeV/c	5°
8 GeV/c	2°

Figure 5: A target station

Technical Problems:
- Beam Power 3 MW
- Radiation Levels & Shielding
- Remotely Installed/Repaired Components

CROSS SECTION OF TARGET AREA SHIELDING

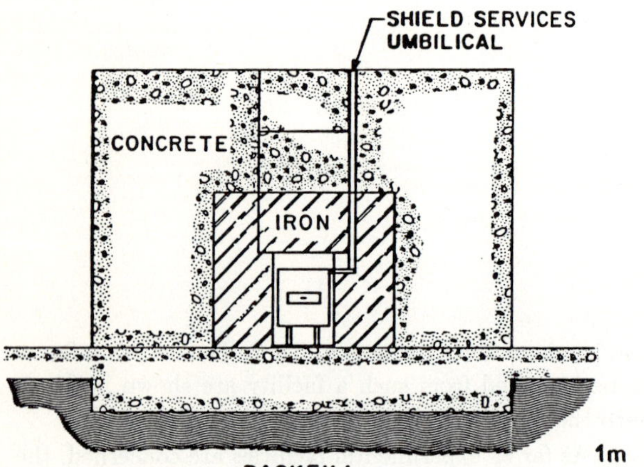

Remote Handling Features
- 'Canyon' Approach
- Vertical Installation of Components
- 'Smokestack' Access for Targets & Collimator
- Repair of Components in Hot Cells

Figure 6: Cross section of target area

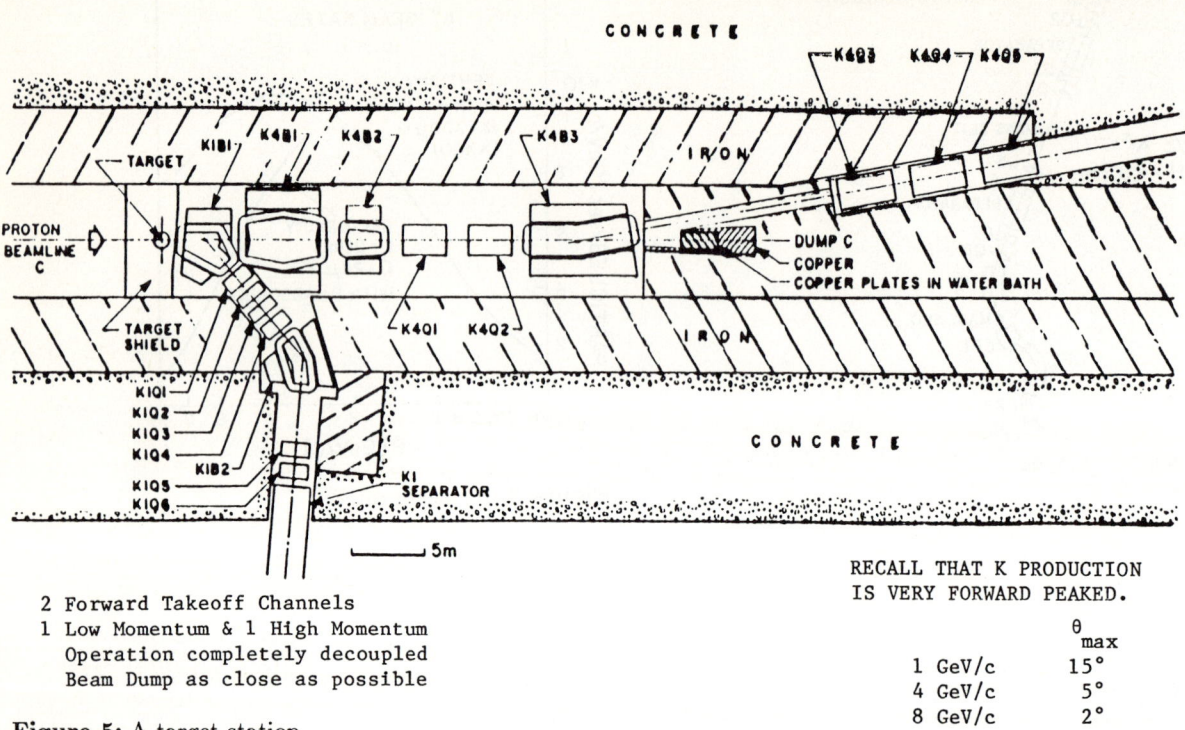

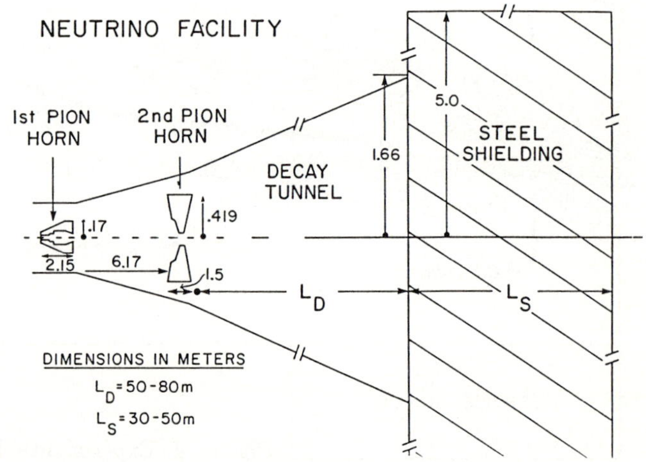

Figure 7: Neutrino facility

Table 4 summarizes the present state of our thinking on beamlines at the Kaon Factory. The major change is the addition of K20, a beamline going from 6 to 20 GeV/c. The need for a high quality separated beam line of at least 15 GeV/c became apparent at our hadron spectroscopy workshop in February, so we are proposing a line with three stages of separation which is 160 m long. The fluxes shown in the table are expected to be quite adequate for a program of spectroscopy with a LASS-like detector. The beamline is shown in fig. 9.

Table 4: Six Separated Beams at KAON
Emphasis on purity: wanted/unwanted ≥ 2
Unseparated beams can have $\sim 10\times$ more intensity
Resolution possible 0.1 – 0.5 %

Channel	Momentum [GeV/c]	Solid Angle [msr]	Momentum Acceptance Δp/p in %	Length [m]	Separation
K20	20 – 6	0.1	1	160	RF(3)
K6	6 – 2.5	0.08 – 0.30	3	110	RF(3)
K2.5	2.5 – 1.25	0.5 – 2.0	4	54	DC(2)
K1.5	1.5 – 0.75	2.0	4	30	DC(2)
K0.80	0.80 – 0.55	6.0	5	18	DC(2)
K0.55	0.55 – 0.40	8.0	6	14	DC(1)

Channel	P [GeV/c]	K^- $\times 10^6$/sec	K^+ $\times 10^6$/sec	π^- $\times 10^9$/sec	π^+ $\times 10^9$/sec	$\bar{p}$ $\times 10^6$/sec	p $\times 10^9$/sec
K20	23	0.04	0.8	0.2	0.2	0.0012	130
	17	1.2	27	0.21	0.9	0.0022	41
	6	2.4	4.8	1.4	2.3	15	3.3
K6	6	15	34	1.9	3.6	23	
	3	2.5	4.5	3.2	5.0	43	
K2.5	2.5	66	119	16	24	110	
	2.0	39	76	21	30	91	
	1.5	14	27	25	36	52	
	1.25	5.4	9.7	27	37	26	
K1.5	1.5	193	366	49	69	81	
	1.2	52	93	36	49	25	
	1.0	18	31	27	36	8.3	
	0.8	3.7	6.3	18	23	1.9	
K0.8	0.8	99	203	87	113	7.1	
	0.65	32	59	63	80	2.6	
	0.55	10	19	44	55	1.0	
K0.55	0.55	41	80	80	101	1.5	
	0.50	21	44	67	82	0.93	
	0.45	9.2	21	50	61	0.53	
	0.40	3.8	9.4	33	44	0.30	

As an example of a low momentum beamline, I would like to tell you about some work we have been doing on redesigning the LESB1 beamline at Brookhaven. Our interest in this particular line stems from the fact that it is the site of experiment 787, a search for the rare decay $K^+ \to \pi^+\nu\bar{\nu}$, which D. Bryman will describe in a later talk. This experiment is a collaboration between TRIUMF, BNL and Princeton, and we have an obvious interest in improving the beamline to increase the beam flux and decrease the π/K ratio. As well as improving our experiment, rebuilding the beamline, perhaps undertaken as a joint BNL/TRIUMF project, would give TRIUMF beam designers and engineers direct hands-on experience in designing, building and commissioning a kaon line. We could contribute our experience in radiation hardening, etc. The project would be a prototype for later Kaon Factory beamlines. J. Doornbos has already done some work at TRIUMF on the design which I will now describe to you.

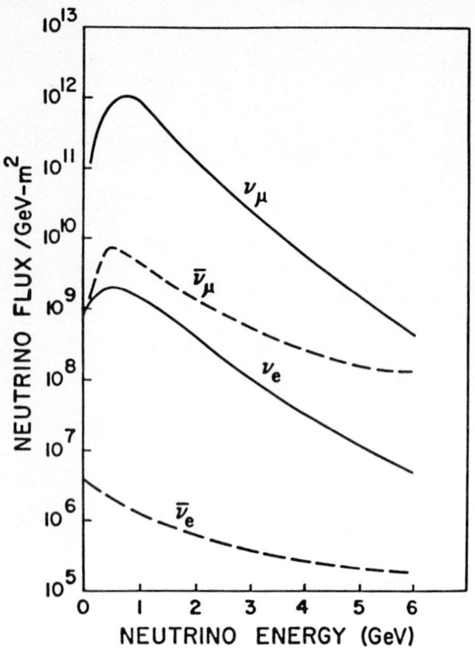

Figure 8: Neutrino fluxes at KAON

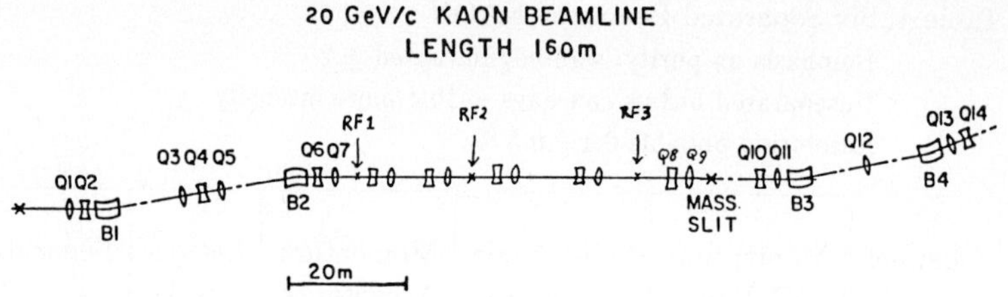

Figure 9: K20 beamline

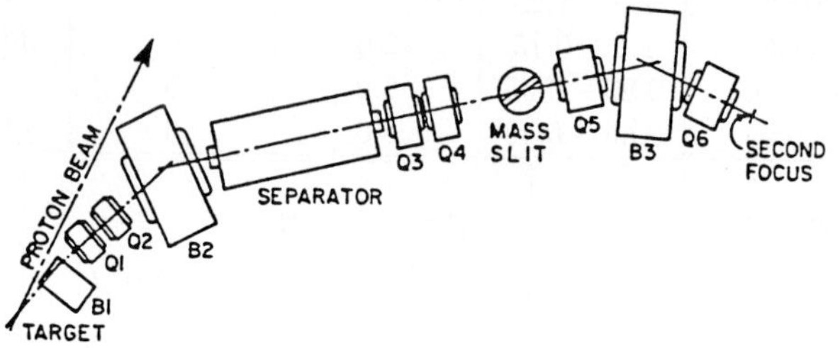

Figure 10: LESB1 beamline at Brookhaven

The present LESB1, shown in fig. 10, is a B-Q-Q-B single stage DC separated beamline with a take-off angle of 10°, solid angle of 2.6 msr and momentum acceptance of 4.5%. Doornbos' new design is for a Q-Q-B-B, two-stage DC separated beamline with a take-off angle of 0°, a solid angle of 12 msr and a momentum acceptance of 6%. It is shown schematically in fig. 11, with a blow-up of the front end shown in fig. 12. Putting the quadrupoles first instead of a dipole increases the solid angle by a factor of 4.5. Adding a second stage of separation decreases the pion contamination dramatically, as shown by Monte Carlo calculations which indicate that the main sources of contamination are (a) cloud pions from K and hyperon decays around the production target, (b) muons from pion decay after B, and (c) scattering from vacuum boxes and poletips. All these sources are reduced by an order of magnitude by the second separator and mass slit in the new design, apparently leaving kaon decay in flight after the second mass slit as the main contamination. If these calculations are to be believed, an upper limit on the pion contamination hitting the experimental target is 25% of the kaon intensity. Of course, the new design is longer (2.6 m) than the old one leading to 35% more loss by decay, but the overall effect is to increase the flux by 2.9 while improving the π/K ratio by an order of magnitude.

We need to do more calculations and detailed design and to persuade BNL and TRIUMF managements that this would be a good project benefitting both laboratories in order to make the design a practical reality. The aim would be to have the new beamline ready for the new BNL booster.

How would all the KAON Factory beamlines be laid out in an experimental hall? I showed you our initial layout for the proposal in fig. 4. For comparison, fig. 13 shows the BNL experimental hall and fig. 14 the new experimental hall being constructed at KEK. In both of these the philosophy is to split the primary proton beam into two or more proton lines with two targets on each. In contrast, the EHF and the LAMPF AHF designs adopted the philosophy of having only a single proton line with multiple targets. Choosing between these two options involves balancing considerations of efficiency of proton use versus flexibility of scheduling experiments, etc. In principle, the single proton line concept makes more effective use of the available protons. Flexibility in varying the split between users would have to be done by varying the target thicknesses. Note, however, that because of absorption of the produced hadrons in the target, only about 35% of the incident protons can be effectively used to produce secondary particles by a single target. Making the target thicker, so more protons interact, does not increase the secondary flux. Fig. 15 shows calculations done at KEK indicating that the maximum useful target thickness is about 6 cm of platinum.

At TRIUMF we have looked at three layouts, shown in fig. 16. The first two assume the primary proton beam is split in two with two targets on each. In (a) there are two beams from each target, while (b) allows three. The third option (c) has a single proton line with three targets and up to three beamlines from each.

Our present thinking is that the first option (a) would be the best and an experimental hall based on this principle is shown in fig. 17. It incorporates two un-

Figure 11: Two stage separated 800 MeV/c kaon beam

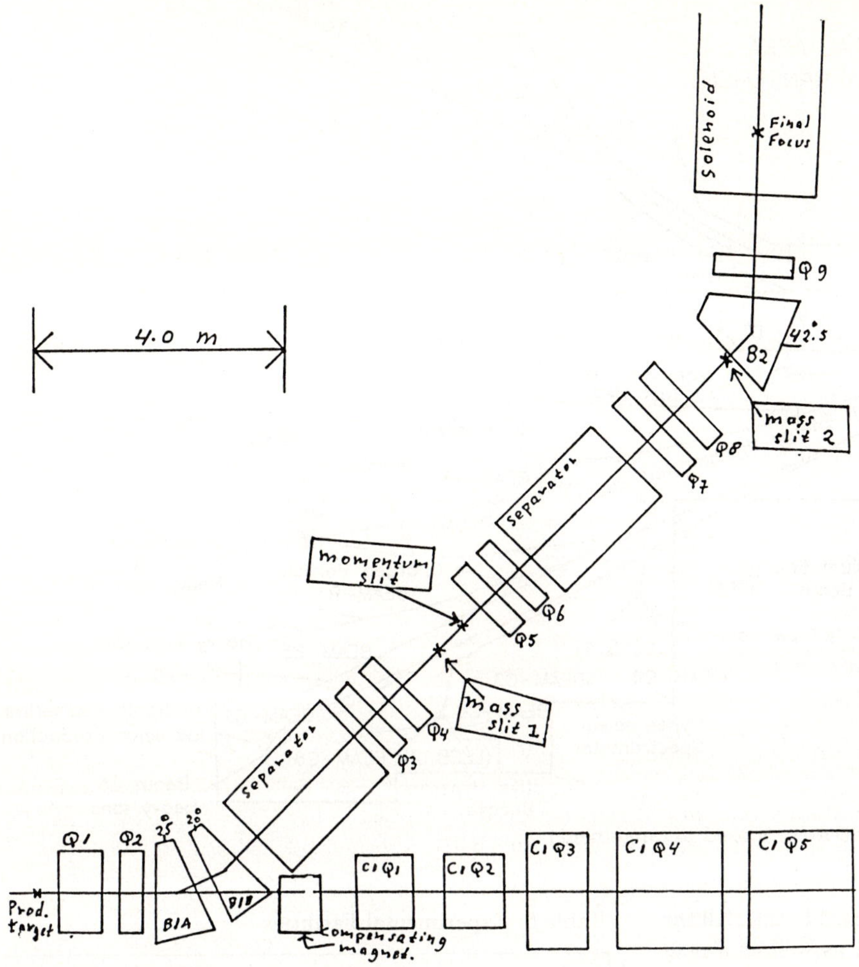

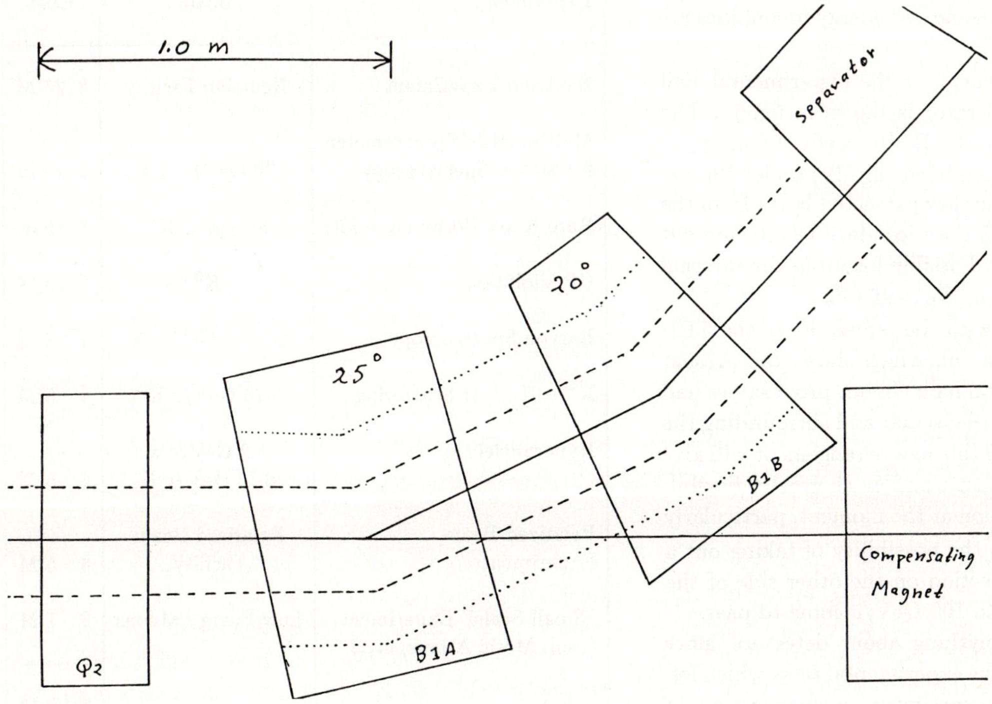

Figure 12: Close-up view of 800 MeV/c kaon beamline

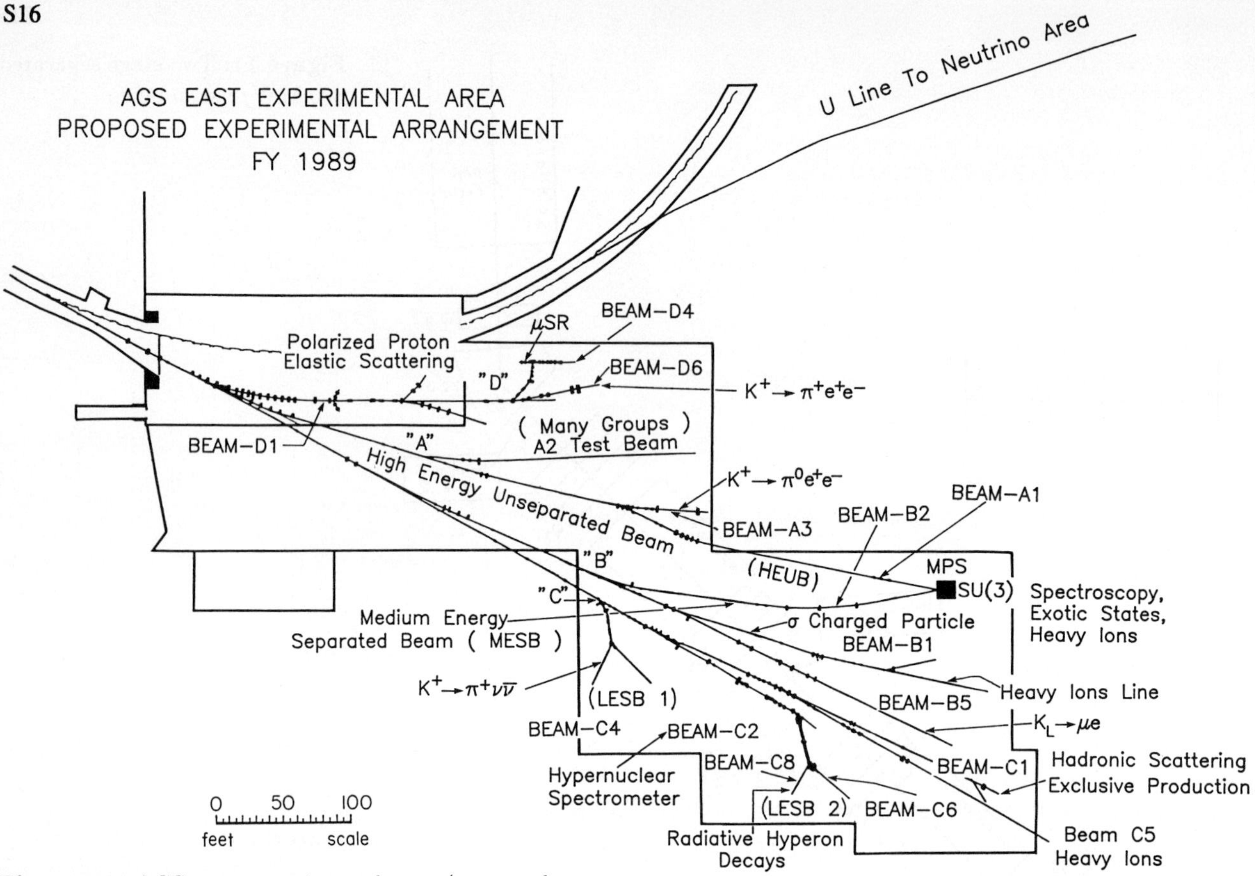

Figure 13: AGS east experimental area (proposed arrangement)

polarized proton beams and two polarized beams. All the secondary beamlines listed in table 4 as well as a neutrino line with active components are located under the same crane coverage and some low energy muon lines are added for μ_{SR}, etc.

A possible arrangement of the experimental hall with respect to the accelerator is shown in fig. 18. The fast extracted beam from the D-ring feeds the neutrino facility, with the target and horn (fig. 19) under the experimental hall crane. The slow extracted beam from the E-ring is split in two with the secondary beams laid out as in the previous figure. Possible locations for internal targets will be discussed in the next talk.

How all this fits in with the remainder of the TRIUMF site is shown in fig. 20, which shows the present cyclotron and experimental halls. In our proposal we had the main accelerator tunnel circular and surrounding the present cyclotron. We feel this new arrangement will give us more room for experiments and much greater flexibility. It is the preferred option at the moment, particularly since it would leave open the possibility of taking out a beam in the opposite direction on the other side of the ring should the upgrade to 100 GeV/c come to pass.

I will say hardly anything about detectors, since they will be discussed in the experimental talks which follow. Table 5 lists a set of experiments such as one might envision being approved early in the experimental pro-

Table 5: Experimental Facilities

Experiment	Beam	Cost
Neutrino Experiment	Neutrino Facility	\$ 25 M
Multiparticle Spectrometer for Meson Spectroscopy	20 GeV/c K,π	\$ 20 M
Rare Kaon Decay (K→ $\nu\bar{\nu}$)	Stopping K^+	\$ 15 M
CP Violation	K^0	\$ 15 M
Baryon Spectroscopy	2.5 – 6 GeV/c K	\$ 3 M
K^+ – Nuclear Scattering	0.75 GeV/c K^+	\$ 5 M
Hypernuclei	1-2 GeV/c π 0.75 GeV/c K	\$ 5 M
Polarized Beam Experiment	Polarized Beam Facility	\$ 5 M
"Small Scale" Experiment (μ_{SR}, Mesic Atoms, etc.)	Low Energy Muons	\$ 7 M
		\$100 M

Figure 14: KEK new developments under construction

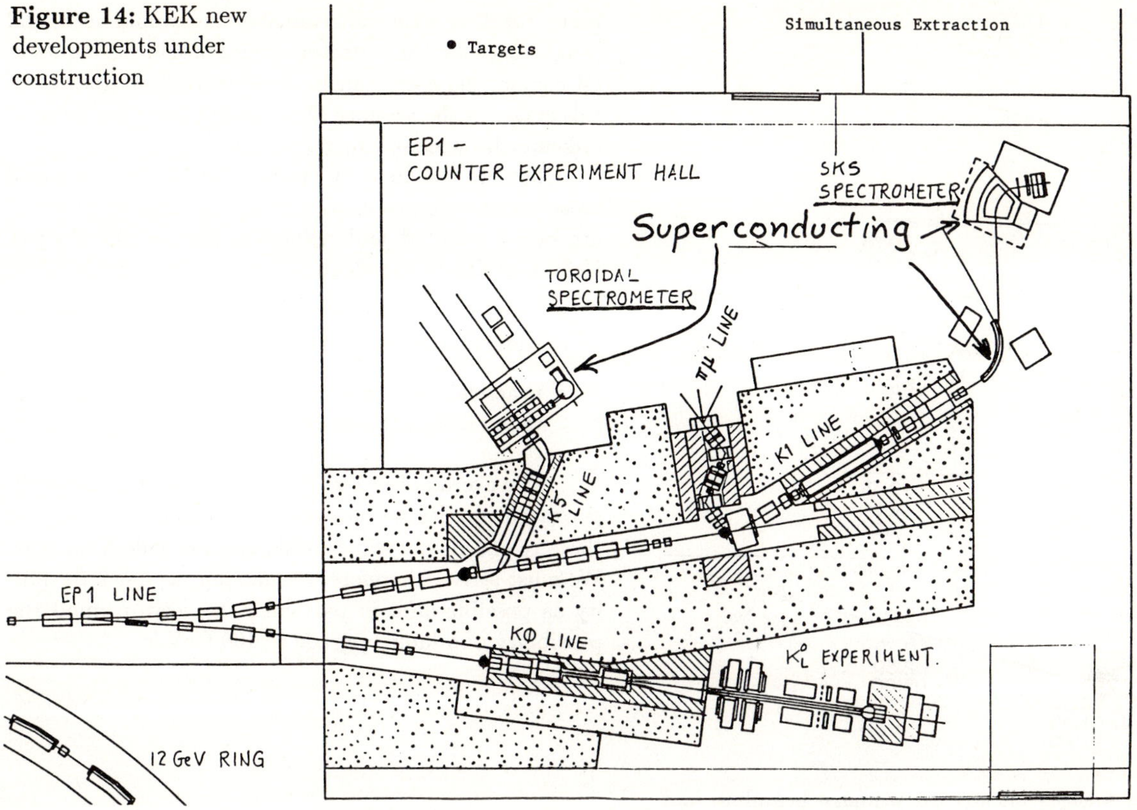

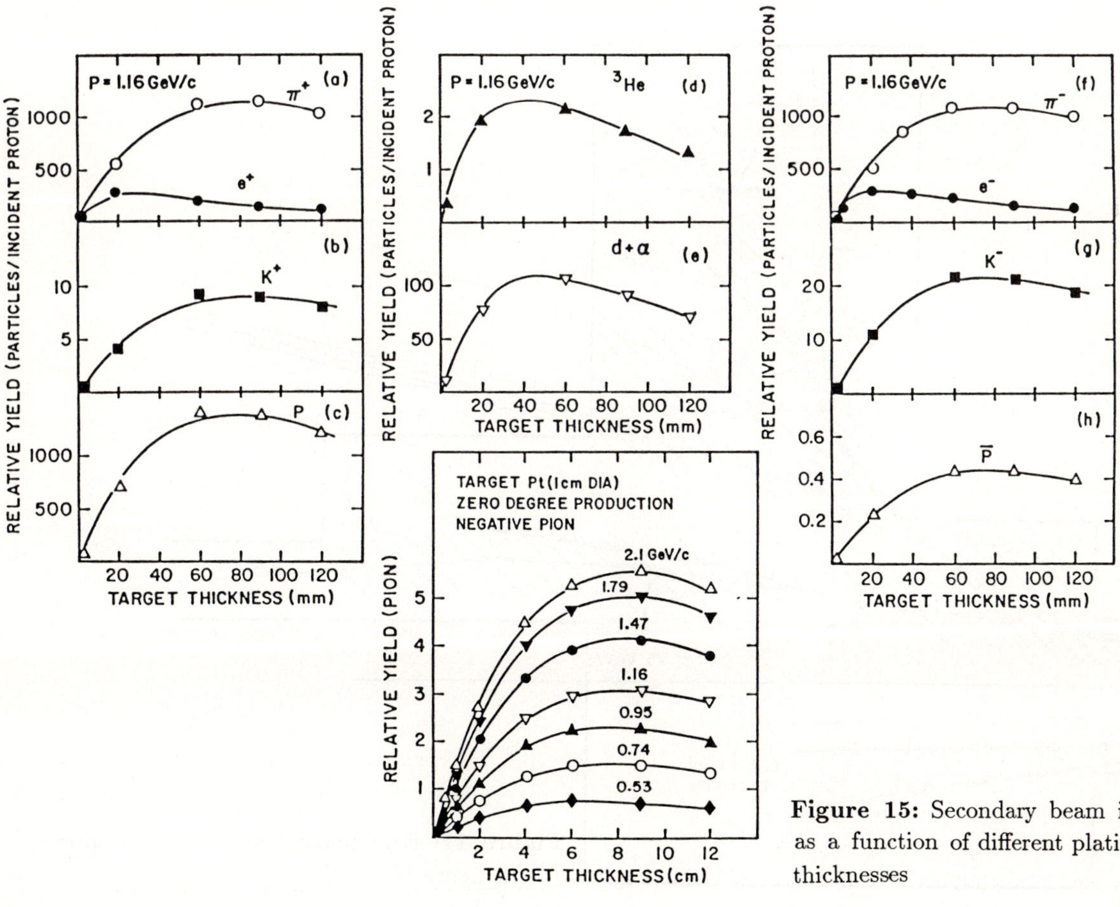

Figure 15: Secondary beam intensities as a function of different platinum target thicknesses

gram, together with the beamlines they would need, and rough costs for the detectors which would be required. Of course, the costs can be little more than guesses at this stage, in the absence of proposals, but they serve to indicate the scale of things.

Let me conclude by saying that the experimental areas and beamlines contained in our original proposal are being updated and refined as part of the Project Definition Study to reflect:

- physics input from the community gathered in workshops such as this, and
- practical and costing considerations as more detailed design takes place.

The overall objective is to provide for the presently perceived physics needs while recognizing that many of these requirements will be different a decade from now. We must strive, therefore, to maintain as much flexibility as possible and, in particular, to keep in mind the possibility of a later upgrade to 100 GeV/c.

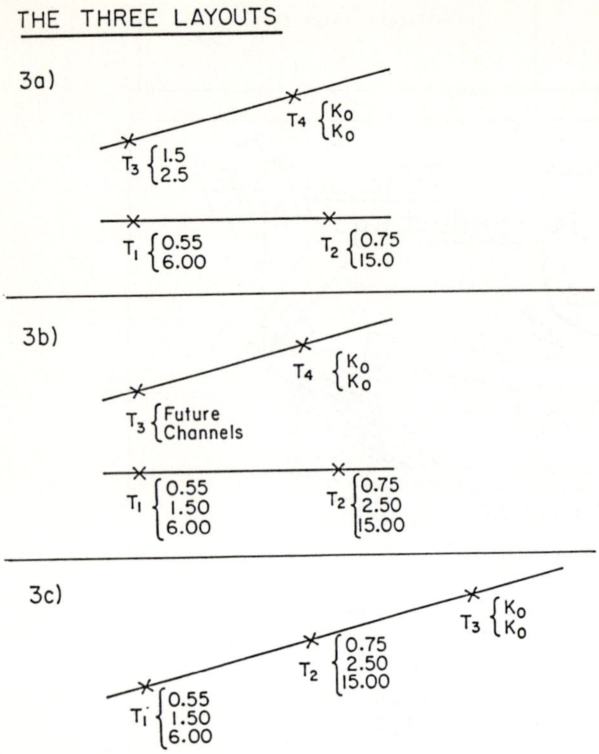

Figure 16: Several layouts for primary beamline and targets.

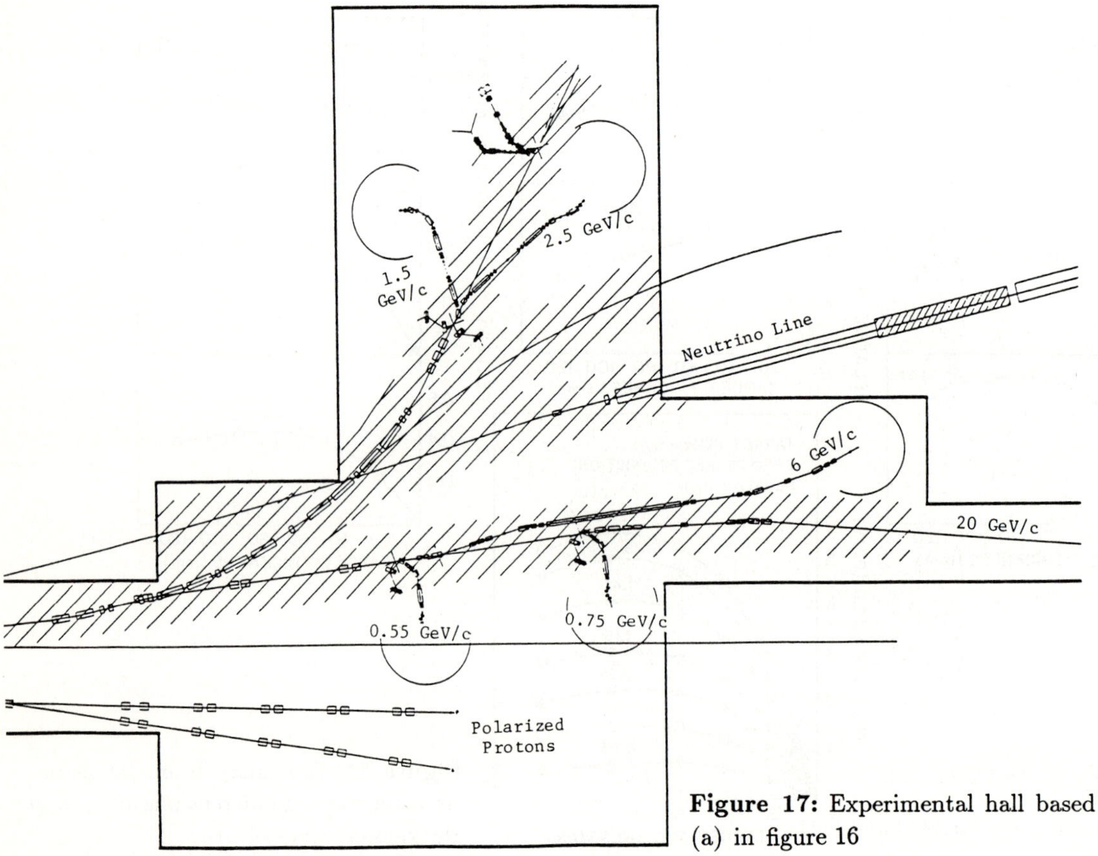

Figure 17: Experimental hall based on option (a) in figure 16

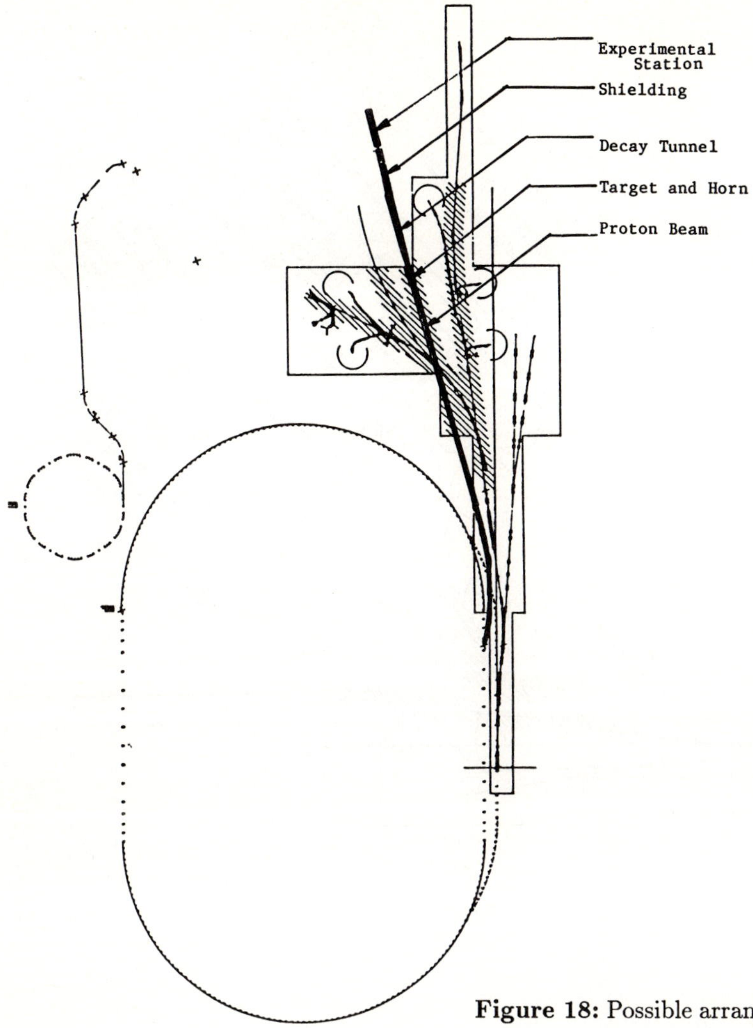

Figure 18: Possible arrangements of experimental halls

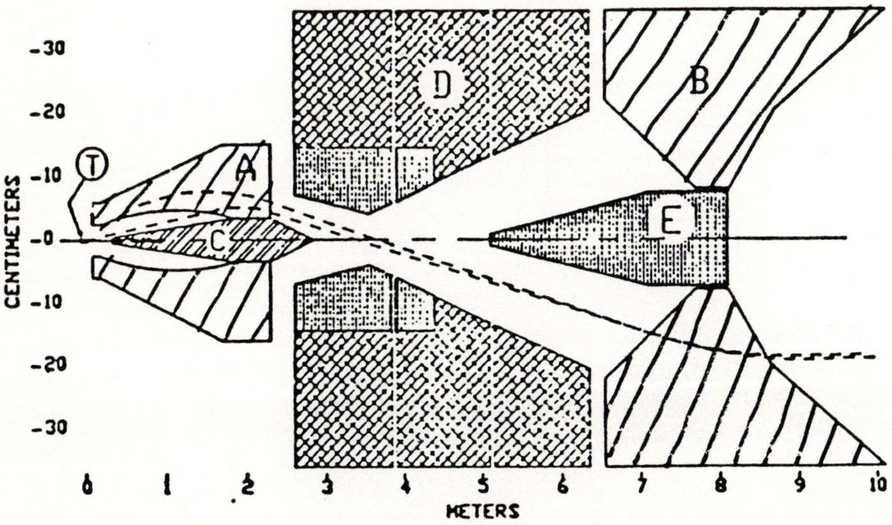

//// System to be used initially at KAON

Figure 19: Brookhaven narrow band horn

Figure 20: Site layout

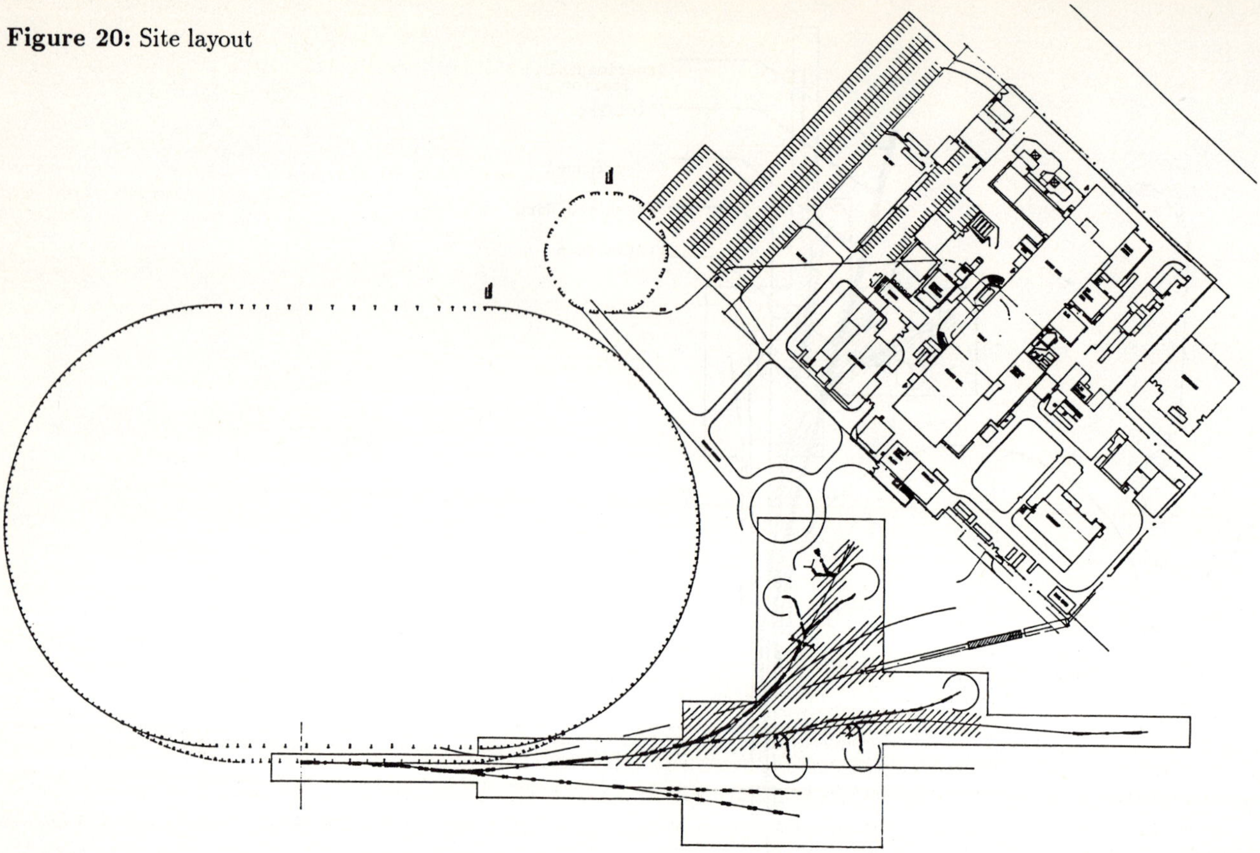

Z. Phys. C – Particles and Fields 46, S21–24 (1990)

Zeitschrift für Physik C Particles and Fields

Polarized internal targets at KAON

C. A. Miller

TRIUMF, 4004 Wesbrook Mall, Vancouver B. C., Canada V6T 2A3

Abstract

We discuss a new proposal that polarized gas jet targets might be installed in the intense polarized proton beam stored in the KAON Extender ring. The resulting luminosity could be competitive with present-day external target experiments while the purity of these targets offers new possibilities for nucleon spin physics at KAON. Since changes in the present building designs would be required, the potential user community is being consulted.

In other presentations at this workshop, particularly that of Alan Krisch, I think we will hear about the surprising spin effects that have been observed in proton-proton elastic scattering in the supposed hard scattering regime ($P_\perp^2 \geq 2\ (\mathrm{GeV}/c)^2$) where such effects were expected to disappear. The experiments that produced these as-yet unexplained data were rather heroic long-term efforts, in large part because they were done at accelerators that were designed without polarized beam in mind. As we have heard from Mike Craddock at this meeting, KAON is being designed to make the acceleration of polarized protons simpler and available much earlier in the life cycle of the facility. It will be especially helpful that high intensity polarized beam is already produced by the existing cyclotron injector.

Another fundamental aspect of the KAON design that could be well suited to nucleon spin physics is the flexible five-ring accelerator design. In particular, the Extender (E) ring is intended to store a circulating beam of 2.8 Amperes. Normally, this is unpolarized beam which is extracted "slowly" in 0.1 seconds and then refreshed by the next Driver (D) cycle. However, there seems to be no fundamental obstacle to storing in the E-ring for much longer, say several seconds, during which time this extremely intense beam could interact with very thin targets installed in the ring. This possibility for the KAON facility is not yet part of the "reference design" and detailed design work directed specifically toward it has only just begun. I am presenting these ideas here as a lobbyist rather than a designer, with the purpose of learning the degree of user interest in spin physics with internal targets. If sufficient interest is expressed, the accelerator ring layout could be modified to eventually accommodate such experiments.

Other aspects of accelerator design would also be affected. As we have heard earlier, the polarization of the beam can be maintained through the acceleration cycle either by jumping the relatively few depolarizing resonances or by installing "Siberian Snakes" in both the D- and E-rings which precess the spin axis by 180 degrees on each pass through the ring. The latter method is much preferred for the purposes of internal target experiments for two reasons. It is expected that the resonance-jumping method may limit the stored beam intensity because use of the full ring acceptance might cause excessive depolarization. More importantly, only with snakes can we hope to provide both transverse-vertical and longitudinal polarization directions at the internal target. This could be achieved via the "adiabatic precession" method in which the E-ring snake is ramped slowly after the beam is stored [1]. Transverse-horizontal beam polarization does not appear to be possible.

In principle, any beam energy between 3 and 30 GeV could be stored but with lower intensity at the lower energies because of the correspondingly larger emittance. For the purposes of spin physics with internal targets, we wish to maximize the intensity of polarized beam stored in the E-ring. In normal unpolarized operation, refilling the E-ring with 2.8 A every 0.1 seconds requires 100 microamperes CW from the injector cyclotron. At present, this cyclotron produces "only" 0.5 microamperes of polarized protons. A new optically pumped ion source is expected to soon raise this to several microamperes but with normal accelerator cycles, this would still not be enough to fill the E-ring to capacity. The flexible 5-ring design offers a possible solution. The extended storage cycle in the E-ring allows time for an extended accumulation cycle in the A-ring which stacks beam from the cyclotron via stripping injection of H^- ions. The accumulated intensity is therefore limited not by phase space conservation but only by spills arising from multiple traversals of the stripping foil. These spills must be kept to the level of about a percent during normal high intensity operation to avoid excessive ring activation. However, the E-ring should need refilling much less often to maintain a stored beam for internal targets. Therefore, larger spills from each extended accumulation cycle could presumably be tolerated. Although this method of concentrating the available polarized beam from the cyclotron has not been studied in detail, it seems possible that the A-ring can compensate for the lower intensity of polarized beam from the injector and make available the full intensity of stored polarized beam in the E-ring (limited by instabilities).

The internal targets would be in the form of gas jets. Such targets can be as thick as 10^{14} atoms per cm^2 [4]. In combination with 2.8 A in the E-ring, the resulting luminosity would be $1.8x10^{33}$ $cm^{-2}s^{-1}$ which corresponds to 10 nA of polarized protons on 6.7 mm of liquid hydrogen, for example. The main advantage of this internal target would be the possibility of detecting low energy recoil particles.

The main physics interest in an internal target facility arises from the possibility of polarized proton targets containing negligible contamination of non-hydrogenous material. This would permit the complete exploration of the spin dependence of processes initiated by two polarized protons. Elastic scattering measurements done up to now at ANL and BNL [2] have employed solid polarized hydrogen targets, which present to the beam mostly non-hydrogeonous background material. This type of target has been useful only for exclusive measurements involving several kinematic over-constraints to discriminate against background. Even in such cases, the background flux in the detectors from the target material will impose a limit on useful luminosity. This limit is already being approached in the current generation of elastic scattering measurements at BNL with a luminosity in the 10^{34} $cm^{-2}s^{-1}$ range. The thermal load on their polarized target is already at the limit of present cryogenic target technology. Also the kinematic constraints needed to cope with the background are expensive and inconvenient: two large magnetic spectrometers are required in coincidence, having necessarily limited acceptance. It seems that if the luminosity available with polarized internal targets continues to improve, this approach will come to dominate spin physics with two polarized protons.

The large spin-dependence already observed in the elastic data from ANL and BNL has not yet been explained in detail by any theoretical calculation but has inspired models which show some promise [3]. A comprehensive data set spanning the 3 to 30 GeV regime would serve to discriminate among such models. On the other hand, QCD-motivated theories are expected to be more applicable to inclusive processes. Measurements of such processes initiated by two polarized protons are presently almost impossible because kinematic over-constraints are not available to discriminate against the overwhelming background from polarized solid targets. This is an area where internal polarized targets could make a unique contribution, especially because large luminosities are not necessary.

Several groups are presently developing the technology of polarized gas targets. The FILTEX collaboration [5] hopes to use such a target to polarize the stored anti-proton beam at LEAR. They plan to use the difference in total $p - \overline{p}$ cross sections for spins parallel versus anti-parallel to preferentially deplete one spin state in the anti-proton beam. Their goal is a target thickness of 10^{14} cm^{-2} in a storage cell which is essentially an open-ended tube co-axial with the beam. The polarized atomic beam from a conventional atomic beam source (with refinements for high output) is injected through a port near the middle of the cell. Because the atoms collide with the cell walls about a hundred times before escaping, this type of cell can enhance the target density as much as two orders of magnitude over that of a simple gas jet.

A Novosibirsk-ANL collaboration [6] has been testing a polarized-deuteron storage cell in the VEPP-3 electron ring at Novosibirsk. They are developing an optically pumped polarized source. It has already produced a flux of $6x10^{16}$ polarized D atoms per second but is expected to yield $4x10^{17}$ s^{-1} at P_z=0.5. This should result in a target thickness of 10^{14} cm^{-2}.

Members of both of these groups are collaborating in a proposal to install polarized targets in the HERA electron storage ring at DESY, in an effort to measure the spin structure function of both the neutron and proton with high precision [7]. The importance of such experiments should ensure that this target technology will continue to be under active development.

The effective thickness produced in a storage cell receiving a given flux of polarized atoms is determined by its conductance; i.e. its length and aperture areas. For example, a cell 40 cm long with 1 cm^2 apertures receiving 10^{17} atoms per second is expected to present a thickness of 10^{14} cm^{-2}. If we are to enjoy the main advantage of internal targets - freedom from background material - we must be concerned about beam halo interacting with cell walls. A problem with installing such

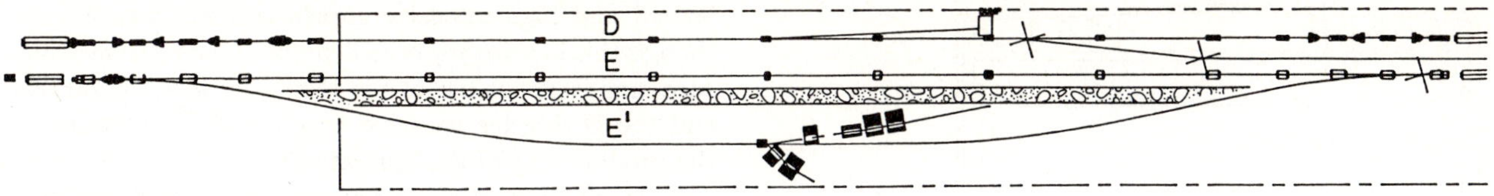

Figure 1: Proposed layout of beamline E′ with experimental area

a cell in the KAON E-ring is that this ring is not designed as a collider ring with minimum beam size at a target location. In fact, the beam emittance must be kept large to avoid beam instabilities at such large circulating currents. Consequently, a storage cell in the E-ring might have apertures larger than this. Since the thickness scales as d^{-3} where d is the aperture diameter, the density would be correspondingly lower. We must bear in mind however, that we are designing a facility with a lifetime measured in decades during which polarized source technology can be expected to advance considerably. Even with today's sources, the luminosity would be useful for interesting physics.

One promising new technology is an ultra-cold well-focussed jet of polarized hydrogen atoms now being developed by a University of Michigan/MIT collaboration [8] for use as an internal target in the 3 TeV ring at UNK. They plan to measure the p-p analyzing power in the 0.4 to 3 TeV range with $P_\perp^2$ varying from 2 to 8 $(GeV/c)^2$. This group believes it may be possible to achieve a target thickness in the 10^{14} cm^{-2} range without using a storage cell. This would indeed be a polarized proton target free of all background material.

Any internal target will impose an additional gas load on the E-ring vacuum system. Judging by the design studies done for other such installations, this can be controlled at an acceptable level by an appropriate differential pumping system. The total pressure bump caused by the target is expected to be comparable in terms of ion generation or multiple scattering to the residual gas load at the normal operating pressure of 10^{-9} Torr. Beam lifetime should be limited only by emittance growth due to multiple scattering. Depolarization of the target by the magnetic field of the beam bunches, a problem at electron rings, should not be a problem at KAON mainly because of the small instantaneous current (short bunch period and long bunches).

How adaptable is the present KAON E-ring design to the needs of internal target experiments? The primary design goal has been achieving high intensity slow extraction from the E-ring with acceptable spills and at minimum cost. Since the intensity is so far beyond present operating experience, slow extraction is expected to be the source of the most serious beam losses. Therefore the present ring layout calls for the straight section of the E-ring in the extraction hall to be physically separated from the other rings by enough shielding to prevent activation of components of the other rings. Although such a separated straight section is just what is needed for "mini-beta" optical sections to provide a small beam size for the internal target, the extraction section is expected to be too crowded and heavily irradiated to be useful for this purpose. The other side of the ring is constrained inside a tunnel to be too close to the other rings.

We are forced to conclude that some change in the present layout would be needed to provide for any non-trivial experiments using internal targets. In considering such changes to the ring layout, the guiding principle must be to provide the space to install such a facility when funds become available, possibly some time after initial operation, while minimizing increases in costs which can't be deferred, especially of buildings. At present, the only practical possibility seems to be to provide an alternate bypass straight section parallel to the extraction section. This new section, which we call E′, is unused during normal slow extraction operation and is activated only for internal target experiments with stored polarized beam. This would be accomplished by exciting two normally-dormant standard E-ring dipoles situated in the first gaps in the E-ring next to the arcs. A total of twelve additional standard dipoles appear to be sufficient to provide 7.8 m separation between the E and E′ beam lines, with 15 m both upstream and downstream of the internal target location for the min-beta quadrupole sections. This proposed layout is shown in figure 1. Detailed study of the optics of this system has begun and a design with reasonable lattice functions has been found[10]. A beam size at the target location of the order of 1mm seems to be possible.The ring separation allows 4.5m for spectrometers on either side of the E′ section, in addition to 1.7m of shielding to prevent activation of components in this E′ experimental area. Of course, this shielding is far from enough to allow personnel access to this area during slow extraction operation.

This proposed E′ layout is compatible with the presently planned extraction hall and overhead crane at ground level. The major impact it would have on undeferrable costs that would have to be included in the initial KAON funding request is due to the need to approximately double the width of the excavation for the extraction hall at beam level in the vicinity of the target location. As shown in figure 2, this 18m wide hall would

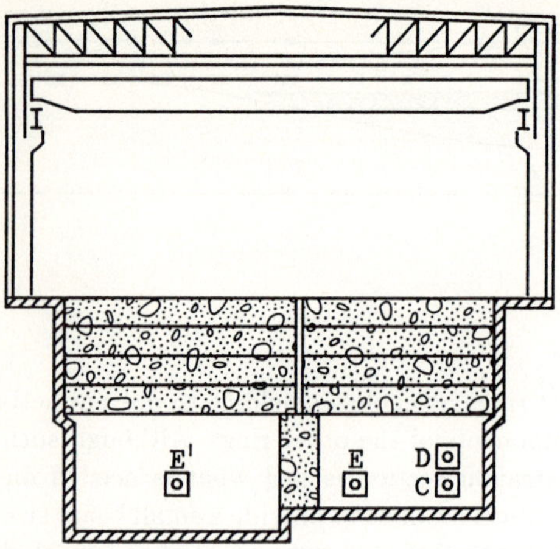

Figure 2: Cross section of extraction hall with beamline E′ in place.

then have to be covered with additional removable concrete shielding beams. All of these beams would have to be in place for any of the rings to operate.

Figure 1 also shows schematically how the proposed experimental area could accommodate a spectrometer system typical of that used for the elastic scattering measurements at BNL. Because of the limited space, the spectrometer dipoles bend the detected particles toward the beam line. Not shown is additional shielding around the detectors. Inclusive measurements could be expected to require somewhat less space transverse to the beam line, typically for a forward spectrometer like that for UA6 and a solenoidal recoil spectrometer enclosing the target [9].

Of course, the E-ring would not be available for secondary beam production during periods of stored polarized beam. However, in principle direct fast extraction from the D-ring could continue during the time intervals not needed for replenishing the E-ring. This time-sharing mode would require quickly switching between the polarized and high intensity unpolarized ion sources. Although this has not yet been demonstrated to be feasible, such simultaneous use of the E-ring for internal targets and the D-ring for neutrino production would enhance the productivity of the laboratory.

If provision of space for spin physics with internal targets is not made in the original accelerator buildings, it will be difficult and expensive to do so in the future. Therefore it is crucial that physicists interested in such experiments think about and express their needs now.

References

[1] K. Steffen, in Proc. Polarized Beams at SSC, Ann Arbor, June 1985, ed. R.G. Lerner, AIP Conf. Proc. No 145, p. 162; T. Roser, in Proc. High Energy Spin Physics, Minneapolis, Sept. 1988, ed. K.J. Heller, AIP Conf. Proc. No 187, p. 1442; L. Teng, in Proc. AHF Int. Workshop, Los Alamos, 1989, ed. H.A. Thiessen

[2] F.Z. Khiari *et al.*, Phys. Rev. D39, (1989) 45

[3] J. Soffer, Nucl. Phys. B279, (1987) 97, and Proc. High Energy Spin Physics, Minneapolis, Sept. 1988, ed. K.J. Heller, AIP Conf. Proc. No 187, p. 728.

[4] Workshop on Internal Targets for COSY, Jülich, April 1987, ed. W. Oelert

[5] G. Graw *et al.*, in Proceedings of the IV LEAR Workshop, Villars, 1987, ed. C Amsler *et al.*, Harwood Acad. Publ., Chur Switzerland (1988), p. 249.

[6] S.I. Mishnev *et al.*, in Proc. High Energy Spin Physics, Minneapolis, Sept. 1988, ed. K.J. Heller, AIP Conf. Proc. No 187, p. 1286.

[7] The HERMES Proposal (draft), DESY, Sept., 1989

[8] T. Roser *et al.*, preprint UM HE 89-7, March 1989

[9] Proposal for a European Hadron Facility, EHF-87-18, Ch. 5

[10] U. Wienands, Private communication

Z. Phys. C – Particles and Fields 46, S25–29 (1990)

Zeitschrift für Physik C Particles and Fields

Hyperons in the bound state approach to the Skyrme model. Magnetic moments of baryons

J. Kunz, P. J. Mulders

National Institute for Nuclear Physics and High Energy Physics,
NIKHEF-K, P.O. Box 41882, NL-1009 DB Amsterdam, The Netherlands

Abstract

We construct the electromagnetic current in the bound state approach to the Skyrme model using the Callan-Klebanov ansatz. The current is used to calculate the magnetic moments of the baryons. The general agreement of the magnetic moments in the bound state approach with the experimental magnetic moments is qualitatively satisfying. We disagree with some of the results of Nyman and Riska.

Already for almost thirty years attemps have been undertaken to understand the structure of hadrons. It has become clear that in order to understand the excitation spectrum of baryons and mesons, their weak and electromagnetic properties and the scattering processes at low energies, nonperturbative aspects of the underlying theory of quantum chromodynamics (QCD) are needed. This has led to the various models for hadrons. Among these, one of the oldest is the Skyrme model which we will use in our approach.

In the large N_c limit, where N_c is the number of colors, QCD becomes a theory with (stable) mesons, in which baryons arise as solitons. The most important mesons are the (light) pseudoscalar mesons, the Goldstone bosons of spontaneously broken chiral symmetry. In the simplest version of the Skyrme model [1, 2], based on a derivative expansion with a second order and a fourth order term (the Skyrme term), these are the only mesons taken into account. The other essential part of the model is the Wess-Zumino (WZ) term. It accounts for a vertex with an odd number of mesons, which is known to exist in QCD if one has three flavors involved. It describes processes like $K^+K^- \longrightarrow \pi^+\pi^0\pi^-$.

The Skyrme model approach to baryons starts from a classical solution of the mesonic lagrangian, a soliton with a nontrivial topological structure. The corresponding topological current is identified with the baryon current.

In the second step one proceeds with the quantization of the classical solution. Zero modes occur because the classical solution is not invariant under spatial rotations and isospin rotations separately, which are good symmetries of the lagrangian. These zero modes are treated via the collective quantization procedure, projecting out states of good isospin and spin. For two flavors, only states with $I = J$ are obtained as a consequence of the symmetry of the classical solution under combined isospin and spatial rotations.

In the three flavor Skyrme model the $SU(3)$ flavor symmetry is strongly broken due to the mass differences between the nonstrange and the strange mesons, notably the mass difference between pions and kaons. Starting with a lagrangian with a massive degenerate pseudoscalar octet of mesons and treating the $SU(3)$ symmetry breaking mass terms in first order perturbation theory leads to too small splittings between the masses of the baryons within flavor multiplets [3]. Some, but not enough, improvement is obtained by the method of "exact diagonalization" [4].

Callan, Hornborstel and Klebanov [5,6] suggested an alternative approach to describe the mass spectrum of the ground state baryons by treating the kaons and pions on a different footing. The classical kaon field vanishes and the kaon fluctuations around the classical hedgehog-like pion field are treated as vibrational modes. Only the $SU(2)$ isospin zero modes are treated via collective variables. The WZ term discriminates between kaons and antikaons. The latter turn out to be bound in certain partial waves. The lowest bound state (with strangeness $S = -1$) has an energy of about $\omega = 0.310\ M_K = 153$ MeV. In the background field of the classical hedgehog solution this bosonic state comes with isospin zero and intrinsic spin-parity $J^P = 1/2^+$.

We follow the procedure of ref. [6] making the ansatz

$$U = \xi\ U_K\ \xi, \tag{1}$$

where $\xi = \sqrt{U_\pi}$ and $U_\pi = \exp(i\vec{\tau}\cdot\vec{\pi}/f_\pi)$ contains the pion fields, while $U_K = \exp(i\lambda_a \cdot K_a/f_\pi)$ with the Gell-Mann $SU(3)$ matrices λ_a (the index a running from 4 to 7) describes the kaon fields in terms of the kaon doublets K,

$$\begin{aligned}\sum_{a=4}^{7}\lambda_a\cdot K_a &= \sqrt{2}\begin{pmatrix} 0 & 0 & K^+ \\ 0 & 0 & K^0 \\ K^- & \bar{K}^0 & 0\end{pmatrix} \\ &= \sqrt{2}\begin{pmatrix} 0 & K \\ K^T & 0\end{pmatrix}.\end{aligned} \tag{2}$$

Inserting the above ansatz for the chiral matrix U in the lagrangian and expanding to second order in the kaon fields leads to

$$\begin{aligned}\mathcal{L} = &\frac{F_\pi^2}{16}\,Tr\left(\partial_\mu U_\pi^\dagger \partial^\mu U_\pi\right) \\ &+\frac{1}{32e^2}\,Tr\left[\partial_\mu U_\pi^\dagger U_\pi, \partial_\nu U_\pi^\dagger U_\pi\right]^2 \\ &+(D_\mu K)^\dagger D^\mu K - m_K^2\ K^\dagger K \\ &-\frac{1}{8}K^\dagger K\left\{Tr\left(\partial_\mu U_\pi^\dagger\partial^\mu U_\pi\right)\right. \\ &\left.+\frac{1}{e^2F_\pi^2}Tr\left[\partial_\mu U_\pi^\dagger U_\pi, \partial_\nu U_\pi^\dagger U_\pi\right]^2\right\} \\ &-\frac{1}{e^2F_\pi^2}\left\{2\ (D_\mu K)^\dagger D_\nu K Tr(A^\mu A^\nu)\right. \\ &+\frac{1}{2}\ (D_\mu K)^\dagger D^\mu K Tr\left(\partial_\nu U_\pi^\dagger\partial^\nu U_\pi\right) \\ &\left.-6\ (D_\mu K)^\dagger[A^\mu, A^\nu]D_\nu K\right\} \\ &-\frac{iN_c}{F_\pi^2}B^\mu\left[K^\dagger D_\mu K - (D_\mu K)^\dagger K\right],\end{aligned} \tag{3}$$

where $F_\pi = 2f_\pi$ and

$$A_\mu = \frac{1}{2}\left(\xi^\dagger\partial_\mu\xi - \xi\partial_\mu\xi^\dagger\right), \tag{4}$$

$$D_\mu K = \partial_\mu K + \frac{1}{2}\left(\xi^\dagger\partial_\mu\xi + \xi\partial_\mu\xi^\dagger\right)K. \tag{5}$$

The last term in eq. (3) originates from the WZ action and is proportional to the number of colors N_c and the baryon current

$$B_\mu = \frac{\epsilon_{\mu\nu\rho\sigma}}{24\pi^2}\ Tr\left(U_\pi^\dagger\partial^\nu U_\pi\ U_\pi^\dagger\partial^\rho U_\pi\ U_\pi^\dagger\partial^\sigma U_\pi\right), \tag{6}$$

where $\epsilon_{0123} = 1$.

To construct the baryons we choose the ansatz $U_\pi(\vec{r}) = \exp(i\vec{\tau}\cdot\hat{r}\ F(r))$ for the classical solution, with the chiral angle $F(r)$ running from $F(0) = \pi$ to $F(\infty) = 0$. The invariance of the background field $U_\pi(\vec{r})$ under combined isospin and spatial rotations enables a partial wave analysis of the kaon field in spinor harmonics Y_{TLT_z}. The lowest solution is found for $L = 1$, $T = 1/2$, and can be written as

$$K(\vec{r}, t) = e^{-i\omega t}\ k(r)\ \vec{\tau}\cdot\hat{r}\ \chi. \tag{7}$$

where χ is a two-component isospinor. Because of the WZ term the spectrum is asymmetric for positive and negative energy solutions. Only the negative energy solutions are bound. The explicit solutions for kaon and antikaon modes, $k(r)\ \chi_i$ (with the index i indicating $\uparrow$ and $\downarrow$ states respectively), are obtained by solving Klein-Gordon type of equations as discussed extensively in refs [6,8,9]. The kaon fields are quantized in the familiar way by replacing the mode amplitudes for kaons and antikaons by particle (a_i) and antiparticle (b_i) creation and annihilation operators. For the quantization of the zeromodes corresponding to the $SU(2)$ isosopin symmetry the semiclassical collective coordinate procedure is followed, which leads to the spin and isospin fine structure in the baryon spectrum. One performs time dependent collective isospin rotations $U_\pi \longrightarrow A(t)U_\pi A^\dagger(t)$ and $K \longrightarrow A(t)\ K = e^{-i\omega t}\ k(r)\ A(t)\ \vec{\tau}\cdot\hat{r}\ A^\dagger(t)\ A(t)\ \chi$. The collective spin operator of the rotator, $\vec{J}_c$, and the total spin operator, $\vec{J}$, are given by

$$\vec{J}_c = -i\Omega\ Tr(A^\dagger\dot{A}\vec{\tau}) - c_K\vec{J}_K - c_{\bar{K}}\vec{J}_{\bar{K}}, \tag{8}$$

$$\vec{J} = \vec{J}_c + \vec{T}, \tag{9}$$

$$\vec{T} = \left(a_i^\dagger(\vec{\tau})_{ij}a_j - b_i(\vec{\tau})_{ij}b_j^\dagger\right)/2 = \vec{J}_K + \vec{J}_{\bar{K}}, \tag{10}$$

where the numbers c_K and $c_{\bar{K}}$ are discussed below. The operator $\vec{T}$ is thus identified as the spin of the kaons. The lowest mode has $T = 1/2$. Note further, that the collective spin equals the isospin, $\vec{J}_c^2 = \vec{I}^2$. The strangeness operator is given by $S = a_i^\dagger a_i - b_i b_i^\dagger$.

The hyperons are constructed by occupying the lowest bound state in the soliton background field $N_{\bar{K}}$ times. Depending on the number of kaons that is bound, the soliton must be quantized with integer or half-integer $I = J$ values. Note, that adding a number of bosonic antikaons with spin 1/2 (i.e. coupling to a symmetric spin state) yields the same quantum numbers as adding a number of fermionic colored s-quarks with spin 1/2. Because of the antisymmetry in color, also in the quark model the spin must couple to a symmetric state. The mass operator for baryons with $N_{\bar{K}}$ bound antikaons is

$$\begin{aligned} M &= M_{cl} + N_{\bar{K}}\omega + \frac{1}{2\Omega}(\vec{J}_c + c\vec{T})^2 \\ &= M_{cl} + N_{\bar{K}}\omega + \frac{c}{2\Omega}\vec{J}^2 + \frac{(1-c)}{2\Omega}\vec{I}^2 + \frac{c(c-1)}{2\Omega}\vec{T}^2, \end{aligned} \tag{11}$$

where we have omitted the index $\bar{K}$ on c. The inertia Ω and the number c are given as integrals involving the chiral angle and the kaon wave function. Choosing the parameters e and f_π to fit the N and Δ masses, implying $M_{cl} = 0.865$ GeV and $1/2\Omega = 98$ MeV [2], the calculation gives $\omega = 153$ MeV and $c = 0.617$, while a best fit to the baryon masses with the mass formula of Eq. (11) yields $M_{cl} = 0.867$ GeV, $1/2\Omega = 99.7$ MeV, $\omega = 215$ MeV and $c = 0.661$. Except for the value of ω, which is too small, the fitted quantities are in excellent agreement with the calculated values. The results are shown in Table 1.

Given the lagrangian it is straightforward to obtain the electromagnetic current from Noethers theorem. In the bound state approach the current can be obtained in two ways. The first way starts with the known flavor currents in the Skyrme model with quadratic, quartic and WZ term [2,10],

Table 1:
Numerical results for the masses of baryons

particle	M_{calc} [GeV]	M_{exp} [GeV]	M_{fit} [GeV]
N	0.939	0.939	0.939
Δ	1.232	1.232	1.232
Λ	1.046	1.116	1.112
Σ	1.122	1.193	1.179
Σ^*	1.303	1.385	1.374
Ξ	1.199	1.318	1.325
Ξ^*	1.381	1.533	1.519
Ω	1.465	1.672	1.671

calculated coefficients: $\omega = 153$ MeV, $c = 0.617$
fitted coefficients: $\omega = 215$ MeV, $c = 0.661$

$$\begin{aligned} J_{a,\mu} =& -i\frac{f_\pi^2}{4}Tr\left(\lambda_a U^\dagger \partial_\mu U\right) \\ &+ \frac{i}{16e^2}Tr\left(\lambda_a [U^\dagger \partial^\nu U, [U^\dagger \partial_\mu U, U^\dagger \partial_\nu U]]\right) \\ &+ \frac{N_c}{96\pi^2}\epsilon_{\mu\nu\rho\sigma}Tr\left(\lambda_a U^\dagger \partial^\nu U\, U^\dagger \partial^\rho U\, U^\dagger \partial^\sigma U\right) \end{aligned} \tag{12}$$

plus the contribution with U and $U^\dagger$ interchanged and in addition a minus sign for the last term. The electromagnetic current $J_\mu^{em} = J_{3,\mu} + (1/\sqrt{3})J_{8,\mu}$ in the bound state approach is then obtained by substituting the ansatz of Eq. (1) including the collective rotations A(t) into these currents. The second way is to apply Noethers theorem directly to the lagrangian in Eq. (3) when the isoscalar current is put in by hand as in the pure $SU(2)$ case [2]. The two methods yield the same result, providing a check on the quite lengthy calculations [11]. The spatial part of the current has terms parallel and perpendicular to $\hat{r}$. In the magnetic moments only the perpendicular current enters, $\vec{J}_\perp = \vec{m} \times \vec{r}$. We obtain for the isoscalar contribution of the electromagnetic current

$$\begin{aligned} m_{I=0}^3(r) =& \frac{-i\, Tr(A^\dagger \dot{A}\tau_3)}{2r^2} \cdot \frac{-\sin^2(F)\, F'}{2\pi^2} \\ &+ \frac{a_i^\dagger(\tau_3)_{ij}a_j + b_i(\tau_3)_{ij}b_j^\dagger}{2\, r^2}\Bigg\{k^2\cos^2(\frac{F}{2}) \\ &+ \frac{1}{e^2F_\pi^2}\left[4k^2\frac{\sin^2 F}{r^2}\cos^2(\frac{F}{2}) \right. \\ &\left. + k^2(F')^2\cos^2(\frac{F}{2}) + 3kk'F'\sin F\right]\Bigg\}, \end{aligned} \tag{13}$$

and for the isovector contribution

$$\begin{aligned} m_{I=1}^3(r) =& \frac{-\, Tr(A^\dagger\tau_3 A\tau_3)}{2\, r^2}\Bigg\{f_\pi^2\, \sin^2(F) \\ &\left[1 + \frac{4}{e^2F_\pi^2}\left((F')^2 + \frac{\sin^2 F}{r^2}\right)\right] \\ &+ (a_i^\dagger a_i + b_i b_i^\dagger)\Bigg\{k^2\, \cos^2(\frac{F}{2})\left(1 - 4\sin^2(\frac{F}{2})\right) \\ &+ \frac{1}{e^2F_\pi^2}\left[4k^2\frac{\sin^2 F}{r^2}\cos^2(\frac{F}{2})\left(3 - 8\sin^2(\frac{F}{2})\right)\right. \\ &+ k^2(F')^2\cos^2(\frac{F}{2})\left(1 - 18\sin^2(\frac{F}{2})\right) \\ &+ 2(k')^2\sin^2(F) \\ &\left. + 3kk'F'\sin(F)\left(3 - 4\sin^2(\frac{F}{2})\right)\right]\Bigg\}\Bigg\}. \end{aligned} \tag{14}$$

In order to evaluate the magnetic moment we have to evaluate the operators in Eqs (13) and (14) between baryon states. The operators entering in the isoscalar part are $-i\ Tr(A^\dagger \dot{A}\vec{\tau}) = (\vec{J}_c + c\vec{J}_K + c\vec{J}_{\bar{K}})/\Omega$ and $(a_i^\dagger(\vec{\tau})_{ij}a_j + b_i(\vec{\tau})_{ij}b_j^\dagger)/2 = \vec{J}_K - \vec{J}_{\bar{K}}$, while for the isovector part $Tr(A^\dagger\tau_3 A\tau_3) = -\gamma(J_c, I)\ J_c^3 I^3$ with certain coefficients $\gamma(J_c, I)$ [11], and $a_i^\dagger a_i + b_i b_i^\dagger = N_K + N_{\bar{K}}$.

Expressed in nuclear magnetons $e/2M_N$ the magnetic moment is given by $\mu = (2M_N/3)\int d^3r\ r^2 m^3(r)$. For a system consisting of a skyrmion with a $N_{\bar{K}}$ bound antikaons we find the magnetic moment operator

$$\mu = a_1\ J_c^3 + (a_2 + a_3 N_{\bar{K}})\gamma(J_c, I) J_c^3 I^3 + a_4 J_{\bar{K}}^3, \quad (15)$$

where a_1 - a_4 are functionals of the chiral angle and the kaon wave function. The resulting magnetic moments of the hyperons in terms of the functionals a_i are given in Table 2.

Table 2: *Matrix elements of the magnetic moment operator for baryons (for highest spin state)*

particle	μ
p	$\frac{1}{2}a_1 + \frac{2}{3}a_2$
n	$\frac{1}{2}a_1 - \frac{2}{3}a_2$
$N \longrightarrow \Delta$	$\frac{2\sqrt{2}}{3}a_2$
Λ	$\frac{1}{2}a_4$
Σ^+	$\frac{2}{3}a_1 + \frac{2}{3}a_2 + \frac{2}{3}a_3 - \frac{1}{6}a_4$
Σ^0	$\frac{2}{3}a_1 - \frac{1}{6}a_4$
Σ^-	$\frac{2}{3}a_1 - \frac{2}{3}a_2 - \frac{2}{3}a_3 - \frac{1}{6}a_4$
$\Lambda \longrightarrow \Sigma$	$-\frac{2}{3}a_2 - \frac{2}{3}a_3$
Ξ^0	$-\frac{1}{6}a_1 - \frac{2}{9}a_2 - \frac{4}{9}a_3 + \frac{2}{3}a_4$
Ξ^-	$-\frac{1}{6}a_1 + \frac{2}{9}a_2 + \frac{4}{9}a_3 + \frac{2}{3}a_4$
Ω^-	$\frac{3}{2}a_4$

In Table 3 the numerical results are given. The calculated magnetic moments of baryons are compared with the experimental magnetic moments. Furthermore the ratios of the magnetic moments to the proton magnetic moment are given. The calculated values are in general too small, by about 30%, just as in the $SU(2)$ Skyrme model. For the ratios, however, the agreement between theory and experiment is remarkably good in view of the fact that they result from a calculation that is parameter-free, except for fitting e and F_π in the nonstrange sector to N and Δ masses. Notably the calculated ratio $\mu_\Lambda/\mu_p = -0.221$ comes out in perfect agreement with the experimental result $\mu_\Lambda/\mu_p = -0.219 \pm 0.002$. In most other cases there is qualitative agreement leaving room for improvements. In the last column of Table 3a the result of a fit to the magnetic moments using the operator in Eq. (15) is given. In this fit a_1, a_2, and a_4 have been determined from the magnetic moments of p, n, and Λ, while a_3 is determined from the other magnetic moments. While the best values for a_1, a_2, and a_4 are about 30% larger,

Table 3a: *Numerical results for the magnetic moments μ of baryons*

particle	μ	μ_{exp}	μ_{fit}
p	1.870	2.793	2.793
n	−1.313	−1.913	−1.913
$N \longrightarrow \Delta$	2.251	~ 3	2.251
Λ	−0.414	-0.613 ± 0.004	−0.613
Σ^+	2.066	2.42 ± 0.05	2.467
Σ^0	0.509	—	0.791
Σ^-	−1.048	-1.157 ± 0.025	−0.885
$\Lambda \longrightarrow \Sigma$	−1.557	-1.61 ± 0.08	−1.6760
Ξ^0	−1.153	-1.250 ± 0.014	−1.297
Ξ^-	−0.138	-0.69 ± 0.04	−0.631
Ω^-	−1.243	—	−1.839

calculated: $a_1 = 0.557$, $a_2 = 2.39$, $a_3 = -0.052$, $a_4 = -0.83$.
fitted: $a_1 = 0.880$, $a_2 = 3.53$, $a_3 = -1.02$, $a_4 = -1.23$.

Table 3b: *Numerical results for the ratios μ/μ_p*

particle	μ/μ_p	$(\mu/\mu_p)_{exp}$
p	1.000	1.000
n	−0.702	−0.685
$N \longrightarrow \Delta$	1.204	~ 1.1
Λ	−0.221	-0.219 ± 0.002
Σ^+	1.105	0.87 ± 0.02
Σ^0	0.272	—
Σ^-	−0.560	-0.414 ± 0.009
$\Lambda \longrightarrow \Sigma$	−0.833	-0.58 ± 0.03
Ξ^0	−0.616	-0.448 ± 0.005
Ξ^-	−0.074	-0.25 ± 0.02
Ω^-	−0.664	—

but have the same ratio as the calculated coefficients, the best value for a_3 is of the same order of magnitude which is considerably ($\sim 20\times$) larger than the calculated coefficient.

Our results differ substantially from those of Nyman and Riska [12]. These differences cannot be attributed to the different ansatz, $U = \sqrt{U_K} U_\pi \sqrt{U_K}$ with which they start. In their derivation of the isoscalar contribution to the magnetic moements, Nyman and Riska have not accounted for the contribution $c_{\bar{K}} J_{\bar{K}}$ in the collective spin operator (see Eq. (8)). This contribution affects the coefficient a_4 of $J_{\bar{K}}$, as can be seen from a comparison of Eqs (13) and (15). Further, Nyman and Riska do include one term, μ_B, which originates from the Wess-Zumino term and which is of higher order in $1/N_c$ compared to the other contributions. We have omitted this term in order to be consistent, since in the derivation of the bound state lagrangian in Eq. (3) one systematically neglects such higher order contributions [5]. As expected, however, μ_B is small. A final difference is the interaction term in the isovector contribution to the magnetic moments, proportional to $n_{\bar{K}} J_c^3 I^3$. This term looks like a kaon contribution to the isovector part of the magnetic moments. We obtain this contribution in both derivations. Its calculated coefficient, a_3, however, is quite small compared to the other

The presence of more than two flavors of quarks has been a fortunate circumstance in hadronic physics, that has greatly contributed to our knowledge. It provides a testing ground for many hypotheses and models that are put forward on the basis of the understanding of the ordinary nonstrange matter that surrounds us. The possibility of having easy access to the strange sector in studying the structure of hadrons is an exciting prospect of KAON.

This work is supported by the Foundation for Fundamental Research (FOM) and the Netherlands Organization for the Advancement of Scientific Research (NWO).

References

[1] T.H.R. Skyrme, Proc. R. Soc. **A260** (1961) 127, Nucl. Phys. **31** (1962) 556

[2] G.S. Adkins, C.R. Nappi and E. Witten, Nucl. Phys. **B228** (1983) 552

[3] P.O. Mazur, M.A. Nowak, M. Praszałowicz, Phys. Lett. **147B** (1984) 137;
M. Praszałowicz, Phys. Lett. **158B** (1985) 264;
M. Chemtob, Nucl. Phys. **B256** (1985) 600

[4] H. Yabu and K. Ando, Nucl. Phys. **B301** (1988) 601

[5] C.G. Callan and I. Klebanov, Nucl. Phys. **B262** (1985) 365

[6] C.G. Callan, K. Hornborstel and I. Klebanov, Phys. Lett. **202B** (1988) 269

[7] N. Scoccola, H. Nadeau, M. Nowak and M. Rho, Phys. Lett. **201B** (1988) 425

[8] J. Kunz and P.J. Mulders, Phys. Lett. **215B** (1988) 449

[9] U. Blom, K. Dannbom and D.O. Riska, Nucl. Phys. **A493** (1989) 384

[10] A. Kanazawa, Prog. Theor. Phys. **77** (1987) 1240

[11] details can be found in J. Kunz and P.J. Mulders, in preparation

[12] E.M. Nyman and D.O. Riska, University of Helsinki preprint HU-TFT-88-50 (November 1989)

Z. Phys. C – Particles and Fields 46, S31–35 (1990)

Zeitschrift für Physik C Particles and Fields

Y* resonances in the mass range 1520–2430 MeV/c²

D. V. Bugg[1], D. Axen[2]

[1] Queen Mary College, London
[2] Physics Department University of British Columbia, Vancouver, B.C., Canada

Abstract

A simple set-up for studying almost all K^-p reactions is discussed. It is also suitable for studying πN and K^+N inelastic scattering.

1. The Proposed Set-up

Until now, most data on πN and KN inelastic processes have been obtained in bubble chambers. Statistics are limited to typically 1000 events/mb at momentum steps of about 25 MeV/c. We show that a simple and economical set-up is capable of covering most of 4π and identifying almost all K^-p reactions, including those with π^0, n or K_2^0 in the final state, from either a liquid hydrogen target or a polarised target. Cross-talk between channels is superior to that in most bubble chamber experiments, and is generally at the level of 1% or so. Even from the polarised target, background levels from carbon are at the level of 5–10%, and can be subtracted accurately.

We propose to survey the spectroscopy of $S=-1$ resonances produced in K^-p reactions with statistics 2 to 3 orders of magnitude higher than bubble chambers and to produce comprehensive polarisation data on all channels. The objective is not only to locate missing states, but also to establish systematically the decay modes of all resonant states, i.e. to establish eigenvectors as well as eigenvalues. The equipment is also capable of surveying $S=0$ resonances produced in $\pi^{\pm}p$ reactions, and also the physics of K^+p inelastic reactions.

The key to the simplicity of the experiment is that a magnet is not required providing one has, (a) detection of γ, K_L^0 and n, and (b) time of flight information on one slow particle. The following table lists the overall number of kinematic constraints of each reaction, assuming neutrals are detected.

$K^-p \rightarrow$	$K^0 n$	3C	(1)
$\rightarrow$	$\Lambda^0\pi^0$ (or η)	3C	(2)
	$\hookrightarrow \pi^-p$		
$\rightarrow$	$\Sigma^+\pi^-$	3C	(3)
	$\hookrightarrow n\pi^+$ or $p\pi^0$		
$\rightarrow$	$\Sigma^-\pi^+$	3C	(4)
	$\hookrightarrow n\pi^-$		
$\rightarrow$	$\Sigma^0\pi^0$(or η)	3C	(5)
$\rightarrow$	$K^-\pi^+n$	2C	(6)
$\rightarrow$	$K^0\pi^0 n$	2C	(7)
$\rightarrow$	$K^-p\pi^0$	2C	(8)
$\rightarrow$	$K^0\pi^-p$	2C	(9)
$\rightarrow$	$\Sigma^\pm\pi^\mp\pi^0$	2C	(10)
$\rightarrow$	Σ^0 (or Λ^0)$\pi^+\pi^-$	2C	(11)
$\rightarrow$	Σ^0 (or Λ^0)$\pi^0\pi^0$	2C	(12)
$\rightarrow$	Ξ^-K^+	3C	(13)
	$\hookrightarrow \Lambda^0\pi^-$		
$\rightarrow$	Ξ^0K^0	3C	(14)
	$\hookrightarrow \Lambda^0\pi^0$		
$\rightarrow$	Λ^0 (or Σ^0)ω	2C	(15)

Consider the first reaction as an example. If the K^0 is detected via the interaction of a K_L^0 and if the time of flight of the slower of the two particles (K_L^0 or n) is measured, there is one unknown momentum, and the event is in the 3C category. If $K^0 \rightarrow \pi^+\pi^-$, there is one additional momentum, but one additional constraint, the K^0 mass, and the event is again in the 3C category. In subsequent reactions, two body decays (eg $\pi^0 \rightarrow \gamma\gamma$) likewise produce an additional unknown momentum, but an additional mass constraint, hence no change in the overall number of constraints.

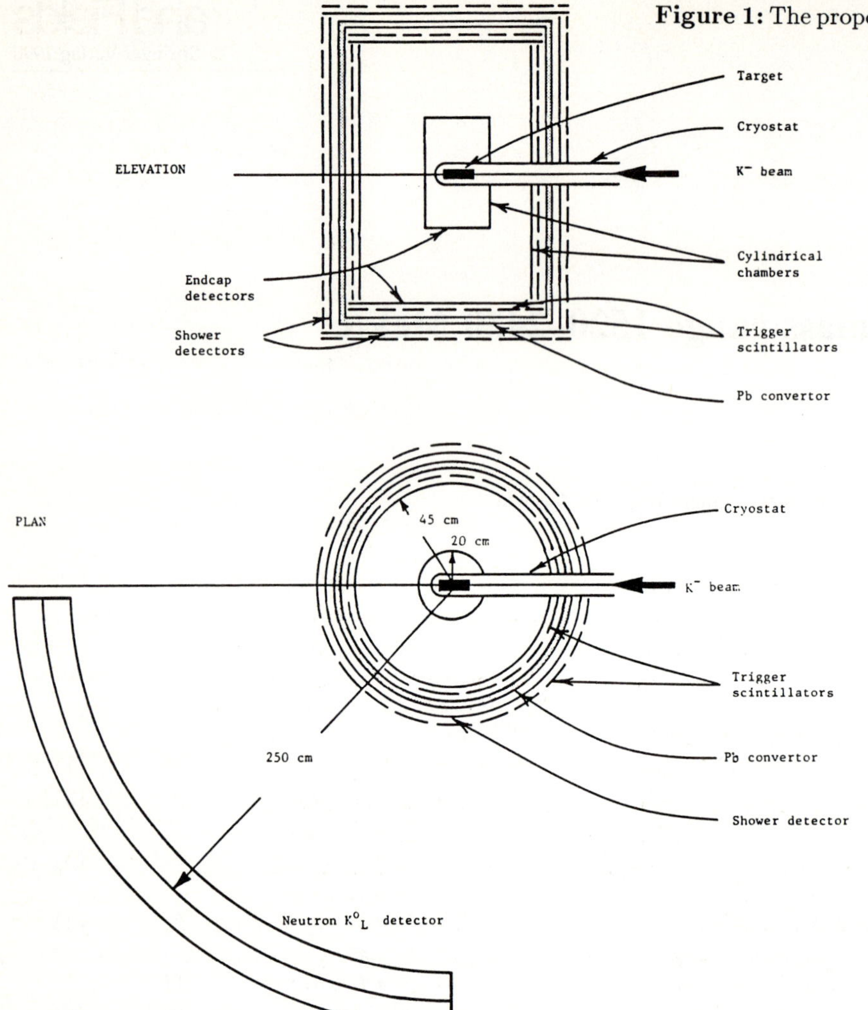

Figure 1: The proposed experimental setup (not to scale)

The equipment is sketched in fig. 1. It consists of three types of detectors, (a) charged particle detectors with nearly 4π coverage, (b) γ detectors with nearly 4π coverage and $\geq 90\%$ detection efficiency, and c) a large detector for K_L^0 and neutron detection, sampling 0-90^0 lab with about 25% efficiency and time of flight recording. Charged particle detection is achieved by cylindrical chambers of radius about 20 and 45 cm, surrounding a target of about 15 cm length, which may be liquid hydrogen or a frozen spin target (using a dilution refrigerator). The top and bottom of the cylindrical detectors are closed by "end-caps" made of plane chambers. The radius of the 20 cm cylindrical detector is a compromise between (i) cost of digitising wires and (ii) small probability for K_S^0 and Λ^0 decaying beyond this chamber; (at 2.5 GeV/c, a K_S^0 has a mean free path for decay of 12.5 cm). Outside this layer of detectors are about 3 radiation lengths of lead, followed by another layer of detectors to identify showers. Finally, neutrons and K_L^0 are detected in a position-sensitive liquid scintillation detector, 2m high at a radius of 250 cm; this detector consists of layers about 10 cm high (hence digitising vertical coordinates ± 5 cm), with photomultipliers on each end of every layer. Times of arrival of light at these two photomultipliers determine the horizontal coordinate of the interaction with an accuracy which, in our experience, is ± 3.5 cm for neutrons and ± 2.5 cm for γ and charged particles. They also determine the time of flight of the neutral particle with an accuracy of about ± 0.5 ns.

All these reactions have been subjected to a Monte Carlo simulation. The conclusions of the Monte Carlo study are:

1) Reactions with integrated cross sections ≥ 1 mb may readily be separated by 2C fits with a level of cross-talk between them and with other reactions generally in the 0.5–2.0% region,

2) The same is true for 3C fits to reactions with integrated cross sections ≥ 10 μb,

3) These levels of cross-talk are comparable with or less than what follows from, (a) K decays in flight,

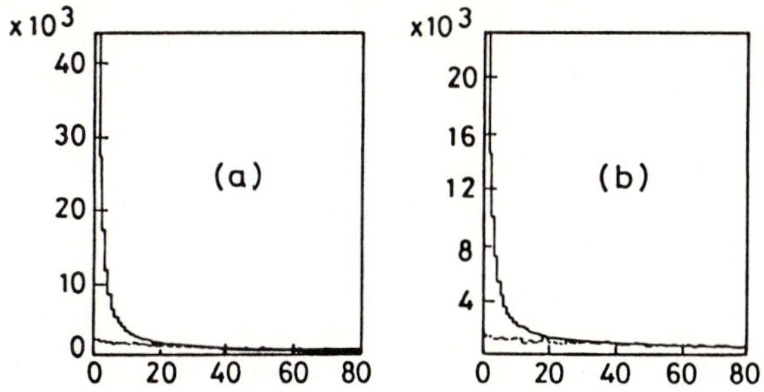

Figure 2: Number of events versus χ^2 for $pp \to n\pi^+$ at (a) 800, (b) 500 MeV [1].

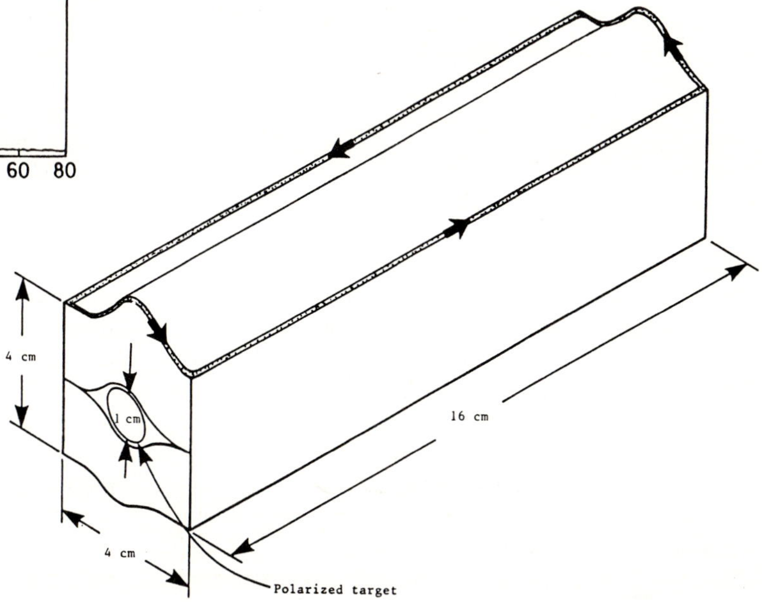

Figure 3: A sketch of the holding coil for the frozen spin target

(b) second scatters of exit particles in the material of the polarised target and in the detectors, i.e. intrinsic experimental limits,

4) Vertices of $\Sigma^\pm$, Ξ and K^0 decays are identified accurately by tracking the charged particles; it is not necessary to locate accurately the vertices of Λ^0 decays, which have a mean free path so short as to be negligible in kinematic reconstruction of γ and n directions,

5) The backgrounds from carbon events (simulated with Fermi momenta of 200 MeV/c) from the polarised target are typically ~10%.

These computer simulations may be backed up by our experience in measuring $pp \to n\pi^+$ with a similar (but more restricted) set-up at TRIUMF and LAMPF, using a polarised beam and polarised target [1]. At 500 MeV, the cross section for this reaction is 2.5 mb out of a total cross section of about 30 mb. However, the reaction was separated extremely cleanly with a 2C fit. Two demonstrations of this are given in fig. 2, which shows the number of events vs. χ^2; there is an integrated background under the signal of typically 10–15% from carbon after the 2C fit. These results are entirely consistent with Monte Carlo simulations of carbon events. In this measurement, the 25 kG field of the polarised target caused considerably greater perturbation than will be the case with the 2.5 kg holding field required for the frozen spin target proposed for the K^-p experiment.

2. The Frozen Spin Target

A frozen spin target of butanol with a volume of 50 cc (4 cm diameter vertical cylinder × 4 cm high) has been constructed at TRIUMF with a vertical cryostat, an operating temperature of 0.06°K, and a holding field of 2.5 kG. Under these conditions, the relaxation time of the target polarisation is about 4 days. The mean polarisation if 85%, the difference from 100% originating largely from non-uniformities in the direction of the holding field.

For the K^-p experiment, we propose a frozen spin target with a horizontal cryostat and a target length of about 15 cm. The required target diameter depends on the quality of the beam optics, but we tentatively assume 1 cm. In this case, the holding field can be provided by a superconducting coil of niobium–titanium, wound inside the cryostat, with a coil shape sketched in fig. 3. Individual turns are capable (conservatively) of carrying 15A in a wire of 10 mil diameter. The holding field of 2.5 kG can be provided with about 1000 turns, wound into a stack about 4 cm high x 1.6 mm thick (1.3% of a radiation length). This thickness is comparable with the thickness of materials of the cryostat, and causes less multiple scattering than the material of the polarised target itself. The stray field of the coil is a nuisance, but its effect on particle trajectories can be calculated easily in an iterative fit to the kinematics of any particular reaction. The beam can be brought in along the axis of the cryostat, with the aid of a pitching magnet upstream.

The polarised target needs to be polarised in a uniform field greatly superior to the holding field. This will be a large diameter C magnet with cobalt steel–tips, with a central field of 25 kG. The cryostat needs to be retracted periodically into this field.

3. Physics

It is already clear that resonances fit qualitatively into octets and decuplets of the quark model with L excitations, i.e. the symmetry $SU(6)\otimes O(3)$. In the $S = -1$ system, there are four missing $L = 1$ $[\mathbf{70}]^-$ members, two with quantum numbers $\Sigma(1/2^-)$ and one each with quantum numbers $\Sigma(3/2^-)$ and $\Lambda(3/2^-)$. Isgur and Karl [2] predict these to be in the mass range 1650-1880 MeV/c^2. At higher energies, the L=2 $[\mathbf{56}]^+$ spectrum is seriously incomplete. One objective of the present experiment is to locate these missing resonances.

Two of the most important polarization data sets in this respect are likely to be $P(K^-p \rightarrow \overline{K^0}n)$ and $P(K^-p \rightarrow \Sigma^-\pi^+)$. There are presently extensive data on differential cross sections for K^-p and K^-n elastic scattering and for $K^-p \rightarrow \overline{K^0}n$. And there are extensive and accurate polarisation data for elastic scattering. The data on $P(K^-p \rightarrow \overline{K^0}n)$ are the missing link offering hope of isolating uniquely the two spin–dependent amplitudes in each of the I = 1 and I = 0 elastic channels. Alston–Garnjost et al. [3] have drawn attention to the sensitivity of phase shift analysis to these data. Likewise, in the $\overline{K}N \rightarrow \Sigma\pi$ channel, there are cross section measurements from bubble chambers for $\Sigma^\pm\pi^\mp$ final states and polarisation data for $\Sigma^+\pi^-$, where the Σ^+ decay analyses its own polarisation. Again, one polarisation measurement $P(K^-p \rightarrow \Sigma^-\pi^+)$ provides the missing link to determine all of the amplitudes. In general terms, polarisation data are very important because of their sensitivity to phases of amplitudes.

An important qualitative feature of the model of Isgur and Karl is the prediction of multiplets $[\mathbf{70,0}]^+$, $[\mathbf{70,2}]^+$ and $[\mathbf{20,1}]^+$ for which there are few, if any, candidates at present. The excuse for their invisibility is that they are weakly coupled to πN and $\overline{K}$N. Thus, high statistics data on other channels are important, particularly sequential decays of the type

$$K^-p \rightarrow \text{ prominent } Y^* \rightarrow \Lambda^*\pi \quad \hookrightarrow \Sigma\pi, \Lambda\pi$$

where Λ^*(or Σ^*) is predicted to have a large branching ratio to both initial and final states. For example, there are six such missing $\Lambda^*(3/2^+)$ states in the mass range 1960-2175 MeV/c^2, which might be accessed via decay of Y_1^* (2250), produced at about 2 GeV/c K^- beam momentum.

There is another fundamental objective to the experiment. Present theories of spectroscopy are guided largely by mass values. But it is well known that energy eigenvalues are readily perturbed by (i) small changes in the potential (ii) inelastic thresholds, without the basic symmetry being destroyed. For example, in atoms, the spectrum of the hydrogen atom is reproduced in Li, Na, etc., but the inner cores of electrons perturb the eigenvalues. Likewise, in the s–d shell in nuclei, one cannot calculate the spectrum of eigenvalues from the free NN interaction; an empirical effective interaction is used instead. In baryon spectroscopy, it is remarkable that many resonances appear at inelastic thresholds (e.g. Nη, Nρ, Nω, $\Sigma\eta$, NΔ?), and one wonders whether these channels do not distort the mass spectrum substantially. Hence, if we are to learn anything quantitative about the Hamiltonian beyond its approximate symmetry, we need to have a complete picture of wave functions, hence branching ratios, which relate directly to eigenvectors of the Hamiltonian. Currently, there is widespread complacency that the spectrum of excitations fits into an L–excitation quark model qualitatively. Quantitatively, there are problems in fitting eigenvalues (e.g. the L-S splitting of $\Lambda(1405)$ and $\Lambda(1520)$). The eigenvectors from current phase shift analyses are a mess, and there are major quantitative discrepancies with all attempts to fit them with quark models.

Although this experiment is aimed principally at K^-p interactions, the extension to K^+p inelastic channels (which offers an interesting comparison with regard to threshold effects) and $\pi^\pm$p resonances (where branching ratios to inelastic channels are sorely needed, particularly above 1 GeV) is straightforward.

4. Momentum Ranges and Event Rates

It is desirable to cover ultimately the momentum range 400 to 2500 MeV/c (L = 1 to L = 3 excitations). Practical experience is that coverage of this range requires three separated beams (a) 400–800 MeV/c, (b) 700–1400 MeV/c, and (c) 1250-2500 MeV/c. Where to start the experiment is an open question. The L = 1 multiplet, peaked at 1 GeV/c, seems a good place in physics terms, and beam intensities and experimental resolution are good. One would therefore probably choose the order (b), followed by (a) where beam intensity is poorer, followed by (c), where the spectrum becomes rather complicated. But the interest may change within the time required to mount the experiment.

Event rates are high. Of the ~40mb K^-p cross section, 20 mb is interesting initially. Our experience is that one can record at least 250 events/s, of which half are from hydrogen in the polarised target. Thus it is quite realistic to record 300K hydrogen events/hour (assuming 77% operating efficiency). If we focus attention on those reactions requiring detection of n or K_L^0 (detection efficiency ~25%), this requires a beam intensity of about 6.4

x 10^4/s. Following the tactics of bubble chamber experiments, one would take data at approximately 25 MeV/c steps of momentum. It is clear that data handling is a major headache, since one could easily accumulate 10^8 triggers in a week's running! The initial strategy should be to record 10^6 good events at each momentum with a loose trigger; nearly all of the events are valuable. Thereafter, one would fall back to more selective triggers using on-line microprocessors to isolate rare events, e.g. ΞK.

5. Calibrations

For cross section measurements, one needs to know the detection efficiency for neutrons and K_L^0. The liquid scintillation detector for these particles needs to be calibrated using tagged K_L^0 and neutrons. This is tedious, but straightforward. At TRIUMF, our experience is that one can calibrate the neutron detection efficiency with an absolute accuracy of 2–3% using tagged neutrons from small-angle np elastic scattering. Tagged K_L^0 may be obtained from the reaction $K^-p \to \overline{K^0}n$ as part of the primary experiment, following the procedures developed by Alston-Garnjost et al. [4]

6. Costs

A detailed costing of the neutron/K_L^0 detector has been done, with the result £170K ≡ \$340K Cdn in 1989 prices. The cost of a frozen-spin target developed from scratch is \$1M Cdn, but there is reasonable hope of using many components from the present TRIUMF target. The cost of a new horizontal cryostat and superconducting coil for the holding field is ~\$250 Cdn. The cost of the chambers themselves is ~\$300 Cdn, and there is again reasonable hope that readout electronics may be taken from existing experiments.

7. Further Experiments

With carbon replacing lead, and an extra layer of charged particle detectors, one has a polarimeter, capable (for example) of measuring R and A parameters in K^+p elastic scattering [6]. For these measurements, the holding field of the polarised target needs to be longitudinal.

References

[1] R. Shypit et al., submitted to Phys. Rev. C.

[2] N. Isgur and G. Karl, Phys. Rev. D18, (1978) 4187; ibid D19, (1979) 2653; ibid D20, (1979) 1191; ibid D21, (1980) 1868.

[3] M. Alston-Garnjost et al., Phys Rev. D18, (1978) 182.

[4] M. Alston-Garnjost et al., Phys Rev. D17, (1978) 2226.

[5] A. Garcia and P. Kielanowski, Phys. Rev. D26, (1982) 1090.

[6] R. A. Arndt, L.D. Roper and P.H. Steinberg, Phys. Rev. D18 (1978) 3283.

Z. Phys. C – Particles and Fields 46, S37–43 (1990)

Zeitschrift für Physik C Particles and Fields

KAON and COSY – common aspects in hadron physics

M. G. Huber, B. Ch. Metsch

Institut für Theoretische Kernphysik, Nußallee 14–16, D-5300 Bonn 1, Fed. Rep. of Germany

Strong Interaction Physics is actually the least understood domain of Microscopic Physics, in particular in view of the precision with which the electromagnetic and weak interactions are known. This applies both to the knowledge of strong interaction phenomena as well as to our theoretical understanding of them. In the present contribution it is sketched how meson factories like KAON and high precision light baryon accelerators like COSY can contribute in a complementary way to a better investigation and a deeper understanding of Strong Interaction Physics.

1. Introduction

In the realm of microscopic physics the phenomena of the strong interaction and the corresponding structure of hadrons, including that of complex hadrons such as nuclear systems, are the least understood. Although it is generally accepted that a fundamental theory of strong interactions, Quantum ChromoDynamics exists, its intrinsic non-linearity together with its non-perturbative nature prohibits a straightforward application to the structure and interactions of hadrons at low energies. Moreover, one should bear in mind, that even if one were able to solve the theory in the relevant domain it might still be impossible to predict or even calculate phenomena that involve the degrees of freedom of many particles, like the phenomena of semi-conductivity and super conductivity in solid state physics.

In the present contribution we will therefore concentrate on a brief review of topics concerning the hadron structure and hadron-hadron interactions. The phenomena can be classified according to the baryon number B:

i) B=0-system: Meson spectroscopy and meson-meson interaction,
ii) B=1-system: Baryon spectroscopy and baryon-meson interaction,
iii) B=2-system: Dibaryon spectroscopy and baryon-baryon interaction,
iv) Systems with B>2: Multibaryon spectroscopy and collective excitation modes, multi-quark configurations.

For all these systems, irrespective of the baryon number B, the appropriate "bottom-up" strategy seems to be to start with an exploration of the dynamical structures ("subnuclear spectroscopy") and an interpretation even in simple (semi-) phenomenological models; in a next step then one ought to try to relate the phenomena to the underlying (quark-gluon) dynamics. One of the most important issues in this respect is certainly the investigation of the confinement mechanism especially in multihadron systems, since it can be expected, that the confinement will be influenced by the presence of other hadrons in the immediate environment.

In order to facilitate the discussion it is perhaps useful to discriminate between the description of the **internal quark dynamics** (i.e. the assumptions made for the quark confinement and the residual quark-quark or quark-antiquark interaction) and of the **external dynamics**, which should take into account that simple quark configurations like $(q\bar{q})$ and (q^3) couple to more complicated configurations by the creation or annihilation of extra $(q\bar{q})$-pairs. Such hadronic coupling vertices are then considered as the fundamental ingredients for the description of the scattering and the decay of hadrons and thus for the structure of the hadronic excitation spectrum itself (see Fig 1).

This paper is organized as follows: In section 2 the relevance of a high precision proton machine like the **CO**oler **SY**nchrotron presently under construction at the KFA-Jülich is indicated by sketching two experiments: one aimes at the investigation of meson-meson interactions using light nuclear fusion as an energy-momentum tag and another experiment that by the use of recoilless kinematics could contribute to the investigation of dibaryonic resonances. Section 3 then contains a discussion of a strategy of how to study the structure of multihadron systems (exotic nuclei) with high intensity meson beams (with flavour changing meson scattering) ,as will be available at the **KAON**-Facility. Furthermore collective excitations and multiquark excitations will be briefly discussed. Section 4 contains some concluding remarks and an outlook.

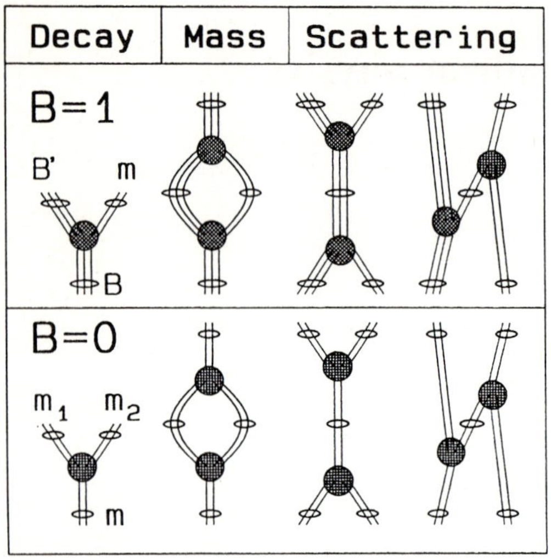

Fig.1. Interrelation of hadronic vertices: the self-energy contribution to the mass, the decay width, and the scattering amplitude for baryons and mesons, respectively

2. High precision Proton-induced Reactions

In Fig. 2 the excitation energy attainable in proton induced reactions with the **CO**oler **SY**nchrotron presently under construction at the KFA Jülich [1] is sketched alongwith the energies of the relevant single-baryonic excitations and the relevant meson-baryon decay channels. As can be seen the COSY machine will be able to explore excitations of baryonic matter upto well above 1 GeV, where also the excitation of gluonic degrees of freedom can be expected. In order to illustrate the special features of COSY, — its flexible and precise beam characteristics as well as a high energy resolution — we will discuss two types of proton induced reactions.

2.1. Tagged meson production and meson-meson interaction.

The prototype for tagged B=0 production is the reaction

$$p + d \longrightarrow {}^{3}He + X \qquad (1)$$

$$\hookrightarrow m$$

$$\hookrightarrow m_1 + m_2$$

This reaction has been used by Plouin et al. [2] at the **SATURNE** facility in Saclay to investigate the production of single mesons in a nuclear

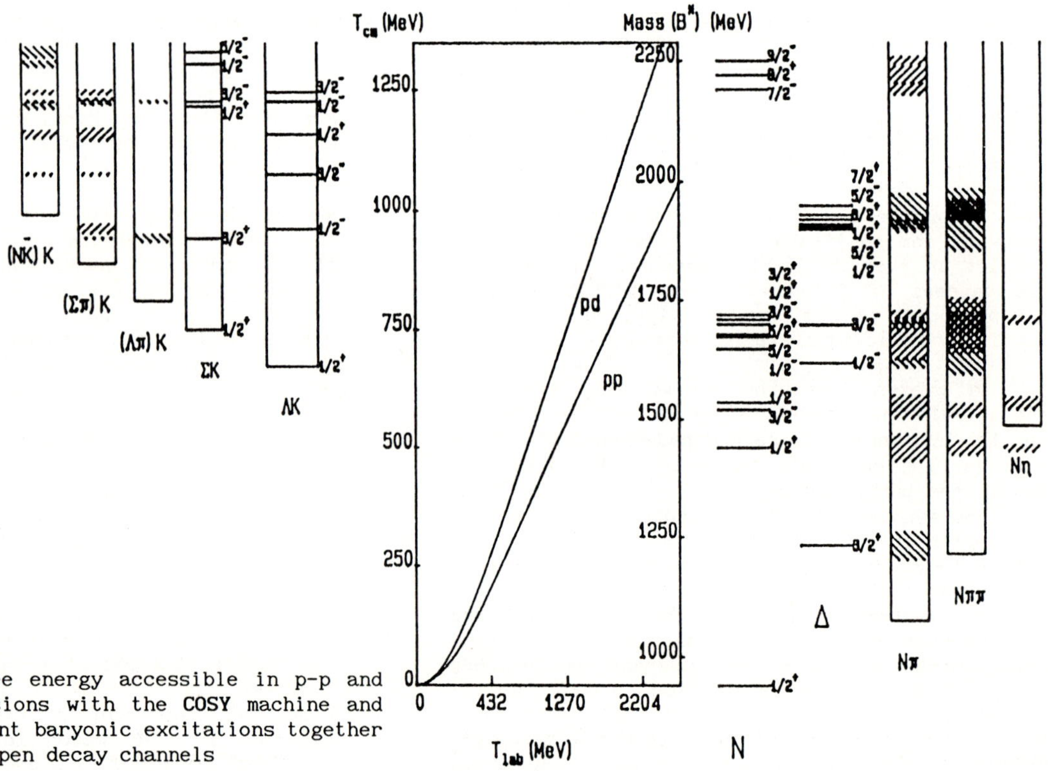

Fig.2. The CM-free energy accessible in p-p and p-d collisions with the **COSY** machine and the relevant baryonic excitations together with the open decay channels

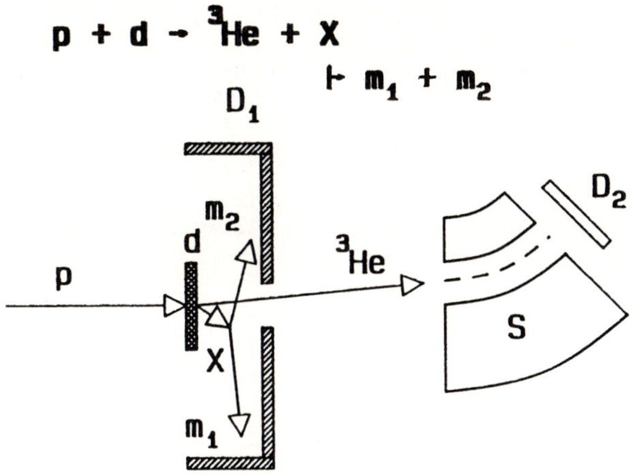

Fig.3. Schematic setup for an exclusive measurement of two meson production, in the p + d → ^{3}He + m + m' reaction, by complete spectroscopic determination of the recoil nucleus and the directions of the charged mesons

fusion reaction . The authors report a so-called threshold excitation curve (see Fig. 1 of Ref[2]) with pronounced structures at the single meson production thresholds as well as in the continuum, which are up-to-now not completely understood.

The same reaction can be used for an investigation of double meson production and the meson-meson interaction by complete identification of the ^{3}He nucleus including its energy [3]. A possible setup is sketched in Fig. 3.

By measuring in addition the directions of the two (charged) mesons the three-body final state is even kinematically overdetermined; this obviously helps to reduce the background. Using Dalitz-plots the kinematical regions where the relative motion between the final nucleus and a meson and between the two mesons is smallest can be determined, thus allowing for

i) a study of the excited intermediate states **(intermediate state spectroscopy)**, where in particular the study of the role of e.g. ΔΔ-excitations in two-meson production processes could be studied, or
ii) a study of meson-meson resonances (**final state spectroscopy)** or, in general, the meson-meson final state interaction in particular near the two-meson production threshold,

respectively. The latter is particularly interesting because the interpretation of the scalar meson resonances, especially of the a_0 and f_0 states is still under debate and a precise measurement should shed some light on such issues as the coexistence and/or mixing of $(q\bar{q})$- and $(q^2\bar{q}^2)$-configurations that, as has been pointed out in the introduction, are intimately related to the internal and external quark-dynamics of mesonic states.

2.2. Dibaryon spectroscopy using recoilfree energy transfer.

The pickup-reaction (p,d) at 0° scattering angle

$$p + X \longrightarrow d + Y \qquad (2)$$

where B(Y) = B(X)-1 can be used to study excitations of the (multi-)baryonic system Y by transferring energy without momentum, and consequently without angular momentum (see Fig 4). The proton and the deuteron having the same momentum **p**, the excitation energy ω of the system Y is given by the difference of the kinetic energies:

$$\omega(\bar{\mathbf{p}}) = \left[\sqrt{(m_p^2+\mathbf{p}^2)} - m_p\right] - \left[\sqrt{(m_d^2+\mathbf{p}^2)} - m_d\right] \qquad (3)$$

varying from 0 upto roughly 1 GeV (the deuteron-proton mass difference) depending on the proton momentum **p**. By a slight variation of the scattering angle the (angular) momentum transfer can be controlled, provided the measurements can be performed extremely precisely. As a consequence one can study subnuclear excitations in a realm of energy-momentum transfer that is in general not accessible in inelastic scattering experiments, see Fig. 5.

As an example we will consider the reaction

$$p + {}^3He \longrightarrow d + X \qquad (4)$$

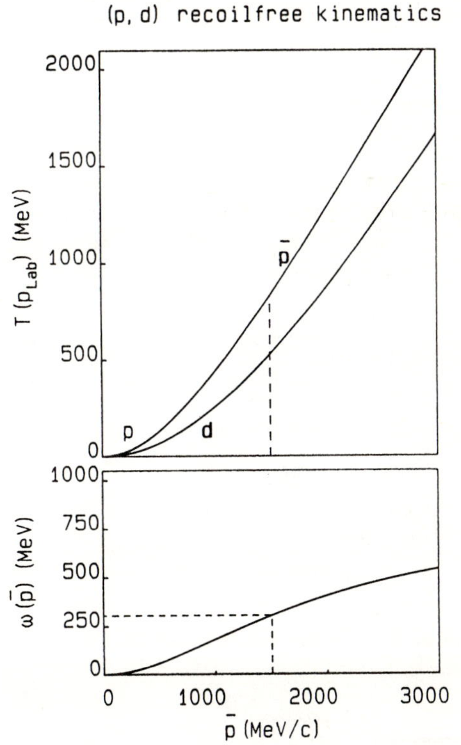

Fig.4. Energy transfer as a function of beam momentum in a (p,d) reaction at 0° scattering angle. The energy transfer is the difference in kinetic energy of the incoming proton and the outgoing deuteron having the same momentum $\bar{p}$

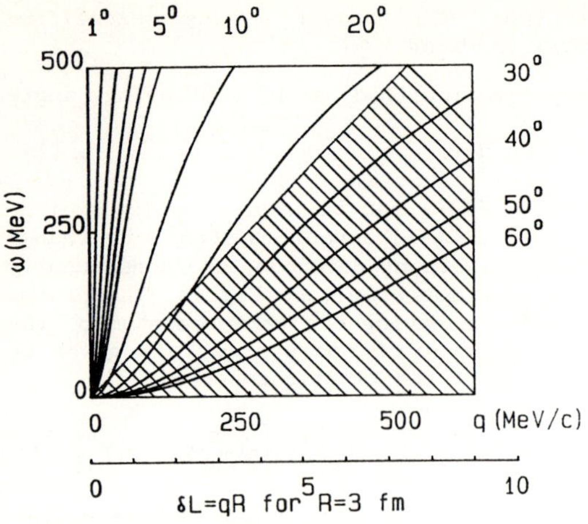

Fig.5. Relation between energy transfer and momentum (or angular momentum) transfer in a (p,d) reaction depending on the deuteron scattering angle. In such a reaction the rest system can be studied at energy-momentum tranfers not accessible e.g. in inelastic electron scattering (shaded area). The relation between linear and angular momentum transfer is indicated δL=qR.

where X has to be a system with baryon number B=2 and isospin T=1. This reaction could e.g. be used to study the dibaryonic structure found by Bock et. al [3] in the reaction

$$\gamma + d \longrightarrow pp + \pi^- \qquad (5)$$

where the pp invariant mass spectrum has been measured. By varying the scattering angle even some information concerning the angular momentum quantum numbers of this proposed system could be obtained. Of course, this resonance can only be seen in reaction (4) if its isospin T=1. If, however, the structure is in the T=2 channel, an associated production of a pion via e.g. the excitation of an intermediate ΔN-state might be employed, see Fig. 6.

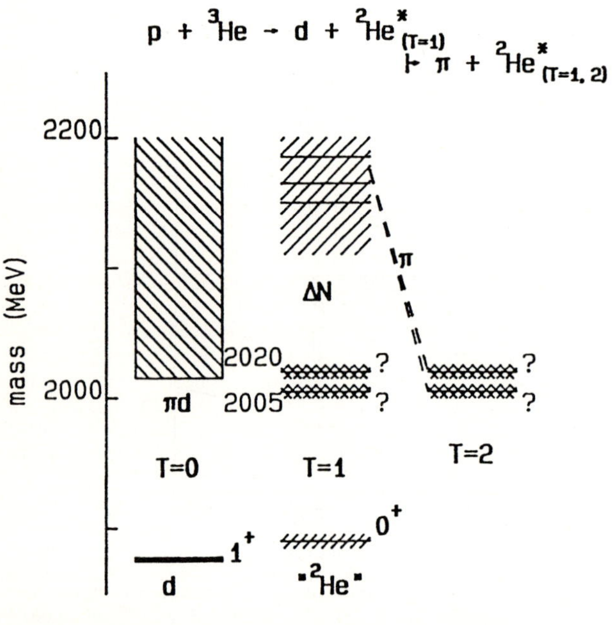

Fig.6. Study of dibaryonic resonances in a $^3He(p,d)^2He^*$ reaction. Dibaryonic resonances with T=2 could be populated via ΔN-intermediate states

3. Subnuclear Spectroscopy.

High intensity pion and kaon beams offer an attractive, interesting and important alternative tool to investigate subnuclear excitations, in particular all possibilities of flavour changing reactions, see Table 1.

Table 1. Flavour transfer in meson scattering[a]

Ejectile \ Projectile	π^-	π^+	K^-	K^+
π^-	I	QQ	S	
π^0	Q	Q	SQ	
π^+	QQ	I	SQQ	
η^0	Q	Q	SQ	
K^-			I	
K^0			Q	
$\bar{K}^0$	SQ	SQ	SSQ	Q
K^+	SQQ	S	SSQQ	I

[a] I = "flavour elastic", no flavour transfer
Q = charge transfer, ΔQ = ±1
S = strangeness transfer ΔS = -1

In Fig. 7 the baryonic excitations accessible in (π,K)-reactions are sketched alongwith the K^--N threshold. Four types of excitations can be discriminated: First of all there are the stable Λ- and Σ-ground states and the "subthreshold" excitations $\Sigma(3/2^+)(1385)$ and $\Lambda(1/2^-)(1405)$, then there is the narrow $\Lambda(3/2^-)(1520)$ state and finally a group of broad overlapping resonances at higher excitation energies. It is to be expected, that each of these strange resonance types will

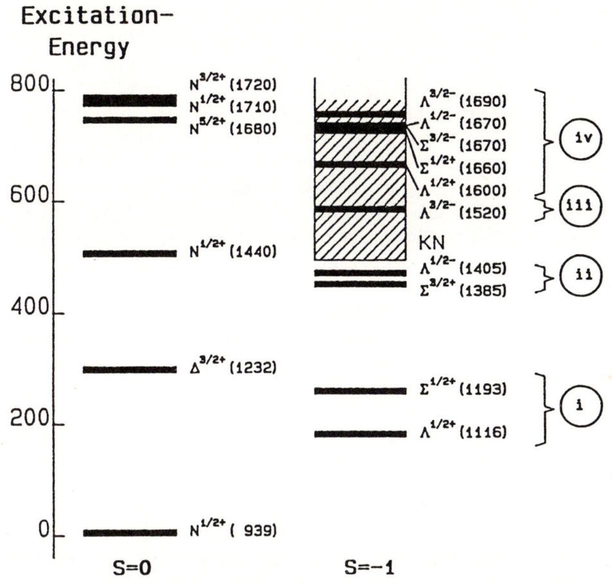

Fig.7. Excitation energies of strange baryon resonances together with the KN decay channel compared to the excitation energies of some nonstrange baryon resonances

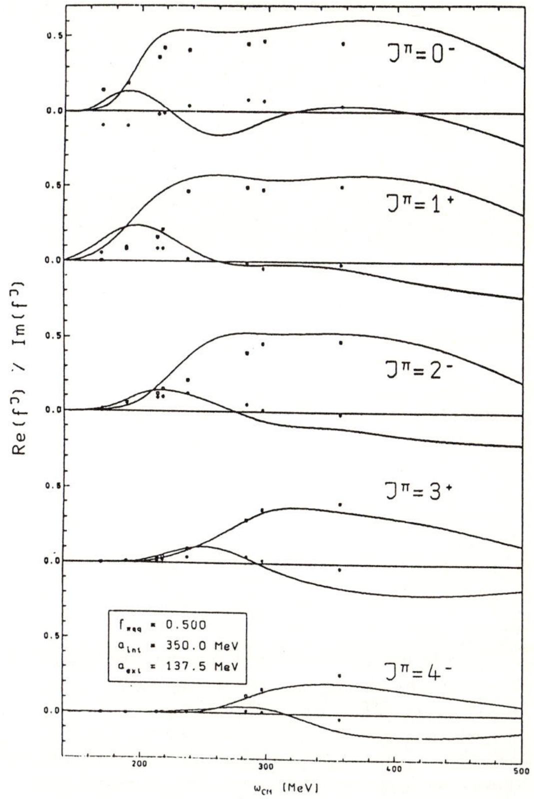

Fig.8. Calculated scattering amplitudes for π-^{12}C scattering [7] compared to the partial wave analysis of [6]

correspond to specific nuclear excitations in strangeness changing meson scattering off nuclei. In order to disentangle these various (strange) nuclear excitation modes, as analogues of the "elementary" N^*-resonances, it will be imperative to aim at a complete partial wave analysis of the meson-nucleus scattering. This together with the question of which kind of issues could be relevant for the underlying quark dynamics will briefly be illustrated for the case of pion-nucleus scattering in the region of the Δ-excitation. This example, however, is expected to be generic for any exclusive meson scattering process.

The subnuclear excitation of a nucleus in the energy region where the nucleon exhibits its resonances, here the Δ-resonance, leads to a number of resonances, that in the so-called Δ-hole model are interpreted as various Δ-N^{-1} configurations coupled to various total angular momentum, parity and isospin quantum numbers. The effects of the residual Δ-N^{-1}-interaction on the position of the resonances in general depends on these quantum numbers, see Fig. 8. The resonant contribution to the pion-nucleus scattering amplitudes at the relevant energies can be written as [4]

$$\langle\pi,A;\mathbf{k}'|\mathcal{T}(E)|\pi,A;\mathbf{k}\rangle = \sum_{J^\pi,T,\mu}\langle\pi,A;\mathbf{k}'|\mathcal{L}^\dagger|\Delta N^{-1};J^\pi T\mu\rangle\left[E-\varepsilon_\mu^{J^\pi T}-i/2.\Gamma_\mu^{J^\pi T}\right]^{-1}\cdot \cdot\langle\Delta N^{-1};J^\pi T\mu|\mathcal{L}|\pi,A;\mathbf{k}\rangle \quad (6)$$

where $\mathcal{L}^\dagger$ denotes the coupling operator to the mesic fields and ε and Γ denote the positions and widths of the various resonances, which — in general — are energy dependent.

These scattering amplitudes are compared to a partial wave analysis in Fig. 8. It was found, that a reasonable description (see also Fig. 9) can be obtained only if the fundamental coupling matrix element $\langle\Delta|\mathcal{L}|\pi N;\mathbf{k}\rangle$ is chosen to be smaller than the free nucleon value, needed to describe π-N scattering at Δ-resonance energies. This result certainly needs a more thorough investigation, but nevertheless it is demonstrated, that observables also of such complex systems as nuclei at higher excitation energies do depend on the assumptions made to describe the internal dynamics of the quarks inside the baryons and mesons, as well the hadronic coupling vertices.

As stated above similar effects are also expected for the excitation of other higher lying resonances in nuclear systems as well as for the excitations of the deuteron (dibaryonic resonances). Since the residual baryon-baryon interaction acts differently in different channels the study of the quark effects along the lines sketched above requires reliable complete partial wave analyses. After all, one should not forget that almost all our knowledge about baryon resonances indeed stems from such analyses, but that the information concerning more complex hadronic systems, like nuclei is up-to-now very fragmentary.

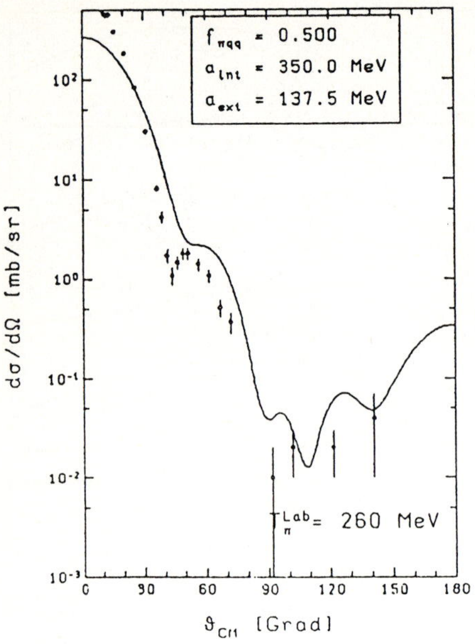

Fig.9. Calculated [7] and experimental (see Ref. [6]) differential cross section for π-^{12}C scattering at T_{π}^{Lab} = 260 MeV.

There is another interesting aspect that arises if one proceeds to investigate the excitation of higher lying baryonic resonances in complex systems. As illustrated for the B=2-system in Fig. 10, the unperturbed positions of the various baryon-baryon combinations are almost degenerate at higher excitation energies. The matrix elements of the residual baryon-baryon interaction are typically several tens of MeV and thus could lead to the interesting phenomenon of the mixing of various baryon-baryon configurations, differing in their internal quark structure due to the interactions mediated by various meson fields. This could lead to a sort of supercollectivity, where it no longer makes sense to discriminate between internal (quark) and external (baryonic) degrees of freedom. In order to probe this mixing of internal and external degrees of freedom the specific decay pattern of some resonances could be used. Especially the η- and K^+-decay are well suited for this purpose, since there are only a few nuclear resonances that couple strongly to these channels (see also Fig. 2).

The very last point touches upon a central issue in the study of intermediate energy excitations of complex hadronic systems: The considerations of the preceeding paragraph only make sense if the notion of spatially isolated hadrons, here baryons interacting via the coupling to meson fields remains actually as meaningful as it is supposed to be for the ground states, i.e. the nuclei. This however need not be the case: At higher excitation energies it might be more appropriate (or at least more efficient) to drop the discrimination of internal and external degrees of freedom and to describe the excitations exclusively in terms of quark degrees of freedom as proposed by Petry et al. [5]. It should be pointed out, that this is of course intimately related to the nature of confinement and the fundamental question whether baryons essentially preserve their identity in a strongly interacting enviroment (local confinement) or that (at least a partial) deconfinement occurs. From the consi-

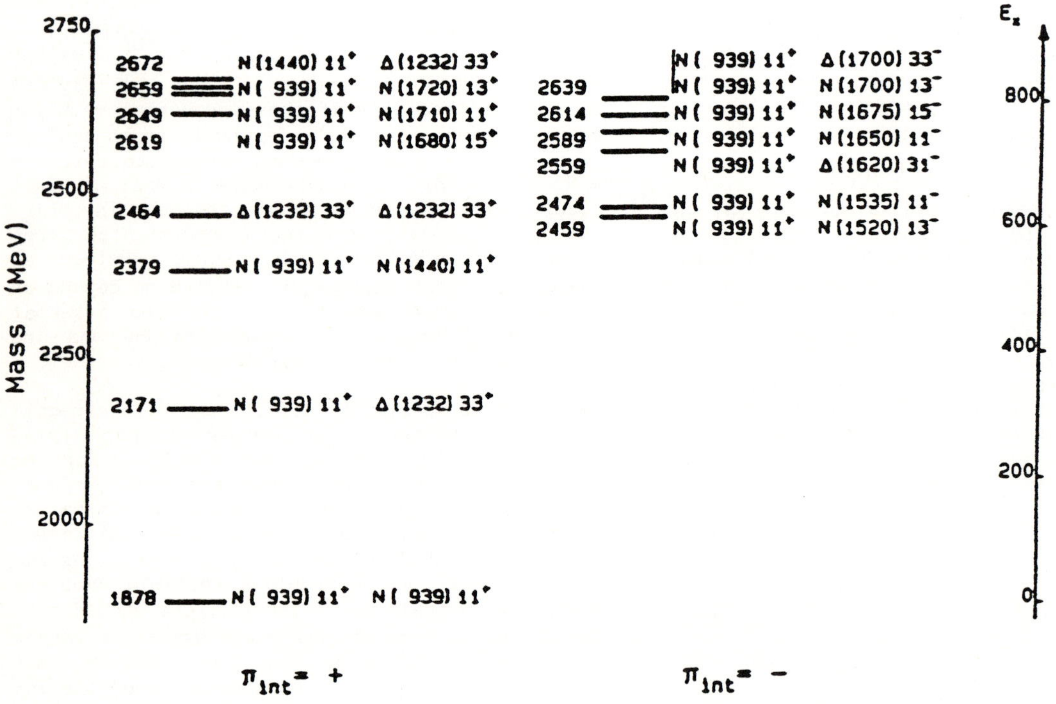

Fig.10. Unperturbed masses of nonstrange baryon-baryon combinations

derations above it might be inferred that this could be not only energy dependent but also state dependent, i.e. depending on the specific baryon-baryon configuration.

4. Summary and Concluding Remarks

In the present contribution it has been pointed out, that a precise and complete investigation of the excitation spectra of simple and complex hadrons via exclusive scattering experiments is an absolute prerequisite for relating the hadronic dynamics to the underlying quark dynamics. In a multihadron picture the role of the interface in this connection is thought to be played by the hadronic vertex matrix elements, that reflect the quark dynamics not only by the mechanism by which additional quark-antiquark pairs are created ("hadronisation at low energies") but also through the hadronic internal quark wave functions, which in turn are again determined by confinement and the residual quark-quark interactions. Of course, if at higher energies quark wave functions do stronger overlap it is more efficient and also physically more relevant, to relate the observables directly to the dynamics in a multiquark treatment.

Whatever specific dynamics might be relevant in a given excitation region will certainly have consequences with respect to the excitation spectrum, in particular of multihadronic systems. It is therefore necessary to be able to perform detailed investigations of this excitation spectrum in a routine fashion. Here the availability of high intensity meson beams like at **KAON** is of crucial importance. High precision proton (and light particle) accelerators like **COSY** are likewise very suited to explore the very same physical issues with complementary tools. Of course one should not forget the information to be gained in electromagnetically induced reactions at stretcher rings such as **ELSA** in Bonn and the **CEBAF** facility presently under construction. In any case a sound infrastructure of various facilities of varying size and at various locations that also can function as a "home base" for university research groups is of essential importance for a successful and efficient use of such a larger scale facility like **KAON** at **TRIUMF**.

References

1. Cooler Synchrotron COSY, KFA Jülich, Jül-Spez (1986) ISSN 034-7639
2. F. Plouin: *"Threshold Productions: η and π^o Beams at SATURNE"* in: Production and Decay of light Mesons, Ed. P. Fleury, pp 114-124, Singapore, New Jersey, London, Hong Kong: World Scientific 1988
3. B. Bock, et al.: Nucl. Phys. **A459** (1986) 573
4. M.G. Huber: Nucl. Phys. **B279** (1987) 249
5. H. Hofestädt, S. Merk, H.R. Petry: Z. Phys. A - Atomic Nuclei **326** (1987) 391
6. O. Dumbrajs, J. Fröhlich, U. Klein, H.G. Schlaile: Phys. Rev. **C29** (1984) 581
7. U. Muschelknautz, Thesis Diploma, Universität Bonn, 1989

Z. Phys. C – Particles and Fields 46, S45–50 (1990)

Zeitschrift für Physik C Particles and Fields

The creation of high energy densities with antimatter beams *

W. R. Gibbs [1], J. W. Kruk [2]

[1] Theoretical Division, Los Alamos National Laboratory, Los Alamos, NM 87545, USA
[2] T. W. Bonner Nuclear Laboratory, Physics Department, Rice University, Houston, TX 77251, USA

Abstract

The use of antiprotons (and antideuterons) for the study of the behavior of nuclear matter at high energy density is considered. It is shown that high temperatures and high energy densities can be achieved for small volumes. Also investigated is the strangeness production in antimatter annihilation. It is found that the high rate of Λ-production seen in a recent experiment is easily understood. The Λ and K_S rapidity distributions are also reproduced by the model considered.

1. Introduction

The use of the antiproton as a nuclear probe provides many opportunities for the study of hadronic physics. One of its outstanding features is the ability to deposit several GeV of energy in a small nuclear volume, thus, creating a region of high energy density in the nucleus [1,2].

While such conditions are of great interest to study the nuclear equation of state, important in themselves for the investigation of stellar structure, they have recently received special attention because of the possibility of a phase change of hadronic matter [3]. Viewed simply, we may expect a (possibly sudden) change in the behavior of a many-body system when the relevant degrees of freedom change. In the nucleus we are accustomed to the normal degrees of freedom being the nucleons, while for the nucleon itself we expect the description to be in terms of quarks and gluons. Note that the meson degrees of freedom have been bypassed in this picture, possibly a serious error.

We can anticipate that the relevant degrees of freedom will change when the energy density of the nucleus becomes equal to that of a single nucleon. In this case the quarks should be able to move freely within the heated region. This simple estimate gives 0.35 GeV/fm^3 for a nucleon radius of 0.86 fm and 1.8 GeV/fm^3 for a radius of 0.5 fm. Recall that the energy density in normal nuclear matter is around 0.16 GeV/fm^3. The scenario just presented is certainly too naive but it sets the scale for the onset of possible new physics.

The excitation of the nucleus is often discussed in terms of temperature, meaning simply the random kinetic energies of the nucleons involved and not implying necessarily that an equilibrium has been reached. The numbers most often quoted for the temperature for such a phase change are around 180 to 200 MeV.

This idea raises several important questions. How might one form such a high energy density region? How do we know that we have indeed created it? What is the signal for the possible phase transition? The second and third questions are related and involve a detailed understanding of hadronic mechanisms in the nucleus and of the properties of the particles created. At this point in time, a much more complete knowledge of the mechanism of production of hadrons in the nucleus, and their evolution in the nuclear medium, is necessary before any high energy density experiment can be interpreted. For the purposes of this talk, let us focus our attention on the first and third questions.

*Los Alamos National Laboratory is operated by the University of California for the U.S. Department of Energy under contract number W-7405-ENG-36. Accordingly, the U.S. Government retains a nonexclusive, royalty-free license to publish or reproduce the published form of this contribution, or allow others to do so, for U.S. Government purposes.

2. Obtaining High Energy Densities

The deposition of a substantial amount of energy in the nucleus is not an easy task. If one tries to do it with high energy proton beams it is seen that the proton tends to make very few collisions with the nucleons in the nucleus. Since the angular distribution for nucleon-nucleon elastic scattering is (to a fair approximation) a function of momentum transfer only the probability distribution of energy transfer is independent of energy, at least for low and moderate bombarding energies. Hence, there is no gain in energy transfer until the energy regime in which most of the collisions are highly inelastic is reached. At these high energies the cascade process (involving the mesons produced) seems to be inhibited. By the use of heavy ions one can mitigate this problem by hitting the nucleus with many nucleons at once. Note the importance of transferring the kinetic energy of the projectile to many nucleons in the nuclear target. If all of the energy is transferred to a single nucleon then the presence of the remainder of the nucleus is useless.

Besides heavy ions, another possible way to create high energy density is to use energetic antiprotons (or antideuterons). The argument is a good deal more subtle than just saying that antiproton annihilation deposits 2 GeV of energy in a relatively small volume. For example, if one considers annihilation of stopped antiprotons only a small fraction of the 2 GeV available is given to the nucleus [4]. Because most of the annihilations take place on the extreme nuclear surface [5] and since the annihilation products are distributed isotropically most of them miss the nucleus completely.

Increasing the antiproton beam energy improves the situation dramatically. First of all, because of the center of mass motion, the annihilation products are pushed forward in the laboratory so that, by 4 – 5 GeV/c $\overline{p}$ momentum essentially all of the mesons enter the nucleus. Thus a significant fraction of the energy is available in the form of mesons. Of course, these mesons (mostly pions) are also carrying more energy since they are the recipients of the kinetic energy of the antiproton as well. Pions do not behave as the nucleons discussed above, but are often absorbed, thus transferring all of their energy (including their mass) to two or more nucleons. The multiplicity of pions in the annihilation process also increases so that it is not unusual if a 20 GeV/c antideuteron annihilates on two nucleons to find 18 pions. Since the average number of nucleons involved in a pion absorption is of the order of 4, one can imagine 72 nucleons being excited in one antideuteron annihilation. Of course all of these nucleons are not independent (i.e. more than one pion can transfer energy to a single nucleon) so that the actual number of nucleons is of the order of half of that, still a very substantial number.

To understand the heating effect of this partition of the initial energy among pions let us suppose that the antiproton momentum is shared among n nucleons in the final state. The problem of following the energy-momentum conservation is very similar to that of coupling a single moving freight car to n freight cars stationary on a track (at least non-relativistically it is). Using the conservation of momentum we find that only $1/n$ of the original kinetic energy remains as directed motion in the final state. The rest of the initial kinetic energy is converted into heat and sound. Thus, antiproton annihilation is a very efficient way to create a high energy density and raise the temperature of the nucleus by converting kinetic energy into thermal degrees of freedom.

3. The Classical Modelling Code

An intra-nuclear cascade code was developed to make quantitative estimates of these effects and since some of the results, as well as a description of the code, have been reported elsewhere [6], I will only mention briefly its workings and the main conclusions. The code has many points of similarity with those developed by Cahay, Cugnon and Vandermeulen [7].

The nuclear target is modelled by a system of A nucleons propagating in Saxon-Woods potential with classical motion. Isotropic (in the CM) NN collisions are governed by an approach distance corresponding to a classical circular cross section of 40 mb. The annihilation and scattering of the antiproton is determined by the antiproton-nucleon cross sections. The products of $\overline{p}$ interactions are taken to be π's, K's, Λ's and $\overline{\Lambda}$'s according to the experimentally measured fractions. The antiproton channels included were:

$$\begin{aligned}
\overline{p}N &\rightarrow \overline{p}N \\
\overline{p}N &\rightarrow \overline{p}N\pi \\
\overline{p}N &\rightarrow \overline{\Lambda}N\pi \\
\overline{p}N &\rightarrow \overline{\Lambda}\,\overline{K}N \\
\overline{p}N &\rightarrow \overline{N}\Lambda K \\
\overline{p}N &\rightarrow \pi's \\
\overline{p}N &\rightarrow \overline{K}K\pi's
\end{aligned}$$

The mesons thus produced are then propagated within the nucleus with no central potential (in contrast to the case for the nucleons) but with the reactions shown below occurring.

$$\begin{array}{llllll}
\pi N \rightarrow & \pi N & \pi N \rightarrow & \Phi N & \overline{K}N \rightarrow & \overline{K}N \\
\pi N \rightarrow & \pi N^* & \pi N \rightarrow & K\Lambda & \overline{K}N \rightarrow & \pi\Lambda \\
\pi N \rightarrow & \pi\Delta & \pi N \rightarrow & K\Sigma & \overline{K}N \rightarrow & \pi\Sigma \\
\pi N \rightarrow & \pi\pi\pi N & \pi N \rightarrow & K^*\Lambda & KN \rightarrow & KN \\
\pi N \rightarrow & \omega N & \pi N \rightarrow & K^*\Sigma & & \\
\pi N \rightarrow & \eta N & & & &
\end{array}$$

The pion induced reactions are especially important since they deposit most of the energy and can produce strangeness as well. The calculation of their reac-

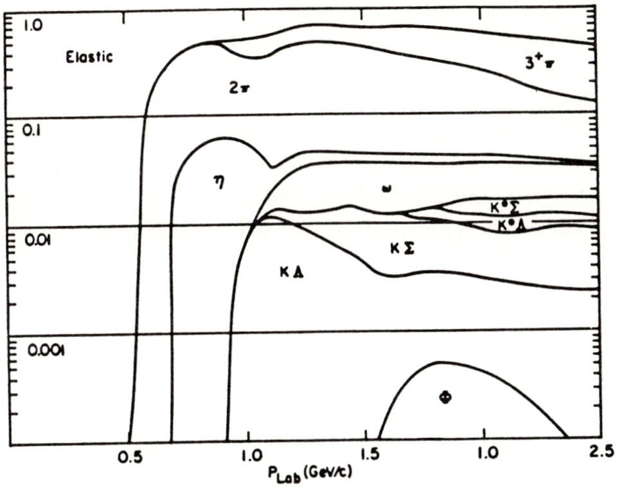

Figure 1: Branching ratios of the pion-nucleon interaction.

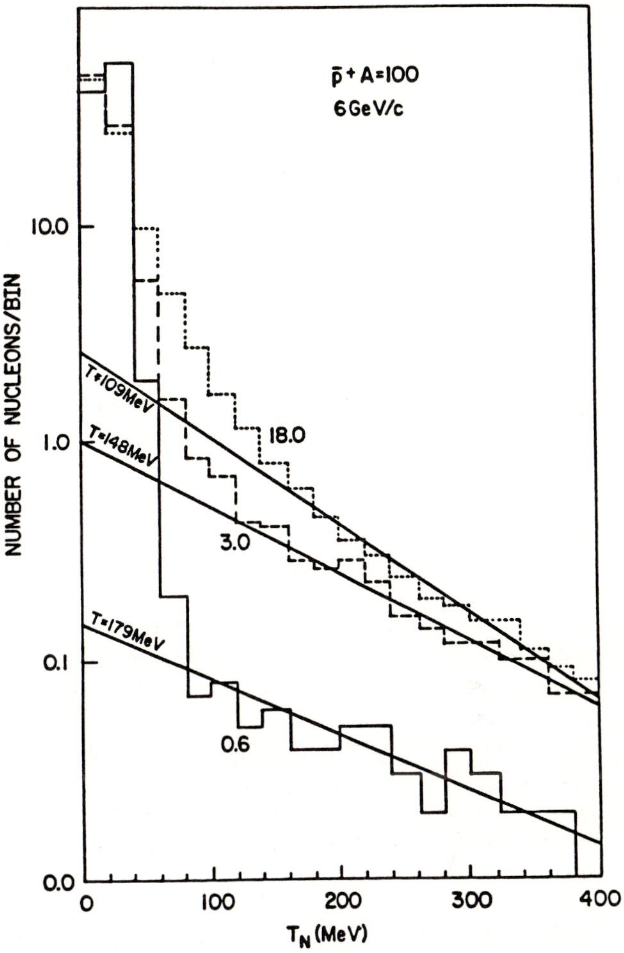

Figure 2: Distribution of the kinetic energies of nucleons after antiproton annihilation in nuclei. Note the "cold" and "hot" distributions.

tions is implemented by first deciding if there is to be a collision based on each pion-nucleon distance compared to the pion-nucleon total cross section in the laboratory. The momentum used for the calculation of the cross section is the momentum that the π-nucleon system would have if the nucleon were at rest, i.e., the Fermi motion of the nucleon is taken into account as if it were on shell. Once it is determined that a π-nucleon collision will take place, a number of branches are possible. These are chosen according to the cumulative probabilities given in figure 1. The logarithmic graph is used so that the small particle production cross sections can be seen. The $(\pi, 2\pi)$ and $(\pi, 3\pi)$ reactions are important for an estimate of energy deposition. They also shadow the strangeness producing reactions. These probabilities were calculated from data taken from the CERN-HERA [8] reports.

The kaon reactions are very important for the transfer of strangeness especially the $\overline{K}$ conversion to Λ's. The $\overline{\Lambda}$ interactions were also included in the following channels:

$$\begin{aligned} \overline{\Lambda}N &\rightarrow \overline{\Lambda}N \\ \overline{\Lambda}N &\rightarrow \overline{\Lambda}N\pi \\ \overline{\Lambda}N &\rightarrow K\pi's \\ \overline{\Lambda}N &\rightarrow K\overline{K}K\pi's \end{aligned}$$

4. Results for Energy Densities

One of the first things one can do, after running a series of cascades, is to look at the distribution of kinetic energies of the nucleons. It is expected that there will be two distinct populations; a cold one corresponding to the nucleons which have not been affected by the annihilation process, and a hot one for the nucleons on which pions have been scattered or absorbed. Indeed, by plotting a graph of the number of nucleons vs. their kinetic energy, the two distributions are easily seen (figure 2). Since the "hot" component is roughly exponential in shape it is natural to determine a "temperature" from it. I put the word in quotation marks since it is only a slope parameter that one extracts, but it does indicate the distribution of kinetic energies of the nucleons in the hot region.

One can follow this temperature to see how long the hot gas is confined as shown in figure 3. Also shown there are estimates by means of a hydrodynamic calculation [1]. It is clear that a high temperature remains for several fm/c, a result that compares favorably with the times in heavy ion collisions.

Passing from this simple measure of excitation, the temperature, to something more precise, the energy density, we see that there are still some ambiguities. If we discuss the energy transfer to nucleons alone, then things are well defined in terms of what energy density is being discussed. However, the deposition of energy among

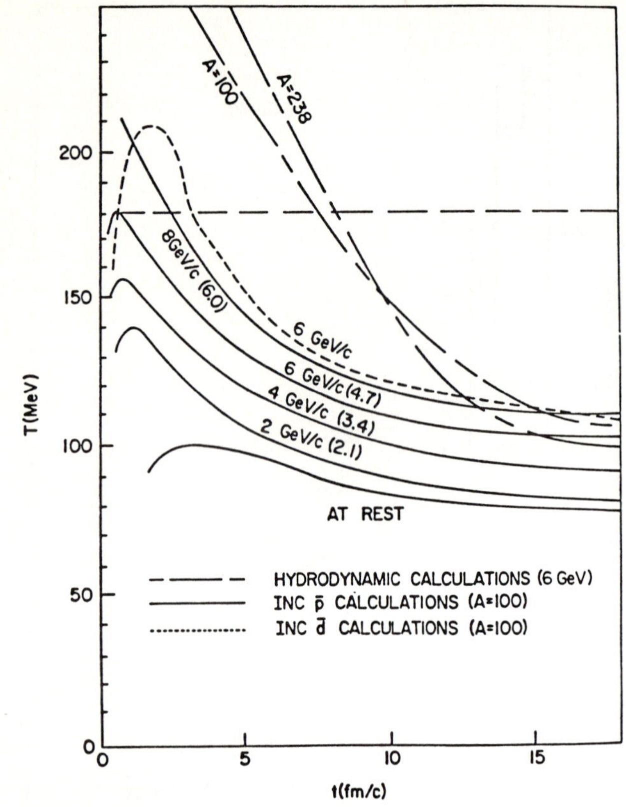

Figure 3: Evolution of the temperature of the hot component as a function of time. One can see the cooling which comes about as the hot part of the gas mixes with the colder surrounding nucleons.

nucleons is quite variable in position in the nucleus from event to event and by following along the axis of annihilation we get a rather pessimistic view of the energy density, of the order of 1 GeV/fm^3.

If we consider the energy density of all particles (including the mesons from annihilation) in a given volume we obtain the result shown in figure 4. This gives a much larger number, of course. Note that it is the energy density of invariant mass which is shown, i.e. the kinetic energy of forward motion has been removed.

5. Strangeness

One of the possible signals that we are dealing with different degrees of freedom is the amount of strangeness created in an annihilation event. We are accustomed to the suppressed production of strange particles due partly to the higher mass of the strange quark and the difficulty of concentrating enough energy to produce $s\bar{s}$ pairs in hadronic interactions. One may well expect the fraction of strange quark production in a quark-gluon plasma to be different from that in a hadronic "soup". Rafelski [9] has discussed the two scenarios. Thus a sudden change in strangeness production as a function of temperature or energy density may be interpreted as a signal of a modification of the structure of the hadronic system.

Strangeness production is normally very small so that only a few percent of strange and antistrange quarks are produced in hadronic reactions. We note that for rapidly occurring reactions, within the nuclear time scale, the strange quarks are essentially stable so we shall

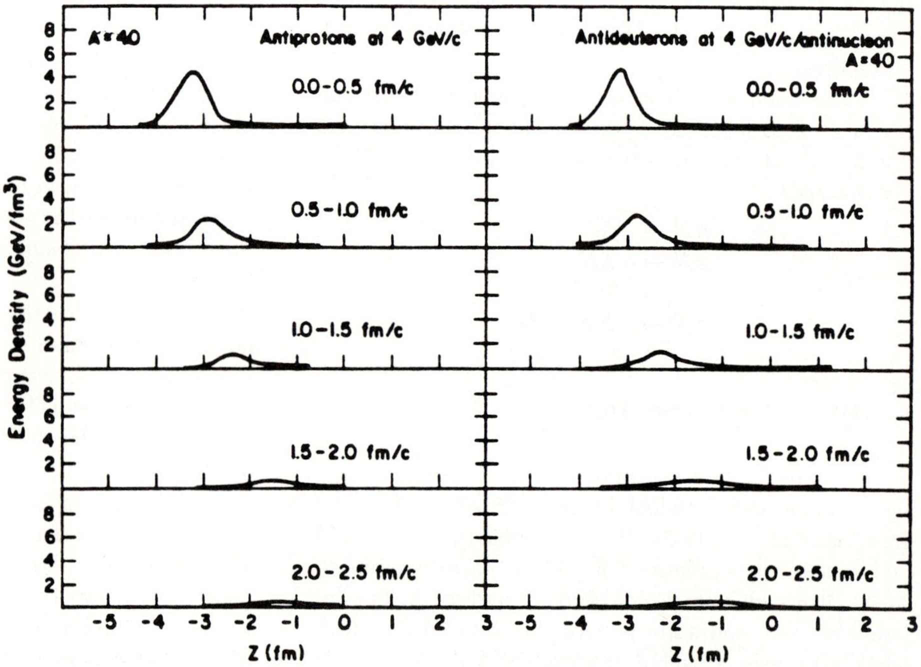

Figure 4: Energy densities along the central axis of annihilation for all particles. Only the density of invariant mass is plotted so as not to include the centre of mass motion.

treat strangeness as a conserved quantity. For this reason there is only pair (strange-antistrange) production. As discussed by Rafelski [9], in a quark-gluon plasma all quarks will be on the same footing, so the number of strange quarks would be equal to the number of light (up and down) quarks. The fraction of strange quarks obtained in this manner is considerably larger than found in ordinary hadronic reactions. He also pointed out that the hot (color singlet) soup will produce an appreciable amount of strangeness by means of the reactions because there are many (non-strange) mesons and, hence, many such reactions take place in the nucleus. This component of strangeness production must be understood, since it is the background for any signal observed indicating the occurrence of a phase transition.

Before entering into the details of how this component can be estimated, we need to discuss the measurement of the strangeness signal itself. The normal convenient experimental variables are the short lived kaons (K_S's), the Λ's and the $\overline{\Lambda}$'s. While the measurement of a Λ gives evidence of the presence of a strange quark and that of an $\overline{\Lambda}$ an antistrange quark, the observation of a K_S only shows, because its observation is by means of a weak decay not conserving strangeness, that there was either a strange or an antistrange quark present. We note that once the strange-antistrange quark pair is produced in the nucleus that both members can be transferred more or less freely between hadrons. For example, $\overline{\Lambda}$'s will annihilate in the nuclear medium, with high probability, leaving their antistrange quark in a K^+ or K^0 meson. The $\overline{K}$ mesons produced in the $\overline{p}N$ annihilation (or otherwise) have a very strong interaction with nucleons and tend to convert, also with high probability, to $\overline{\Lambda}$'s by the $\overline{K}N \rightarrow \pi\Lambda$ reaction. Thus, conclusions drawn from the observation of a single type of strange particle can be very misleading.

As we just remarked we do not have, because of the ambiguity of the K_S detection, a direct measure of the strange (or antistrange) quarks contained in mesons. However, if we use the additional constraint that the number of strange quarks is equal to the number of antistrange quarks, we can obtain a measure of the strangeness produced. We simply compute the number of strange quarks plus the number of antistrange quarks and divide by two. The number of hyperons plus the number of antihyperons gives us the part of this sum contained in the baryonic sector. For the K_S's we must take into account that they signal the existence of either a strange or an antistrange quark. In addition, we must also include unobserved events inferred from these observed particles. First of all, the K_S's represent only one half of the neutral kaons produced, the other half being long lived and hence not observed in the experiment. Since it is known that half of the neutral kaons decay by the 2π mode we can simply multiply by two to obtain the total number of neutral kaons. In many experiments the charged kaons are not observed. If this is the case then we expect, by isospin symmetry, that there are as many charged as neutral kaons. (We neglect the small correction due to the possibility that the nucleus may not have an equal number of neutrons and protons.) Thus, included in the measure of the total number of strange plus antistrange quarks should be a factor of 4 times the number of K_S observed. Thus, the total number of strange (or antistrange) quarks produced in a reaction is:

$$S = (4N_S + N_Y + N_{\overline{Y}})/2$$

where N_S is the number of K_S, N_Y is the number of hyperons and $N_{\overline{Y}}$ is the number of antihyperons observed in each case per annihilation. For a signal of the phase transition one could plot S as a function of temperature or energy density or even simply energy deposition to see if there is an abrupt increase at some given critical value.

Let us take as an example the experiment of Miyano et al. [10] which comes the closest to providing the conditions necessary to produce the putative phase transition. In this experiment a large enhancement in the number of Λ was observed but a large decrease in the number of $\overline{\Lambda}$ and a sizeable decrease in the number of K_S was also seen, compared to the $\overline{p}N$ annihilation in free space. Computing the amount of strangeness, S, in the $\overline{p}p$ system at 4 GeV/c gives:

$$S_0 = (4 \cdot 0.079 + 0.021 + 0.021)/2 = 0.179.$$

Calculating the same quantity for the Miyano experiment we find:

$$S = (4 \cdot 0.055 + 0.129 + 0.003)/2 = 0.176.$$

Since the code indicates that associated production reactions occurring during the cascade process contribute about 0.018 to S the experiment observes less strangeness than expected, but the discrepancy is within the uncertainties. We note that for a signal of excess strangeness production it is the amount above the value of S in $\overline{p}p$ collisions which matters. Thus one should measure $S - S_0$.

By running the classical nuclear modelling code described above we obtain results for the various strangeness channels in good agreement with those seen by Miyano et al. and in agreement with the simple formula just derived. This last is expected since the formula is general. The real question to be answered by the code is "how much strangeness is produced in the πN reactions?" This is the amount that the nuclear production of S should exceed the $\overline{p}p$ production. Table 1 shows the strangeness balance from intra-nuclear reactions. Note that, since all channels have not been included in the code, there are slight differences in input from the general formula. For example, the total percentage of Λ production is 2.1 while the two largest mechanisms (the only two included in the code) add up to 1.5.

Table 1: Frequency of occurrence of various channels, expressed as a percentage of the annihilation rate.

			$\Lambda(\%)$	$K_S(\%)$	$\overline{\Lambda}(\%)$
$\overline{p}N$	$\rightarrow$	$\Lambda\overline{\Lambda}$	0.99	0.00	0.99
$\overline{p}N$	$\rightarrow$	$\Lambda K\overline{N}$	0.52	0.13	0.00
$\overline{p}N$	$\rightarrow$	$\overline{\Lambda}\,\overline{K}N$	0.00	0.14	0.55
$\overline{p}N$	$\rightarrow$	$K\overline{K}\pi's$	0.00	7.02	0.00
$\overline{\Lambda}N$	$\rightarrow$	$K(K\overline{K})\pi's$	0.00	0.33	−1.07
$\overline{K}N$	$\rightarrow$	$\pi\Lambda, \pi\Sigma$	8.48	−2.70	0.00
πN	$\rightarrow$	$K\Lambda$ or $K\Sigma$	1.57	0.51	0.00
		sum	11.56	5.43	0.47
		Exp.	12.88 ±0.80	5.47 ±0.40	0.25 ±0.13

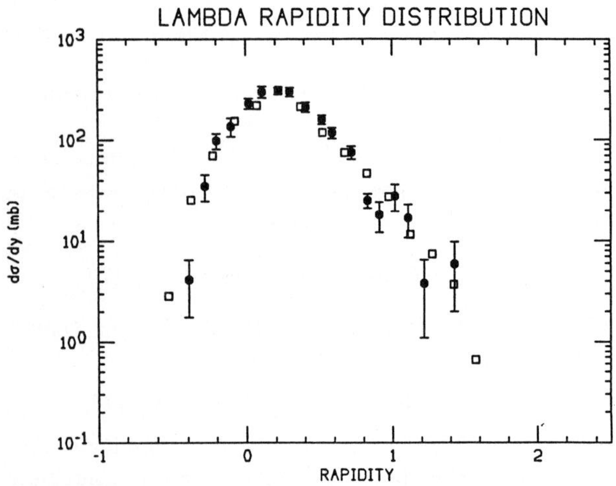

Figure 5: Rapidity distribution of Λ's produced in an antiproton annihilation in Tantalum at 4 GeV/c.

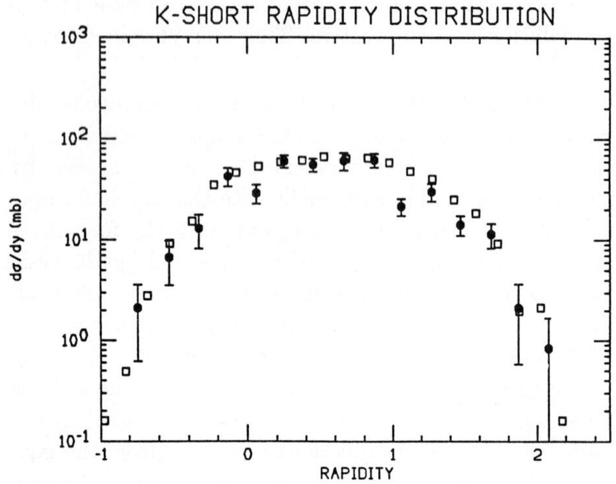

Figure 6: Rapidity distribution of K_S's produced in an antiproton annihilation in Tantalum at 4 GeV/c.

The code has also been used to calculate rapidity distributions. The Λ distribution is shown in figure 5 and the K_S distribution in figure 6. Both distributions are in excellent agreement with the data. Thus, it seems that no exotic effects are needed to explain the data [11], and that there is evidence against the formation of a quark-gluon plasma in the reaction considered.

6. Conclusions

Analysis of the strangeness produced in the recent antiproton-nucleus experiment by Miyano et al. shows it to be completely consistent with no enhanced strangeness production both from the detailed model calculation and from general strangeness conservation considerations. The rapidity distributions are in excellent agreement with the data as well even though they depend on more detailed assumptions of the model.

This work was supported in part by the U. S. Department of Energy.

References

[1] D. Strottman and W. R. Gibbs, Phys. Lett. 149B, (1984) 288

[2] J. Rafelski, Phys. Lett. 91B, (1980) 281

[3] J. Kogut, M. Stone, H. W. Wyld, W. R. Gibbs, J. Shigemitsu, S. H. Shenker and D. K. Sinclair, Phys. Rev. Lett. 50, (1983) 393; J. B. Kogut, E. V. E. Kovacs and D. K. Sinclair, Nucl. Phys. B290, (1987) 431

[4] Armstrong et al. Z. für Phys. A332, (1989) 467

[5] W. B. Kaufmann and H. Pilkuhn, Phys. Lett 62B, (1976) 165

[6] W. R. Gibbs and D. Strottman, Proc. of the "International Conference on Antinucleon- and Nucleon-Nucleon Interactions", Telluride, CO March 18-21,1985. Plenum Press, New York; W. R. Gibbs, Proc. of the "Fermilab Workshop on Low-Energy Antiproton Physics"; W. R. Gibbs, Proc. of the "Intersections Between Particle and Nuclear Physics" Lake Louise, Canada 1986 AIP conference proceedings 150, p. 505.

[7] M. Cahay, J. Cugnon and J. Vandermeulen, Nucl. Phys. A393, (1983) 237

[8] CERN-HERA Reports 83-01,83-02,84-01

[9] J. Rafelski, Nucl. Phys. A418, (1984) 215c

[10] Miyano et al., Phys. Rev. Lett. 53, (1984) 1725; Phys. Rev. C38, (1988) 2788

[11] J. Rafelski, Phys. Lett. 207B, (1988) 371

Z. Phys. C – Particles and Fields 46, S51–56 (1990)

Zeitschrift für Physik C Particles and Fields

The future of Σ – hypernuclei

S. Paul

Max-Planck-Insitut für Kernphysik, D-6900 Heidelberg, Fed. Rep. of Germany

Abstract: A new high rate Σ – hypernuclei formation experiment is suggested for the new KAON facility at TRIUMF. Owing to the experience of past experiments, the set up aims at a (K^-,π) in flight experiment. An implementation of the targets into a TPC is proposed which should allow to detect short lived secondaries from the decay of Σ – hypernuclei. The use of radiation hard high resolution tracking devices and hardwired processors should make possible a production of 50000 recoilless produced hypernuclei/day.

1. Introduction

Most of what we know about the low energy Λ – Nucleus interaction comes from experiments studying the formation and decay of Λ – hypernuclei. Information extracted from these experiments was used to determine the Λ – Nucleus potential and the very weak spin – orbit coupling. Recently, the observation of narrow Λ – hypernuclear groundstates in heavy nuclei showed us the validity of the nuclear shell model even for deeply bound states. On the other hand, the knowledge of the Σ – Nucleus interaction is much less profound. Only a handful of data exists for Σ – hypernuclei taken on light targets only, which by themselves are statistically not overwhelming. They were used to extract the Σ – Nucleus potential depth ($V_C(\Sigma) = 20$ MeV), but no unique information has yet been obtained concerning the spin – orbit interaction or the pure existence of the observed narrow Σ – hypernuclear states. The KAON facility at TRIUMF, however, offers a unique possibility to improve this lack of understanding.

Before I will describe a possible set – up for one of the future high intensity kaon beamlines, I will first summarize the status of Σ – hypernuclei and shortly describe the two basic experimental approaches to the problem.

2. The Status of Σ – hypernuclei

2.1 In Flight Experiments

Narrow Σ – hypernuclear states were first observed at CERN 1978 with a 700 MeV/c Kaon beam using the (K^-,π^-) reaction in flight on a beryllium target [1]. Fig. 1 shows the missing mass spectra for this reaction in units of the hyperon binding energy. For fig.

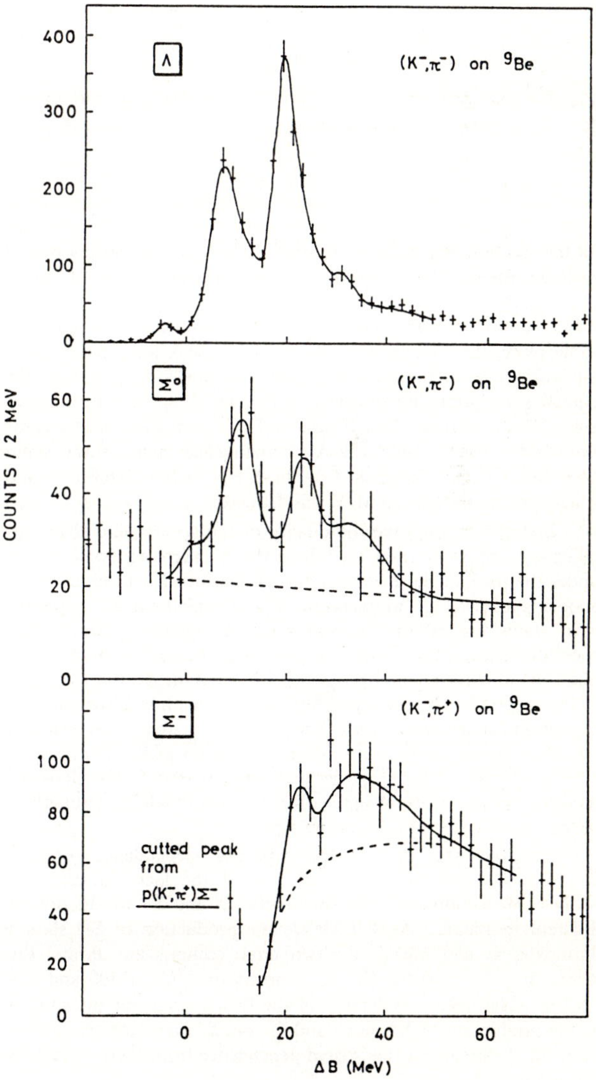

Fig. 1 : The K^-,π reaction on beryllium producing a) Λ, b) Σ^0, c) Σ^- hypernuclei using a 700 MeV/c K^- beam at CERN. The spectra a plotted as a function of the hyperon binding energy.

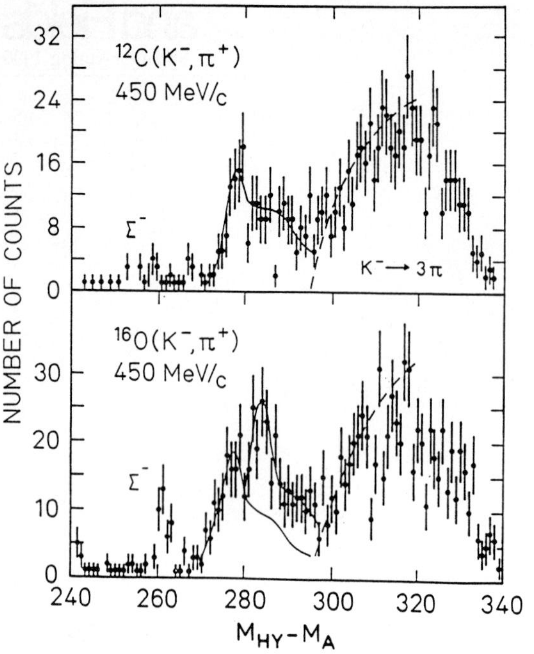

Fig. 2 : The (K^-,π^+) reaction on ^{12}C and ^{16}O with 450 MeV/c K^-. The spectra are shown in units of the missing mass $M_{Hyp} - M_A$

1a the reaction was assumed to produce a Λ on the loosely bound neutron, for fig. 1b a Σ^0 production was assumed. Fig. 1c shows the equivalent reaction (K^-,π^+) for the production of Σ^- on a deeply bound proton. The peak positions are all at roughly the same place, the shift between the Λ and Σ peak indicates a different interaction of these hyperon with one nucleon. Although the statistics are poor, the similarities of Λ and Σ spectra are impressive. The existence of such narrow $\Sigma-$hypernuclear states came unexpected since, unlike the $\Lambda-$hypernuclear states, these states were thought to be quenched by the $\Sigma + N \to \Lambda + N$ reaction and a width of more than 20 MeV was deduced.

In order to improve the quality of the spectra and their significance one soon tried to reduce the background (well visible under the peak of the $\Sigma-$state) due to quasifree $\Sigma-$production by choosing a lower K^- momentum. Fig. 2 depicts the Σ^- hypernuclear states formed in ^{16}O and ^{12}C. As expected, the oxygen spectrum shows two peaks corresponding to the production on nucleons of the $p_{1/2}$ and $p_{3/2}$ shell [2]. Again, it's the similarity of the spectra to the corresponding Λ ones which is convincing although the statistics in each of the spectra is scarce. These spectra served to determine the spin−orbit interaction of the $\Sigma-$Nucleus system. Unlike the $\Lambda-$Nucleus interaction, the $\Sigma-$Nucleus interaction creates a strong spin−orbit splitting of similar strength as in the N−N case but of opposite sign.

In order to deduce the ispospin dependence of the $\Sigma-$Nucleus interaction one made use of the fact that the free K^-N cross sections are different for the Σ^0 and Σ^+ production at different momenta. At 400 MeV/c the production of Σ^+ should dominate, at 450 MeV/c the two cross sections are almost the same. Fig. 3 shows the (K^-,π^-) spectra on ^{12}C for 400 and 450 MeV/c. The distinct difference of the two spectra was interpreted as the production of Σ^+ alone and Σ^0 and Σ^+, respectively. However, the deduction of the isospin dependence from these spectra is not unique.

At BNL another group investigated Σ^-−hypernuclei at higher kaon momenta. Using a 700 MeV/c K^- beam they measured the (K^-,π^+) reaction on ^{12}O, ^{12}C, ^6Li and ^7Li [3]. Although two states similar to the ones for Λ were found for ^6Li, no such peak could be obtained for the other targets. Due to the high momentum of the K^-, causing a large momentum mismatch between the original nucleon and the Σ created, the background due to quasi elastic Σ production was overwhelming. Only using special 'tagging' technics, trying to suppress the quasifree production, they were able to obtain a small signal in the case of the ^{12}C which, however, has not been interpreted by the authors as a narrow peak.

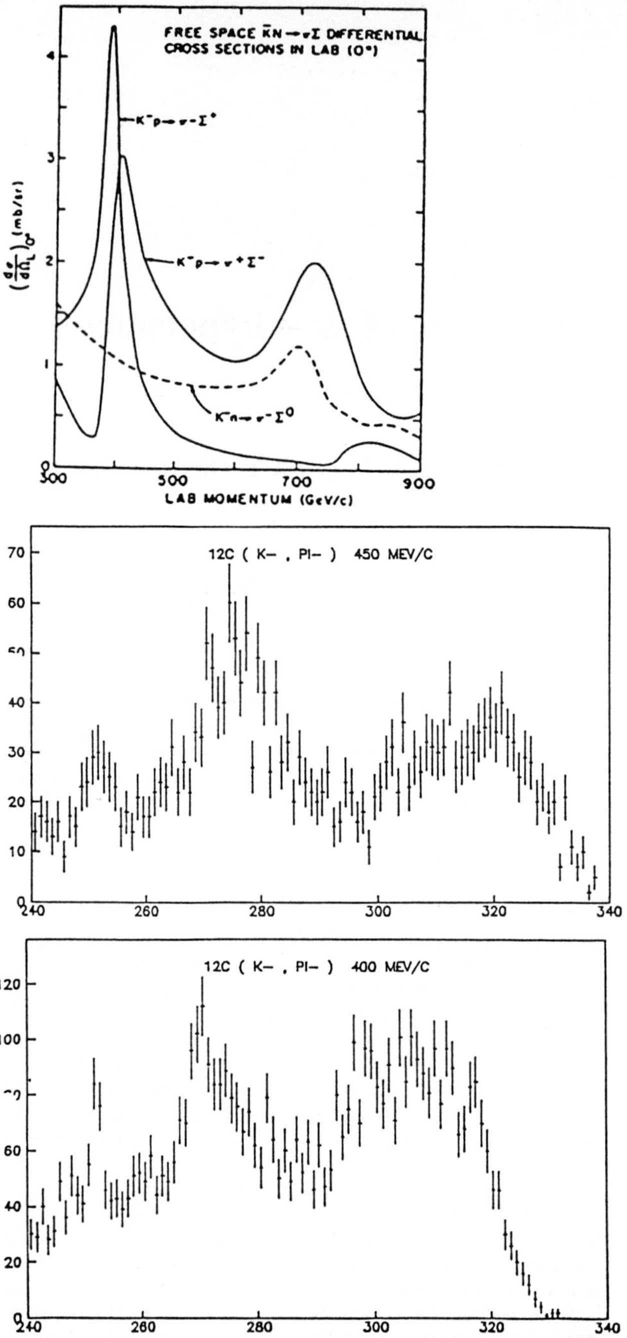

Fig. 3 : Study of the isospin dependence of the $\Sigma-$nucleus interaction. The free (K^-,π) cross−section is shown in a) as a function of the K^- momentum. b) and c) depict the resulting missing mass spectra for the (K^-,π^-) reaction on ^{12}C at 450 and 400 MeV/c.

2.2 Experiments with stopped K^-

Another approach to the study of Σ-hypernuclei is the use of stopped K^- as it was done at CERN for the production of Λ-hypernuclear states in 1973. After this method had been abandoned for more than 10 years, 1983 a group working at KEK employed that method producing spectra in the Σ-hypernuclear mass region in ^{12}C within a few day's only [4]. However, also here special 'tagging' technics were developped to enhance the signal over a dominating background from quasi-free production. Three peaks had been published (instead of only one as at CERN) and the interpretation was dubious. Nevertheless, the production rate seemed high enough that starting in 1986, a collaboration between the MPI-Heidelberg and the University of Tokyo investigated the production of hypernuclei stopping K^- in ^{12}C, ^{16}O, ^{6}Li and ^{40}Ca. No narrow Σ-hypernuclear peaks could be observed in inclusive spectra and only ^{12}C showed two clear Λ-hypernucler states.

In order to reduce background again, a special 'tagging' technique was developped leading to the observation of a peak in the Σ-hypernuclear mass spectrum at practically the same position as the one obeserved previously by the CERN group [5]. Fig. 4a shows the inclusive spectrum for the reaction (K^-,π^+) using a multilayer scintillator target. Asking for only the π leaving the interaction layer in forward direction reduced the background as is shown in fig. 4b. Further requiring a secondary interaction (either from a neutron or the decay of a Λ emitted in parallel with the π) within 2 to 4 layers upstream of the interaction point lead to the spectrum shown in fig. 5 a,b with two selection criteria on the energy deposited in the interaction. Again, the statistical significance is low. Fig. 6, however, summarizes the various experimental efforts searching for Σ^--hypernuclei in ^{12}C. A surprising agreement can be found in the position of the observed hypernulear states. As can clearly be seen, only the spectra obtained with a 450 MeV/c K^- beam show this peak with no background.

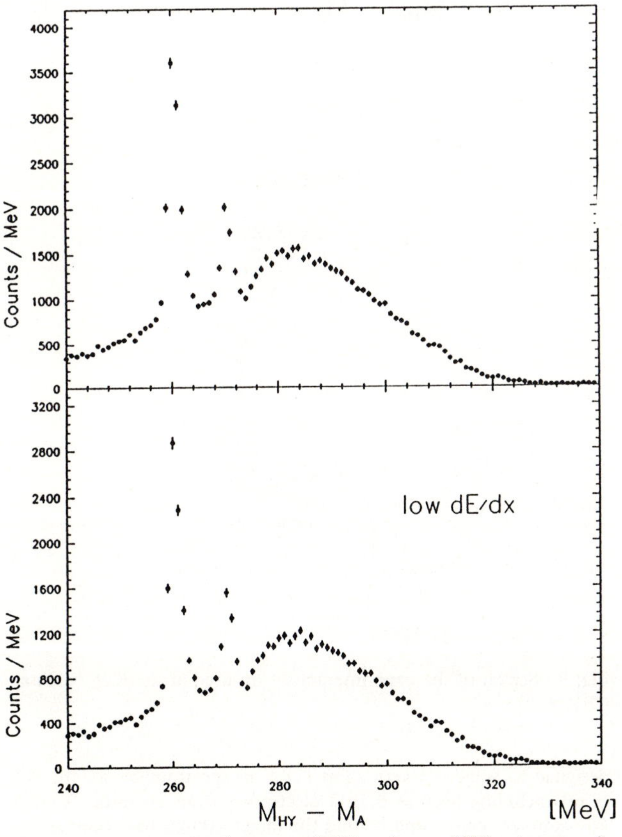

Fig. 4 : K^- stop experiment at KEK using a $(CH)_n$ target. a) shows the inclusive spectrum, b) asks for only one charged particle leaving the intercation point in forward direction.

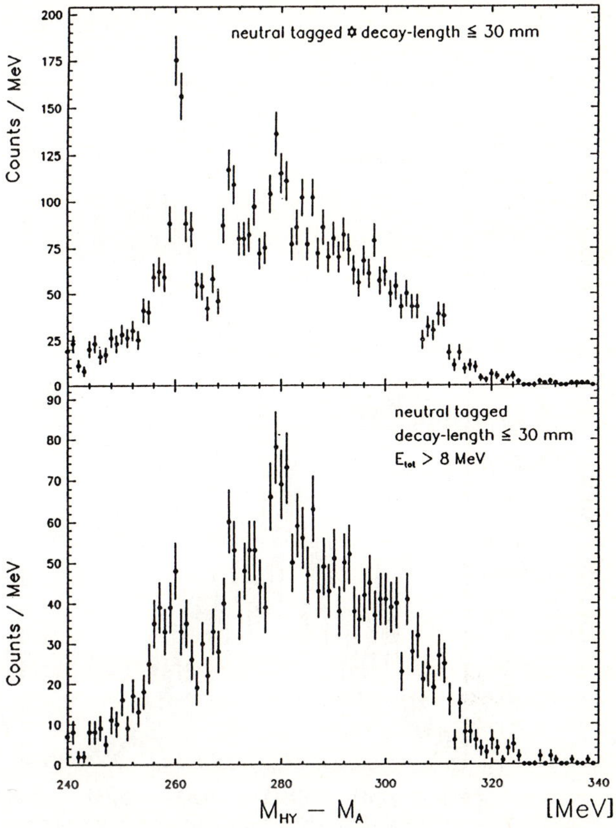

Fig. 5 : Same as fig. 4 but imposing a) lifetime cut on a secondary neutral particle b) leaving more than 8 MeV when interacting upstream of the primary interaction point.

2.3 Experimental methods

Before turning to the 'new' beamlines I first want to compare the two approaches and remind you of the set-up used previously at CERN and KEK. As stated already before, the biggest source of 'background' for the population of substitutional hypernuclear states is the quasi free Σ production. The production probability of these two processes is strongly governed by the recoil momentum transferred to the hyperon produced in the nucleus. If the recoil momentum is low (smaller than the Fermi momentum of the nucleons) the overlap of the wave functions of nucleon and hyperon is large. An increase of the recoil momentum increases the quasi free production. However, the latter process is necessary to gain access to transitions with $\Delta l > 0$ (e.g. population of the hypernuclear ground state). In general, the recoil is fixed by the reaction kinematics and for any given K momentum is minimal if the π is detected in forward direction. Fig. 7 shows the recoil momentum as a function of the momentum of the incoming K^- for the π being detected under zero angle. It is obvious that a K^- beam momentum of 350 MeV/c would be optimal. However, as can be seen from fig. 6, already at 450 MeV/c the quasi free process becomes very important. This explains the large background to be fought at in the case of stopped K^- or at 700 MeV/c.

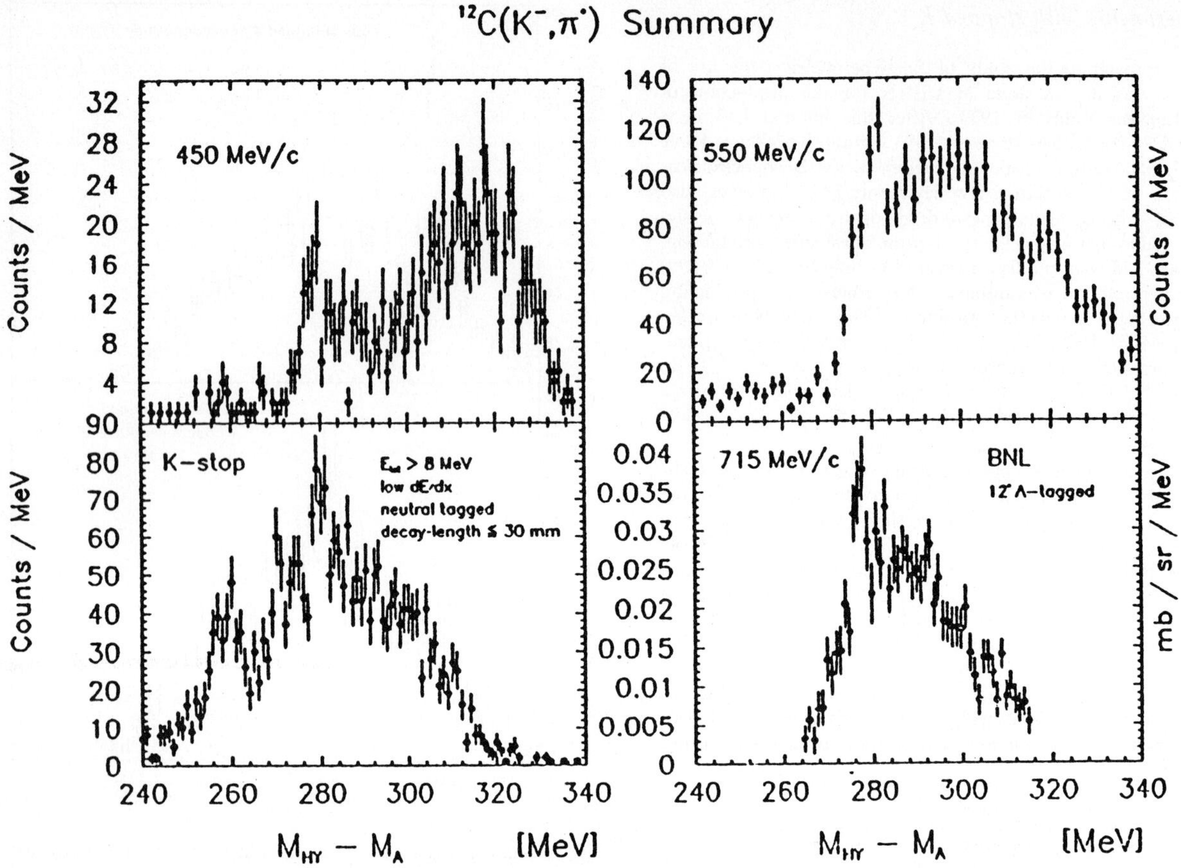

Fig. 6 : Summary of all (K^-,π^+) experiments on ^{12}C at CERN, KEK and BNL.

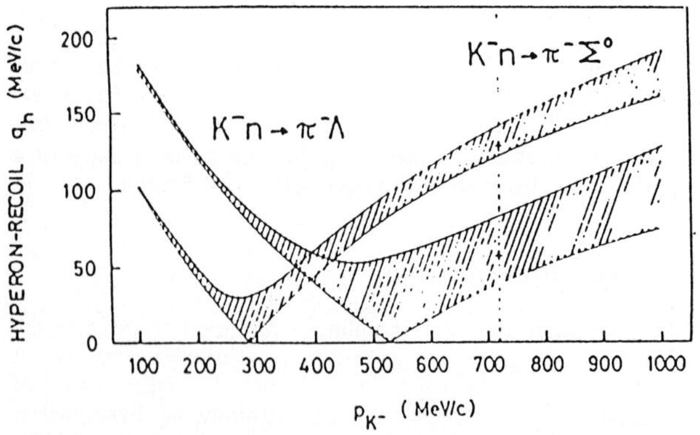

Fig. 7 : Recoil momentum in the reaction K^- n $\to$ π^- Hyperon shown as a function of K^- momentum with the π^- detected in forward direction.

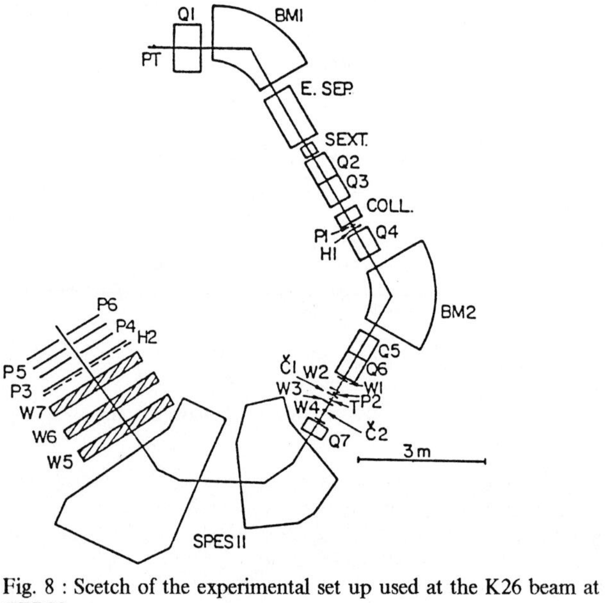

Fig. 8 : Scetch of the experimental set up used at the K26 beam at CERN.

The underlaying ideas of the 'tagging' explained above are based on the assumption (and hope) that the decay process for Σ-hypernuclear continuum 'states' are different from the narrow states observed. In the latter case, the decay should predominantly proceed by the quenching reaction while for the first reaction the Σ has larger chance to escape from the nucleus to either decay freely or to interact again with another nucleus. In the experiments mentioned above such an identification of final state particles was limited by solid angle.

Fig. 8 depicts the experimental set up used at CERN. A very special beamline including higher order correction lenses had been designed to achieve a very short (11.5 m) spectrometer for the K^- beam including a crossed field electromagnetic separator. Cerenkov counters before and behind the target assured the occurence of a $(K^-,\pi^{\pm})$ reaction. The π was detected under 0^0. Due to the finite solid angle (20 msr) the spectrometer accepted particles up to $\pm$ 5^0. The emphasis of the set up used was an online selection of

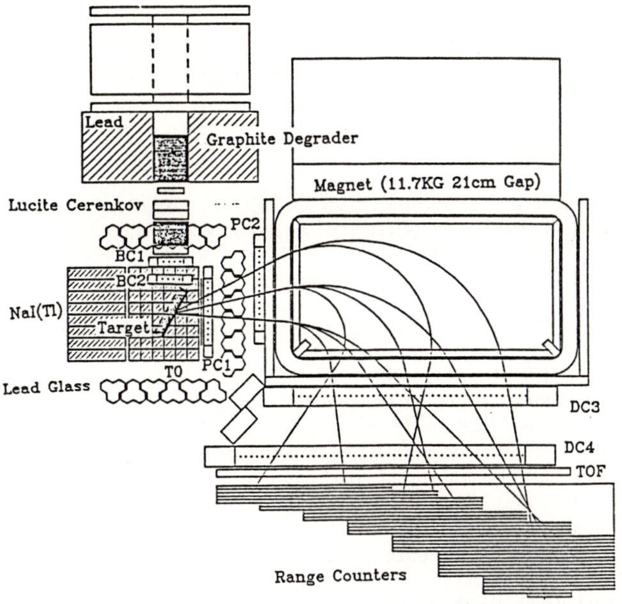

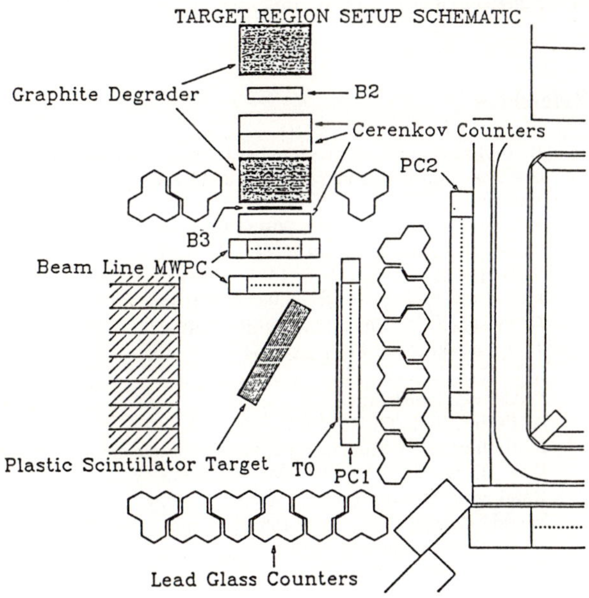

Fig. 9 : Scetch of the experimental set up used for the K$^-$ – stop experiment at KEK.

the reaction process. No counters were present to detect any secondaries from the production or decay of Σ – hypernuclei.

In the case of the KEK experiment (fig. 9) the emphasis was more towards a detection of secondary reaction products, particulary γ from the decay of excited states or π^0 decays. However, the detection probability for π^0 was small (5%). The K$^-$ beam was extracted at 650 MeV/c and then degraded to assure a full stopping of the beam in the target. The π spectrometer (non – focusing dipole magnet) was set perpendicular to the incoming beam. Incoming K$^-$ were identified by means of cerenkov counters. Particle identification for the outgoing particles was performed offline by TOF and range counters. A detection of short lived secondaries was only possible during the runs with a scintillator target and was by no means clear.

3. Σ – hypernuclei at KAON

In the following I will present a possibility of an improved Σ – hypernuclear experiment. The idea behind is to combine and improve the set – ups described above.

A priori it is not evident wether to follow the paths of a K – stop experiment or to continue the in flight reaction. Since a high intensity K beamline reduces the necessity of obtaining a high counting rate by choosing a reaction with high cross – section this advantage of K – stop experiments can be discarded. The necessity of permitting non $\Delta L = 0$ transitions to occur if a large momentum transfer is choosen can also be satisfied in the in flight experiments if the π are detected under a larger angle. In view of the much cleaner spectra which can be obtained with a low momentum K$^-$ beam (see fig. 6) a continuation of the in flight experiments seems desirable in spite of the more demanding K – beam line (K – spectrometer) required. I will therefore restrict myself to the dicussion of an in flight experiment.

3.1 Experimental Set Up

Fig. 10 shows a scetch of a possible experimental set up.

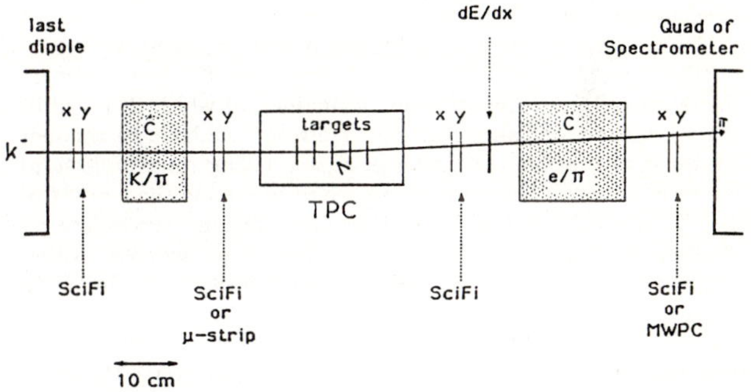

Fig. 10 : Scetch of a possible layout for a Σ – hypernuclear experiment at KAON.

3.1.1 The beamline

Following the ideas presented in the proposal for the kaon factory a low momentum K – beam line similar to the one at CERN can be built. A design making a K – beam of 350 MeV/c feasable would however be desirable, probably requiring a shorter beam line as envisaged in the proposal (18 m) to reduce π – background. The beam particle definition can again be achieved by TOF and Cerenkov counters. To measure the momentum of the incoming particle, scintillating fibre hodoscopes should be used in front and behind the bending magnet to assure good momentum resolution, high rate capability and fast online reconstruction [6].

3.1.2 The Target Region

In order to learn more about the decay of Σ – hypernuclei, short lived secondaries (Σ,Λ) should be observed. This can only be achieved using a thin target (500 μm) implemented into a TPC. First Monte Carlo simulation showed that the lifetime of quasi free produced Σ is about 2 mm compared to 3 cm for Λ coming from the Σ + N → Λ + N. Spatial resolutions better then 150 μm can be achieved. A TPC volume of ≈ 20 · 20 · 10 cm^3 should be sufficient which could house five target slabs spaced by 3 cm. Owing to the high particle flux traversing the TPC, this device should be gated by a first level trigger ([7]). The signal in a TPC is usually read out by wires and a system of pads of which the size partly determines the spacial resolution obtained. Since the multiplicity of an event is low, the two track resolution is not critical allowing more freedom in the size and form of the pads.

3.1.3 Spectrometer

Identification of the outgoing particle can be performed as in the CERN set−up using Cerenkov counters and TOF. The momentum of the particle can be measured in a focussing spectrometer of the SPES 2 type. Such a spectrometer enables a fast online momentum determination if a fast MWPC read out is used (see FASTRO at Ω or PCOS system) (using superconducting magnets a shorter spectrometer could be built reducing μ background from π decay). The detection of μ from K and π decay can be achieved by a range telescope mounted at the end of the spectrometer.

As in the case of the CERN spectrometer a (hardwired) processor can be used to perform online vertex reconstruction and momentum determination strongly reducing the trigger rate.

3.1.4 Count Rates and Trigger

Owing to the use of wire chambers in the set up the maximum particle rate traversing the detector should be kept below 10^6/second. Using a repetition rate of 10 Hz and a π/K ratio of 10 at 400 MeV/c (π/K was 50 at CERN using a shorter beam line but a factor 5 could be achieved using a better channel acceptance with slits and two−stage electrostatic separators ([8]) the number of K/burst is restricted to 10^4 giving a factor 100 higher kaon flux as at the CERN experiment.

If we were to use five ^{12}C targets of 500 μm length we would obtain a trigger rate of 50/burst resulting from K^- interaction in the target (50% are due to elastic scattering), of which 2% lead to the strangeness exchange reaction. However, we have to account for material surrounding the target region adding roughly 10% to the reaction rate. These numbers have to be seen in the context of 10^3 K^- decaying in the 30 cm long target region. An online reduction of at least a factor 200−500 has to be achieved within 100ms to limit the dead time of the data acquisition system (using Fastbus and Spillbuffer systems [6]) requiring an average processing speed of 20−50μsec/event (if the beam can be operated in a continous mode, these time estimates are less rigorous) However, 65% of the K^- decay with μ in the final state which can be rejected by a multiplicity check of the range telescope. π^- from the $K^- \rightarrow \pi^-\pi^0$ decay mode will fall outside the momentum acceptance of the spectrometer. A fast vertex reconstruction should then reduce most of the other background coming from K^- decay. A detailed Monte Carlo simulation has to be used to estimate the final trigger efficiency. Using the reaction (K^-,π^+), the background from K^- decay does not constitute a serious problem.

If we assume the feasability of the set up described above we should write $5 \cdot 10^5$ strangeness exchange reactions to tape/day. If we take the cross−sections for recoilless production obtained from the CERN group into account, ≈50000 events/day should be obtained in the Σ−hyernuclear states. Due to the thin target used, a mass resolution superior to all other experiments should also be obtained.

4. Conclusion

Using a low momentum high intensity K^- beam at the new KAON facility at TRIUMF a better understanding of the Σ−Nucleus interaction should be obtained possibly giving the answer to the puzzle of the existence of narrow Σ−hypernuclear states, their decay and the Σ−Nucleus spin−orbit coupling. Using the results from various experiments summarized in fig. 5 it seems advantageous to perform the $(K^-,\pi^\pm)$ reaction in flight using a 400 MeV/c (possibly 350 MeV/c) K^- beam. The implementation of a TPC, making visible short lived secondaries should help to shed light on the decay of these states. New developments in particle detector and data aquisition systems available should finally solve the problem of unsatisfying statistical significance.

5. Acknowledgements

I would like to thank Prof. B. Povh for the many fruitful discussions concerning the physics of hypernuclei. I gratefully acknowlegde the discussions with Drs. W. Brückner and R. Richter on the newly proposed experimental set up.

6. References

[1] R. Bertini et al., Talk presented at the 8th international conference on high energy physics and nuclear structure, Vancouver, 1979
R. Bertini et al. Phys. Lett. 90B (1980) 375
[2] R. Bertini et al. Phys. Lett. 136B (1984) 29
R. Bertini et al., Phys. Lett. 158B (1985) 19.
[3] L.G. Tang et al., Phys. Rev. 38C (1988) 846
H. Piekarz et al., Phys. Lett. 110B (1982) 428−432
[4] T. Yamazaki et al., Phys. Lett. 54 (1985) 102
[5] S. Paul et al., Nucl. Phys. A479 (1988) 137c−160c
[6] WA89−collaboration
A. Forino et al., CERN SPSC−P−233 (1987)
[7] R. Richter, MPI München, private communication
[8] E.W. Blackmore, The TRIUMF K−factory proposal
Rand workshop on $\bar{p}$ science and technology
World Scientific 1988, 155−168
ed. by B.W. Augenstein et al.

Z. Phys. C – Particles and Fields 46, S57–63 (1990)

Zeitschrift für Physik C Particles and Fields

Direct CP violation observed in decays of neutral K_L mesons

K. Kleinknecht

Institut für Physik, Johannes-Gutenberg-Universität, Postfach 3980, D-6500 Mainz, Fed. Rep. of Germany

Abstract Recent results on CP violation in K^0 decays are reviewed. The measurement of ε'/ε of the NA31 Collaboration indicates for the first time that direct CP violation is present in the weak $\Delta S=1$ transition of $K_L \to \pi\pi$. This result agrees with models of CP violation through weak quark mixing in the Kobayashi-Maskawa scheme, but is at variance with the "superweak" model of Wolfenstein.

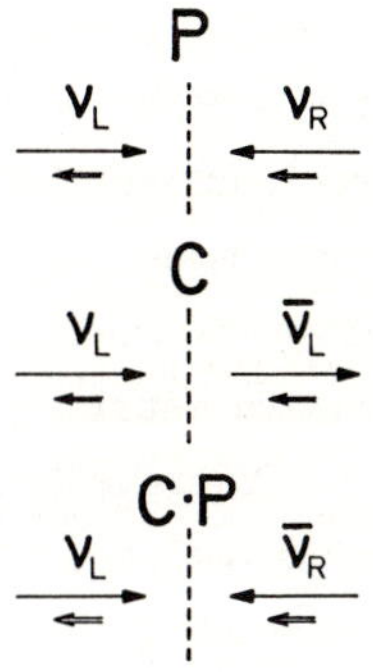

1 Introduction

Parity is violated in weak interactions. This can be visualized by looking at a left-handed neutrino emitted in β^+ decay: its mirror image is a right-handed neutrino, a state which has never been observed. Similarly, the charge conjugation C transforms this left-handed neutrino into a left-handed antineutrino, also a state never observed. However, Landau [1] realized that the combined operation C x P transforms a left-handed neutrino into a right-handed antineutrino, thus connecting two physical states. CP invariance therefore was considered to be replacing the separate P and C invariance of weak interactions.

One consequence of this postulated CP invariance for the neutral K mesons was predicted by Gell-Mann and Pais [2]: there should be a long-lived partner to the known V^0 (K_1^0) particle of short lifetime (10^{-10}sec). According to this proposal these two particles are mixtures of two strangeness eigenstates, $K^0(S = +1)$ and $\bar{K}^0(S = -1)$ produced in strong interactions. Weak interactions do not conserve strangeness and the physical particles should be eigenstates of CP if the weak interactions are CP invariant. These eigenstates are (with $\bar{K}^0 = CP\,K^0$)

$$CP\,K_1 = CP(K^0+\bar{K}^0)/\sqrt{2} = (\bar{K}^0+K^0)/\sqrt{2} = K_1$$

$$CP\,K_2 = CP(K^0-\bar{K}^0)/\sqrt{2} = (\bar{K}^0-K^0)/\sqrt{2} = -K_2\ .$$

Because of $CP(\pi^+\pi^-) = (\pi^+\pi^-)$ for π mesons in a state with angular momentum zero, the decay into $\pi^+\pi^-$ is allowed for the K_1 but forbidden for the K_2; hence the longer lifetime of K_2, which was indeed confirmed when the K_2 was observed.

In 1964, however, Christenson, Cronin, Fitch and Turlay [3] discovered the long-lived neutral K meson also decays to $\pi^+\pi^-$ with a branching ratio of $\sim 2 \times 10^{-3}$. From then on the long-lived state was called K_L because it was no longer identical to the CP eigenstate K_2; similarly, the short-lived state was called K_S. The CP violation that manifested itself by the decay $K_L \to \pi^+\pi^-$ was confirmed by subsequent discoveries of the decay $K_L \to \pi^o\pi^o$ [4], and of a charge asymmetry in the decays $K_L \to \pi^{\pm}e^{+}\nu$ and $K_L \to \pi^{\pm}\mu^{+}\nu$ [5].

The question whether this CP violation (and T violation through the CPT theorem) is due to a fifth force, the "superweak" interaction [6], to a T-violating part of the weak interaction, or to interference of the T-invariant weak interaction with a T-violating part in the electromagnetic or strong interactions, has been studied by many subsequent experiments. These experiments [7] show that only two explanations remain as possible candidates: the superweak model - which however would lead to no other observable CP violating effects outside the K^o system - and the Kobayashi-Maskawa model [8] describing the observed CP violation as due to a complex phase in the six-quark mixing matrix.

At the time of the discovery of CP violation, only 3 quarks were known, and there was no possibility of explaining CP violation as a genuine phenomenon of weak interactions. This situation remained unchanged with the fourth quark because the 2 x 2 mixing matrix has only one free parameter, the Cabibbo angle, and no non-trivial complex phase. However, as remarked by Kobayashi and Maskawa, the picture changes if six quarks are present. Then the 3 x 3 mixing matrix naturally contains a phase, apart from the three mixing angles. It is then possible to construct CP violating weak amplitudes from "box-diagrams" of the form:

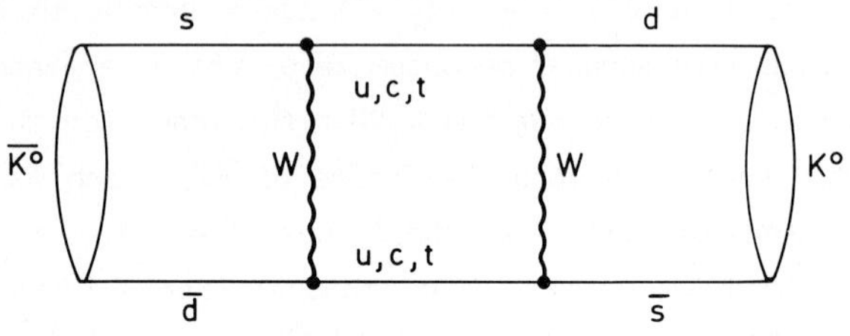

A necessary consequence of this model of CP violation is the non-equality of the relative decay rates for $K_L \to \pi^+\pi^-$ and $K_L \to \pi^o\pi^o$. This "direct" CP violation is due to "Penguin diagrams" of the form

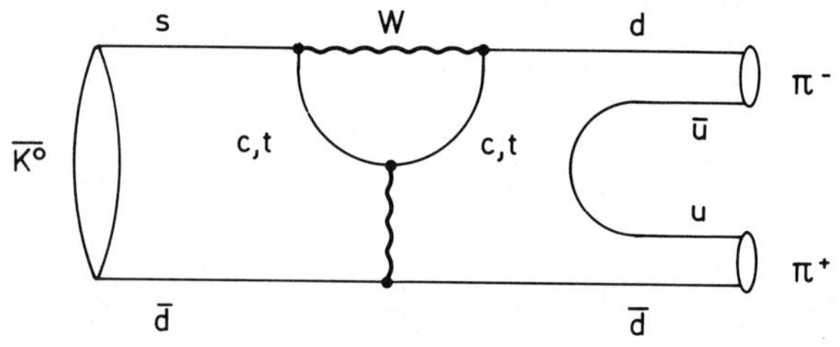

which lead to CP violating $\Delta S=1$ weak decay amplitudes (called ε' in the following section). The search for this direct weak CP violation is currently pursued in experiments at Fermilab, BNL, and CERN.

2 Phenomenology

The experimentally accessible quantities are the ratios of amplitudes of CP violating and CP conserving decays, viz.

$$\eta_{+-} = \frac{A(K_L \to \pi^+\pi^-)}{A(K_S \to \pi^+\pi^-)}$$

$$\eta_{oo} = \frac{A(K_L \to \pi^o\pi^o)}{A(K_S \to \pi^o\pi^o)}$$

It can be shown that these amplitude ratios consist of a contribution from CP violation in the K^o-$\bar{K}^o$ mixing (box diagrams above), called ε, and another one from CP violation in the weak $K \to 2\pi$ amplitudes (penguin diagrams above), called ε':

$$\eta_{+-} = \varepsilon + \varepsilon'$$

$$\eta_{oo} = \varepsilon - 2\varepsilon' \; .$$

In this way η_{+-}, η_{oo} and $3\varepsilon'$ form a triangle in the complex plane. While the modulus and phase of ε are known experimentally quite well [7], the existence of "direct CP violation" through ε' and the magnitude of ε' are to be tested by the "experimentum crucis" for CP violation. It is clear that $\left|\frac{\varepsilon'}{\varepsilon}\right| \ll 1$.

Such experiments on ε' can be done by measuring

$$\left|\frac{\eta_{oo}}{\eta_{+-}}\right|^2 = 1 - 6 \, \mathrm{Re} \, \frac{\varepsilon'}{\varepsilon} \; .$$

3 New experiments on CP violation

Several important experiments on this ratio are now being done.

This, via the relation

$$|\eta_{oo}|^2/|\eta_{+-}|^2 = 1 - 6 \text{ Re } \frac{\varepsilon'}{\varepsilon}$$

gives a measurement of the ε' parameter of direct CP violation. As mentioned in sect.1, ε'/ε can be calculated if CP violation is due to a complex phase in the quark mixing matrix for 6 quarks. Fig.1 shows the result of the calculation as a function of the top quark mass m_t, and of the ratio of the decay rates $\bar{R}=\Gamma(b\to u)/\Gamma(b\to c)$. The prediction also depends on the weak transition matrix element M from K^0 to $\bar{K}^0$. The ratio of M to the value M_{vac} calculated by inserting as an intermediate state the vacuum is called $B_K = M/M_{vac}$. Presently, calculations of M on the lattice point towards a value of $B_K = 0.84 \pm 0.20$ [9], while a "$1/N_c$ expansion" yields $B_K = 0.67 \pm 0.10$ [10] and computations [11] using QCD sum rules give $B_K = 0.55 \pm 0.15$ and $B_K = 0.84 \pm 0.08$. For $m_t \sim 50$GeV and B = 0.67 a value of $\varepsilon'/\varepsilon \sim (1.7 \pm 1.0)\ 10^{-3}$ is expected. In order to establish such a small effect, the ratio $R = |\eta_{oo}|^2/|\eta_{+-}|^2$ has to be measured with a precision of 0.5% or better.

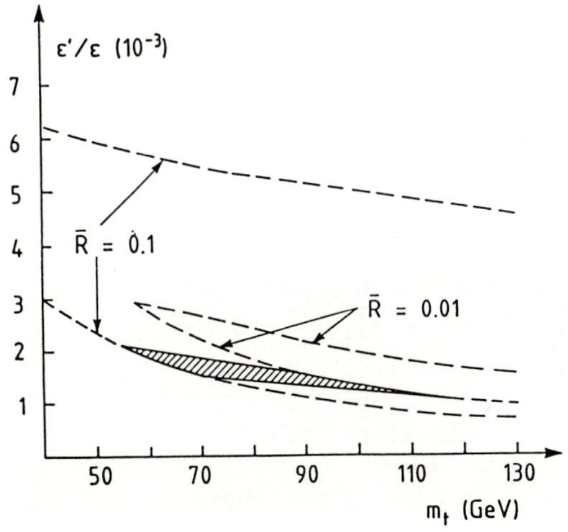

Figure 1 Predicted values for ε'/ε in the six quark model of milli-weak CP violation [10], as a function of the top-quark mass M_t. $\bar{R}$ is the ratio of decay rates of b quarks, $\bar{R}=\Gamma(b\to u)/\Gamma(b\to c)$.

The experiment has been done at three places: by a Chicago-Saclay collaboration at FNAL [13 , 16], by a Yale-BNL group at BNL [14], and by a CERN-Dortmund-Edinburgh-Mainz-Orsay-Pisa-Siegen (NA31) Collaboration at CERN [15]. In addition, a new experiment is prepared at CERN [17].

These experiments employ different methods to measure the double ratio

$$R = \frac{\Gamma(K_L \to 2\pi^0)}{\Gamma(K_S \to 2\pi^0)} \cdot \frac{\Gamma(K_S \to \pi^+\pi^-)}{\Gamma(K_L \to \pi^+\pi^-)}$$

Method I consists in measuring K_L and K_S decays simultaneously by using a split neutral beam with half the area covered by a regenerator producing K_S mesons. Only one decay mode (neutral or charged) at a time is recorded. This method is employed by the Chicago-Saclay collaboration and by its successor, the Chicago-Elmhurst-Fermilab-Princeton-Saclay collaboration.

Method II both decay modes are recorded simultaneously first in a K_L beam, then in a K_S beam. One variant of this method, employed by the Yale-BNL collaboration, uses a regenerated K_S beam, the other (CERN NA31) a K_S beam near the proton target.

Method III aims at measuring all 4 decay modes together from a tagged K^0 or $\bar{K}^0$ source from $\bar{p}p$ - annihilations at rest. This has been adopted by the CERN-LEAR collaboration (PS195) [17].

As an example of a detector for K^0 decays, Fig.2 shows a side view of the NA31 detector. The K_L and K_S decays occur in a vacuum tank, 100 m long and with 2.4 m diameter. The decay products and the neutral beam enter from the left side. On their way, they encounter two drift chambers, the electromagnetic calorimeter with Pb plates in liquid argon, and the hadron calorimeter made of iron and scintillator. Charged pions are detected in the chambers and the calorimeters, the invariant mass of $K \to \pi^+\pi^-$ decays is calculated from the vector momenta of both pions measured this way. The 4 photons from $K \to \pi^0\pi^0$ decay are detected in the electromagnetic calorimeter with an energy resolution of $\Delta E_\gamma/E_\gamma = 7\%/\sqrt{E_\gamma}$(GeV). Two of them are combined to give a neutral pion. Of the three possible

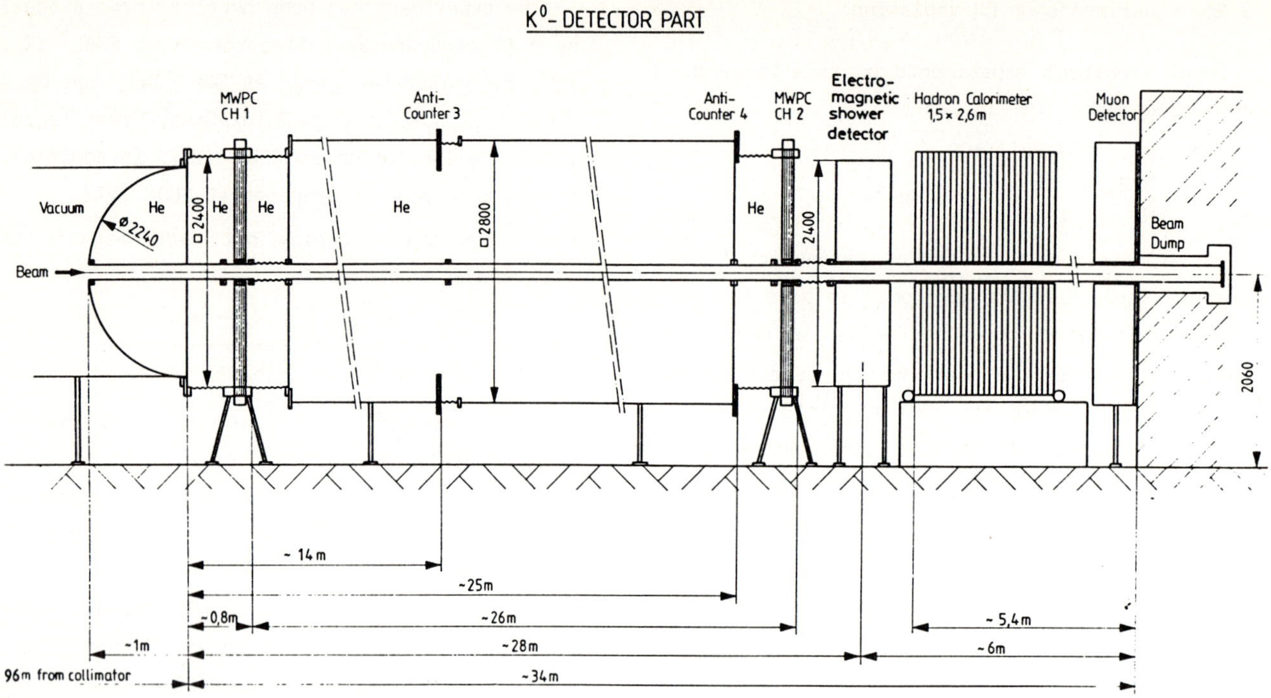

Figure 2 Detector of the CERN-Dortmund-Edinburgh-Mainz-Orsay-Pisa-Siegen collaboration NA31 for the measurement of ε/ε' in Kaon decays.

combinations, the one fitting best the $2\pi^0$ hypothesis is chosen.

Fig.3 shows a lego scatter plot in the plane of two invariant $\gamma\gamma$-masses, for events in the short-lived kaon beam, i.e. $K_S \rightarrow 2\pi^0$. A similar plot is obtained for $K_L \rightarrow 2\pi^0$ decays. Here, the background from $K_L \rightarrow 3\pi^0$ decays (branching ratio 21.5% compared to 0.094% for $K_L \rightarrow 2\pi^0$) shows up as combinations of two pairs of photons outside the π^0 mass peak, see Fig.4.

The ellipsoidal rings around the peak from $K_L \rightarrow 2\pi^0$ events are drawn to contain equal numbers of $K_L \rightarrow 3\pi^0$ events, as obtained from Monte Carlo calculations. The background below that peak is then calculated from a flat extrapolation in the number of events per ring, and amounts to 4%.

For charged decays, $K^0 \rightarrow \pi^+\pi^-$, the directions of the pions are measured in the two drift chambers, and their energies in the electromagnetic and hadronic (iron-scintillator) calorimeters. From these vector momenta the invariant K mass and the distance d_T of the decay plane from the production target are calculated. For $K_S \rightarrow \pi^+\pi^-$, the background is at the 10^{-3} level.

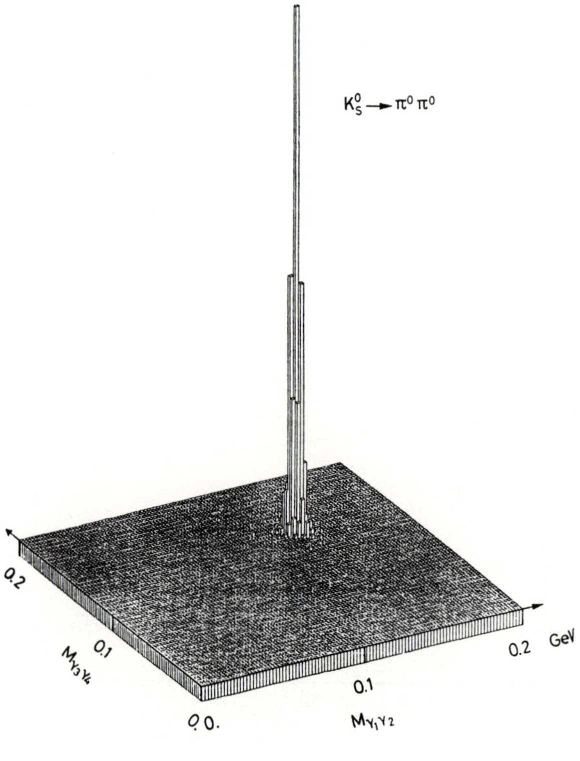

Figure 3 Scatter plot of two $(\gamma\gamma)$ invariant masses in $K_S \rightarrow 2\pi^0$ events from NA31.

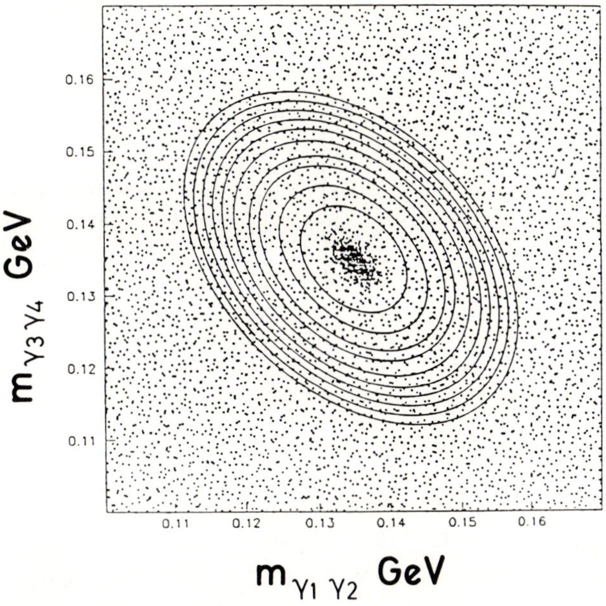

Figure 4 Scatter plot of two ($\gamma\gamma$) invariant masses in $K_L \to 2\pi^0$ and of misidentified $K_L \to 3\pi^0$ decays.

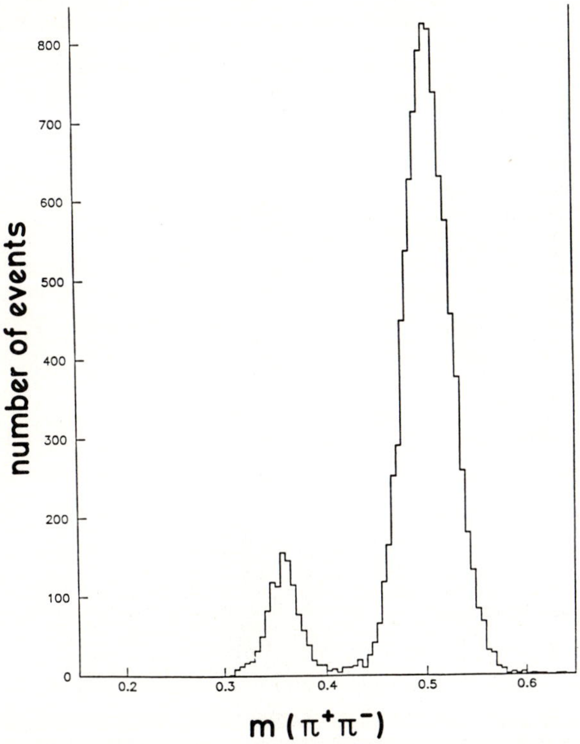

Figure 5 Invariant $\pi\pi$ mass distribution of K_L decay events.

For $K_L \to \pi^+\pi^-$, three-body decays constitute the background. In the invariant $\pi\pi$ mass distribution, Fig.5, the small peak below the K^0 mass comes from $K_L \to \pi^+\pi^-\pi^0$. After a cut selecting

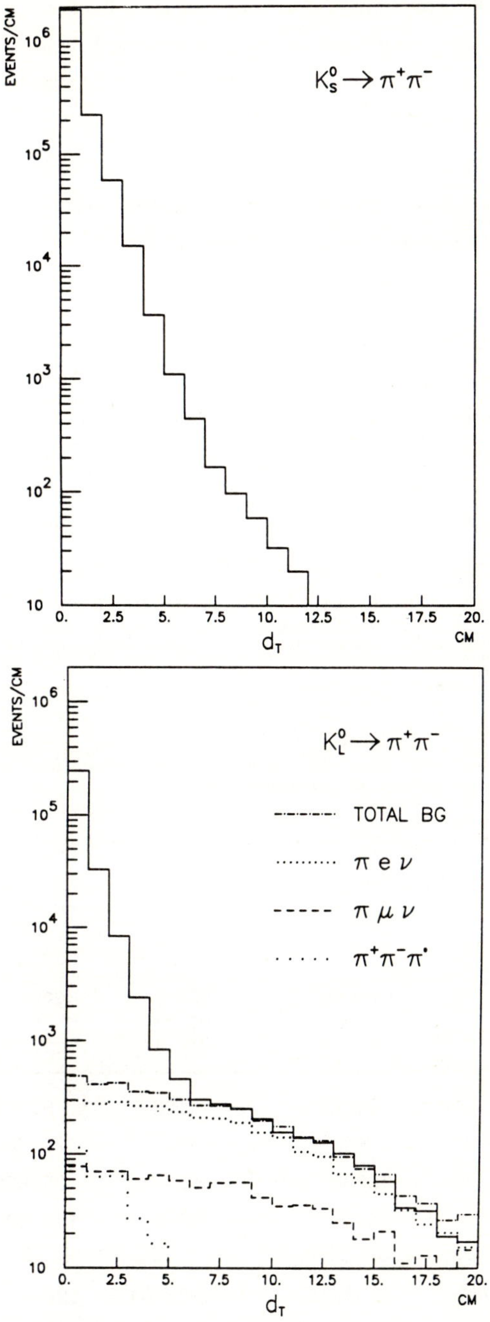

Figure 6 Event distribution for neutral kaon decays into $\pi^+\pi^-$, as a function of the distance d_T between the decay plane and the production target, for K_S and K_L decays, indicating the background composition (Ref.15).

the K mass, the remaining background is obtained by considering the distribution of events in the distance d_T. While two-body decays come from the target, $d_T \sim 0$, three-body decays show a broad distribution. The background below the peak around $d_T=0$ is then obtained from an extrapolation of the observed off-target background sample below the peak using a Monte-Carlo-calculation for the shape of the background distribution (see Fig.6).

The number of events from this experiment NA31 is shown in Table 1, together with data from other experiments. The result is

$$\varepsilon'/\varepsilon = (3.3 \pm 1.1)\ 10^{-3}\ .$$

This result is the first evidence for $\varepsilon' \neq 0$, i.e. the first observation of direct violation, and therefore against the superweak model. The value of ε'/ε is however in agreement with the numbers calculated in the standard model. This new result is compared to previous experimental measurements of this quantity in Fig.7. A confirmation of this result can be expected by 1989 from the continuation of the experiment E731 at Fermilab.

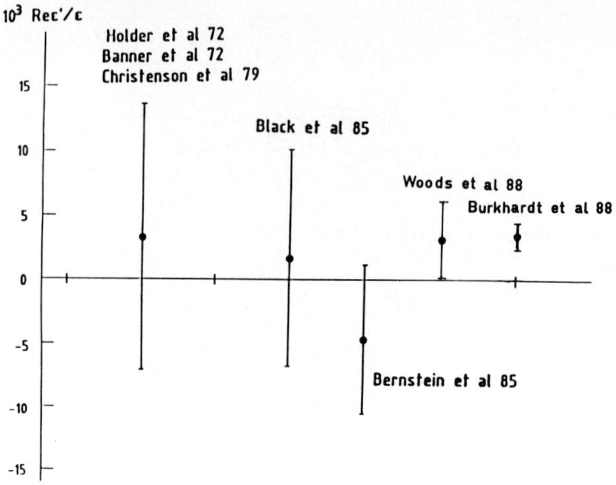

Figure 7 Experimental results on $Re(\varepsilon'/\varepsilon)$ from Banner et al. [18], Holder et al. [19], Christenson et al. [20], Black et al. [14], Bernstein et al. [13], Woods et al. [16], and Burkhardt et al. [15].

Table 1 Recent experiments on ε'/ε

Collaboration	Chicago-Saclay	Yale-BNL	E731 Chicago-Elmhurst-FNAL-Princeton-Saclay	NA31 CERN-Dortmund-Edinburgh-Mainz-Orsay-Pisa-Siegen
	Bernstein et al. Ref.13	Black et al. Ref.14	Woods et al. Ref.16	Burkhardt et al. Ref.15
$K_L \to 2\pi^0$	3152	1122	6.747	109.000
$K_S \to 2\pi^0$	5663	3317	21.788	932.000
$K_L \to \pi^+\pi^-$	10638	85-06	35.838	295.000
$K_S \to \pi^+\pi^-$	25751	20960	130.025	2.300.000
Data taken	1983-84	1983-84	1986	1986
Result	1985	1985	1988	1988
R	1.028 ± 0.032 ± 0.014	0.990 ± 0.048 ± 0.026		0.980 ± 0.004 ± 0.005
$\varepsilon'/\varepsilon \times 10^3$	- 4.6 ± 5.3 ± 2.4	1.7 ± 8.2	3.2 ± 2.8 ± 1.2	3.3 ± 0.66 ± 0.8

References

1. L.D. Landau, Nucl.Phys. 3(1957)127.
2. M. Gell-Mann and A. Pais, Phys.Rev. 97(1955)1387
3. J. Cristenson, J. Cronin, V. Fitch and R. Turlay, Phys.Rev.Lett. 13(1964)138.
4. J.M. Gaillard et al., Phys.Rev.Lett. 18(1967)20; J.W. Cronin et al., Phys.Rev.Lett. 18(1967)25.
5. S. Bennett et al., Phys.Rev.Lett. 19(1967)993.
6. L. Wolfenstein, Phys.Rev.Lett. 13(1964)562.
7. K. Kleinknecht, Ann.Rev.Nucl.Sci. 26(1976)1.
8. M. Kobayashi and T. Maskawa, Progr.Theor. Phys. 49(1973)652.
9. G. Martinelli, paper given at "Workshop on Hadronic Matrix Elements", Ringberg, April 1988.
10. W.A. Bardeen, A.J. Buras and J.M. Gerard, Nucl.Phys. B293(1987)787; A.J. Buras and J.M. Gerard, Phys.Lett. 203(1988)272.
11. R. Decker, Nucl.Phys. B277(1986)660; L.J. Reinders and S. Yazaki, Nucl.Phys. B288(1987)789.
12. F.J. Gilman and J.S. Hagelin, Phys.Lett. 133B(1983)443.
13. R.H. Bernstein et al., Phys.Rev.Lett. 54(1985)1631.
14. J.K. Black et al., Phys.Rev.Lett. 54(1985)1628.
15. "First Evidence for Direct CP Violation", CERN-Dortmund-Edinburgh-Mainz-Orsay-Pisa-Siegen Collaboration: H.Burkhardt, P.Clarke, D.Coward, D.Cundy, N.Doble, L.Gatignon, V.Gibson, R.Hagelberg, G.Kesseler, J.van der Lans, I.Mannelli, T.Miczaika, A.C.Schaffer, J.Steinberger, H.Taureg, H.Wahl and C.Youngman (CERN), G.Dietrich, W.Heinen (Univ. Dortmund), R.Black, D.J.Candlin, J.Muir, K.J.Peach, B.Pijlgoms, I.P.Shipsey, W.Stephenson (Univ. of Edinburgh), H.Blümer, M.Kasemann, K.Kleinknecht, B.Panzer, B.Renk (Univ. Mainz), E.Augé, R.L.Chase, M.Corti, D.Fournier, P.Heusse, L.Iconomidou-Fayard, A.M. Lutz, H.G.Sander (Univ. Paris-Sud), A.Bigi, M.Calvetti, R.Carosi, R.Casali, C.Cerri, G.Gargani, E.Massa, A.Nappi, G.M.Pierazzini (INFN Pisa), C.Becker, D.Heyland, M.Holder, G.Quast, M.Rost, W.Weihs, G.Zech (Univ. Siegen); Phys.Lett. B206(1988)169.
16. "First Results on a New Measurement of ε'/ε in the Neutral Kaon System", Chicago-Elmhurst-Fermilab-Princeton-Saclay Collaboration, M.Woods et al., Phys.Rev.Lett. 60(1988)1695.
17. L.Tauscher, paper given at "Workshop on Rare Meson Decays", Max-Planck-Institut Heidelberg, Dec. 2-3, 1986, Athens-Basel-CERN-Fribourg-Liverpool-Saclay-SIN-Stockholm-Thessaloniki-Zürich-Collaboration, CERN-Proposal PSCC/85-6/P82.
18. M.Banner et al., Phys.Rev.Lett. 28(1972)1597
19. M. Holder et al., Phys.Lett. B40(1972)141
20. J.Christenson et al., Phys.Rev.Lett. 43(1979)1209.

Z. Phys. C – Particles and Fields 46, S65–72 (1990)

Zeitschrift für Physik C Particles and Fields

Rare decays at the kaon factory*

D. Bryman

TRIUMF, 4004 Wesbrook Mall, Vancouver B.C., Canada V6T 2A3

Abstract

Extensive experimental work on rare kaon decays is currently being performed at BNL, KEK and FNAL to search for exotic physics and to examine standard model predictions in unique detail. The Kaon Factory at TRIUMF will produce beams with a hundred-fold increase in intensity over existing machines in the 30 GeV region and will allow even higher precision and higher sensitivity experiments on rare decays to be done. A sample of kaon decay experiments and CP and T violating studies which might benefit from intense kaon factory beams is discussed.

1. INTRODUCTION

The prospects for particle physics experiments at the Kaon Factory are varied and the emphasis will be on attacking many of the same issues addressed at the energy frontier of the high energy colliders. The experiments to be done at the 30 GeV, 100 μA proton synchrotron proposed at TRIUMF will make use of the increase in current over existing machines in this energy range to achieve high precision or high sensitivity to carefully probe predictions of the standard model and to search for new effects. Unprecedented fluxes of kaons, antiprotons, hyperons, neutrinos, pions and muons will all be available to open up new possibilities for experimentation. Some particular areas of interest at the kaon factory include rare kaon decays, CP and T violation, neutrino properties and reactions, and hadron spectroscopy. In the following, a few examples of the physics opportunities generated by the advent of the kaon factory will be discussed.

2. RARE DECAYS AND CP VIOLATION

Rare decays of mesons and leptons play a significant role in challenging the standard model and in searching for effects which could indicate new directions. Kaon decays have been a rich and often surprising source of information at every stage in the development of the present picture of fundamental particles and their interactions. Parity violation, CP violation, neutral currents and the existence of charm are all effects in which kaon decays exhibited crucial or unique features. Kaon decays remain in the forefront of modern high precision attempts to test the accuracy of standard model predictions, to define the nature of CP violation, and to search for neutral flavor changing currents and lepton number violation among other new interactions and particles. For recent reviews of rare kaon decays see ref. 1 and 2.

A. $K^+ \to \pi^+\nu\bar{\nu}$

Reactions which are allowed in the standard model can provide important, detailed information, and can also herald the presence of new effects. The process $K^+ \to \pi^+\nu\bar{\nu}$ offers a prime example of the unique opportunities available in the study of rare kaon decays because a reliable higher order calculation assuming three generations can be confronted by experiment. Nonconformity with the standard model prediction could imply new physics in the form of extra generations or entirely new types of particles or interactions. The rate for $K^+ \to \pi^+\nu\bar{\nu}$ depends on parameters of the Cabbibo-

*Adapted from "Particle Physics Prospects at the Kaon Factory", presented at the Workshop on Intensity Frontier Physics, KEK, Tsukuba, Japan, April 3, 1989.

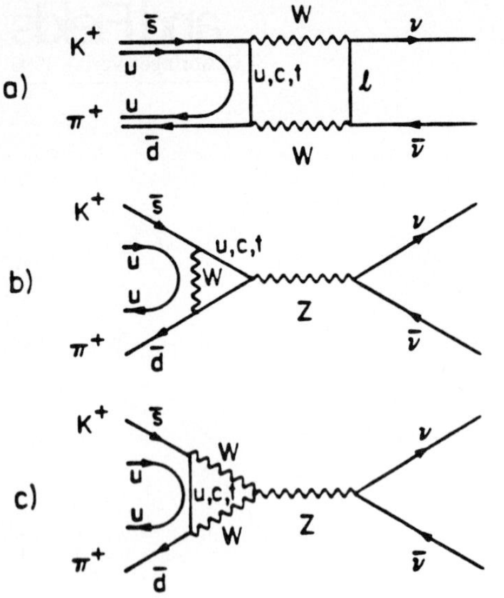

Figure 1: Second order weak diagrams for $K^+ \to \pi^+\nu\overline{\nu}$.

Kobayashi-Maskawa (CKM) matrix as evidenced by the diagrams in figure 1. Constraints on the CKM mixing parameters $V_{ts}^{*}V_{td}$ have been derived from semileptonic B-meson decays, from the measured b-quark lifetime and from the large observed $B_d^0 - \overline{B}_d^0$ mixing which, for example, fixes V_{td} (although with considerable uncertainty at present). The $K^+ \to \pi^+\nu\overline{\nu}$ branching ratio as a function of the t-quark mass with the dependence on uncertainties of B-meson decay observables lies in the region 1 to 7×10^{-10} for m_t in the range 50 to 200 GeV/c^2 [2]. Ellis and Hagelin [3] calculated radiative QCD effects indicating that if the mixing angles and t-quark mass were known a firm prediction for the $K^+ \to \pi^+\nu\overline{\nu}$ branching ratio could be made. Conversely, a measurement of the branching ratio would be significant in constraining these parameters and would allow a direct test of higher order weak corrections in the standard model which is not significantly constrained by uncertain long distance effects [4] as in calculations of $K_L^0 \to \mu\mu$ and the $K_L^0 - K_S^0$ mass difference.

A precise standard model prediction for the $K^+ \to \pi^+\nu\overline{\nu}$ branching ratio allows the reaction to be used to search for new physics. The least exotic addition to the present picture would involve additional generations of quarks and leptons. Since experiments measuring $K^+ \to \pi^+\nu\overline{\nu}$ do not observe the weakly interacting decay products, it is possible that this reaction is accompanied by $K^+ \to \pi^+xx'$ or $K^+ \to \pi^+x$, which occur at comparable or even much higher rates. The window for exotic effects appearing in the reaction $K^+ \to \pi^+xx'$ extends two orders of magnitude from the current limit $B(K^+ \to \pi^+xx') < 1.4 \times 10^{-7}$ [5], to the upper level of the standard model value $B(K^+ \to \pi^+\nu\overline{\nu}) \sim 10^{-9}$. In supersymmetric theories a variety of new particles are hypothesized including the supersymmetric partners of the photon ($\tilde{\gamma}$), the Higgs particle ($\tilde{H}$), the leptons and the quarks. These could contribute to the rate for $K^+ \to \pi^+xx'$, if the masses are sufficiently small. Schrock [6] estimated that, if tree level graphs dominate in the decay $K^+ \to \pi^+\tilde{\gamma}\tilde{\gamma}$, then the branching ratio could be as large as 10^{-7}, near the current limit. Other possibilities for exotic reactions $K^+ \to \pi^+xx'$ and $K^+ \to \pi^+x$ involving scalar or pseudoscalar particles have been suggested. The Majoran (a massless Nambu-Goldstone boson), the axion, light Higgs particles, the familon and hyperphotons are all potential candidates for x above.

An experiment is now in progress at Brookhaven National Laboratory (BNL) to measure the process $K^+ \to \pi^+\nu\overline{\nu}$[7]. The apparatus for BNL 787, a BNL-Princeton-TRIUMF collaboration, pictured in figure 2, is a state-of-the-art detector which builds upon earlier work (such as ref. 4) and points the way to future efforts at the kaon factory. The 787 detector has a large geometrical acceptance (2π sr) for the $K^+ \to \pi^+\nu\overline{\nu}$ decay mode and has been designed to maximize the rejection of background processes such as $K^+ \to \pi^+\pi^0$ ($K_{\pi 2}$), $K^+ \to \mu^+\nu_\mu$ ($K_{\mu 2}$), $K^+ \to \mu^+\nu\gamma$, and others. Sensitivity for identification of unaccompanied pions from $K^+ \to \pi^+\nu\overline{\nu}$ is accomplished through measurements of momentum, kinetic energy, range, decay sequence $\pi \to \mu \to e$, and nearly 4π coverage for detection of photons. The 800 MeV/c K^+ beam is brought to rest in a 10 cm diameter target consisting of groupings of scintillating fibers 2 mm in diameter viewed by photomultiplier tubes. The decay pions pass through a cylindrical drift chamber which measures their momenta in a 1 T magnetic field. The pions then stop in a plastic scintillator range stack which also contains multiwire proportional chambers. Each range stack counter is viewed from both ends by 5 cm phototubes read out by 500 MHz transient digitizers, so that the decay chain $\pi \to \mu \to e$ can be observed for particle identification. The total energy of the decay pions is measured by summing the pulse heights of the target and range array elements. The pion detector is completely surrounded by a 15 radiation length Pb-scintillator gamma veto (1 mm Pb, 5 mm scintillator). Figure 3a shows an example of a calibration event of the type $K^+ \to \pi^+\pi^0$. A blow-up of the segmented target is shown in figure 3b. Energy and time for each target element are available at present from an ADC and a TDC, respectively, so that the incident kaon and outgoing pion elements can be identified. The momentum calculated from the track in the drift chamber is 198 MeV/c, determined with resolution $\sigma_p = 2.5\%$. The pion track energy is found by summing the range stack and target energies to be 97 MeV with a resolution of $\sigma_E = 3\%$ and the range is 31 g/cm^2 with a resolution of $\sigma_R = 3\%$. Correlation of range, energy and momentum are used to verify that the particle is a pion. In addition, the $\pi \to \mu\nu$ decay pulse is observed using the transient digitizer [8] (TD) in the last

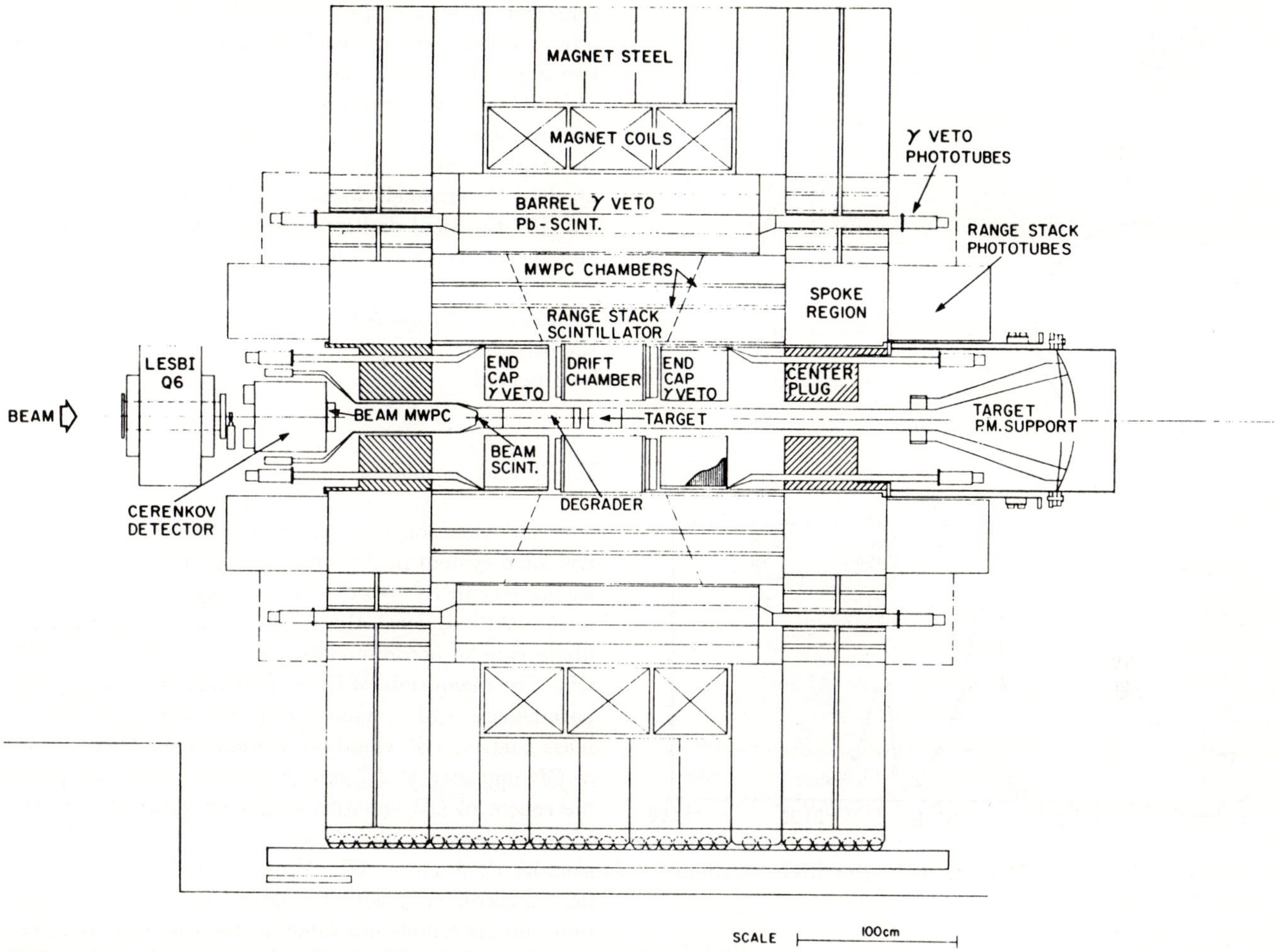

Figure 2: Apparatus for BNL-787 measurement of $K^+ \to \pi^+\nu\bar{\nu}$.

range stack counter hit as shown in figure 3c. The energy and timing of the 4 MeV muon pulse can be obtained and checked for consistency of position using the two ends of the counter. The $\mu \to e\nu\nu$ decay is also observed with the TD during an inspection period of 5 μs. In this event, the two photons from π^0 decay are both observed. We have determined from data that the inefficiency of the photon veto system is $\bar{\epsilon}_{\pi 0} < 4 \times 10^{-6}$ for π^0's from $K_{\pi 2}$ which is consistent with expectations of Monte Carlo calculations.

The 787 experiment had an engineering run in 1988 and has just completed a 10 week run. Some initial (preliminary) results from the 1988 data include new limits on the branching ratio for $K^+ \to \pi^+\nu\bar{\nu}$ (or $K^+ \to \pi^+xx$)

$$\frac{\Gamma(K^+ \to \pi^+\nu\bar{\nu})}{\Gamma(K^+ \to \text{all})} < 3 \times 10^{-8} ,$$

and for $K^+ \to \pi^+a$

$$\frac{\Gamma(K^+ \to \pi^+a)}{\Gamma(K^+ \to \text{all})} < 6 \times 10^{-9} ,$$

where a represents any light, non-interacting particle such as an axion or familon. Present indications are that the experiment may be limited by the available flux of kaons rather than by background processes for the region of phase space (above the $K_{\pi 2}$ peak) being examined. If the standard model prediction is valid then at most a few events from $K^+ \to \pi^+\nu\bar{\nu}$ can be expected even allowing for the increased intensity expected with the AGS booster which is under construction. To examine the spectrum for consistency with the standard model, to investigate any new phenomena which might eventually turn up or to continue the search for this unique process will require the higher intensity available at the kaon factory.

A conceptual design for a kaon factory detector for $K^+ \to \pi^+\nu\bar{\nu}$ capable of operating in a flux of one to two orders of magnitude greater than presently available is shown in figure 4. The basic configuration is similar to that of 787, although the magnetic field strength is 3 T, three times stronger. The primary motivations for the high field are improvement of the momentum resolution, improvement of the pion range stack tracking using greater segmentation (e.g. scintillation fibers) and

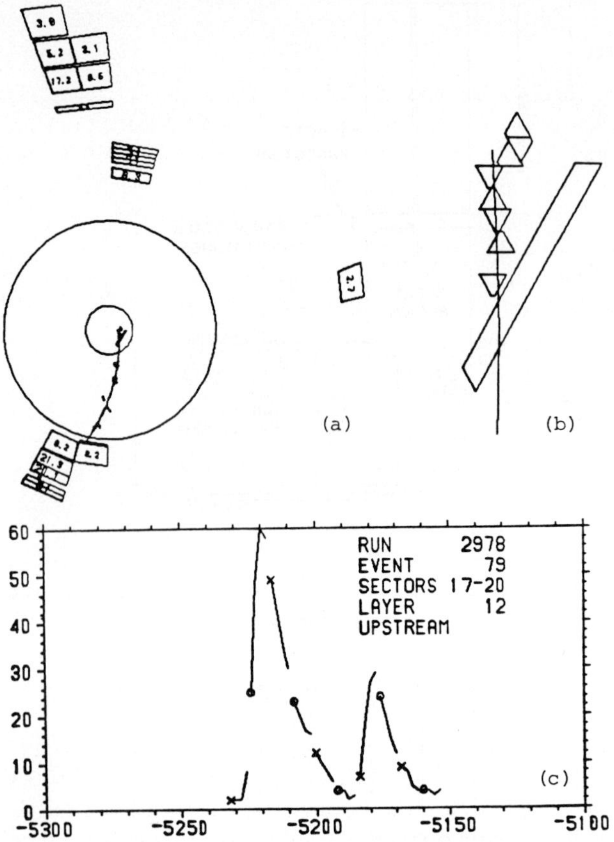

Figure 3: $K^+ \to \pi^+\pi^0$ event in the BNL-787 detector (see text).

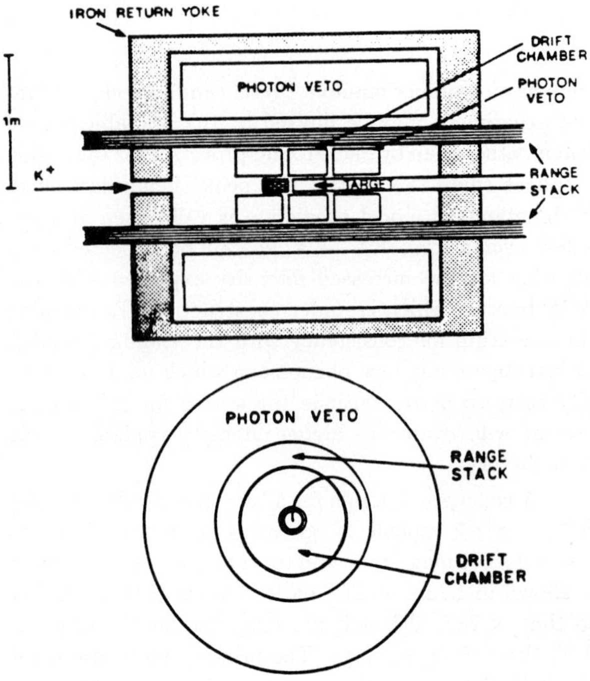

Figure 4: Conceptual design of a kaon factory experiment to measure $K^+ \to \pi^+\nu\bar{\nu}$.

improvement of the photon veto efficiency by the use of a fully active detection medium such as BaF_2. The later two improvements are possible due to the reduced size of the high field tracking apparatus which allows significant reduction in the numbers of channels needed and in the overall volume. To handle the high rates anticipated, all detector channels would be instrumented with 0.5–1.0 GHz transient digitizers like the ones used in 787 [8] and the GaAs CCD's under development at TRIUMF. A sensitivity of $< 4 \times 10^{-12}$ could be achieved with background levels estimated to be < 1 event. Alternatively, if the branching ratio is 5×10^{-10}, 200 to 300 events would be observed.

B. CP and T violation

CP violation has only been observed in the neutral kaon system in $K_L^0 \to 2\pi$ decays and in the charge asymmetry in $K_L^0 \to \pi e^{\pm}\nu$ (K_{l3}^0) decays. In the standard model with at least three generations a CP-violating phase can be accommodated in the quark-mixing matrix. The magnitude of CP violation is indicated by the parameter $\epsilon \sim 10^{-3}$, which has as its source the K^0, $\overline{K}^0$ mass matrix. CP violation is manifested by the level of CP impurity of K_L^0 and K_S^0 states. A second possible source of CP violation originates directly from the $K \to 2\pi$ decay amplitude and is represented by the parameter ϵ'. A recent CERN experiment [9] (NA31) reported consistency with the CKM picture of CP violation, finding a non-zero value (at the three-standard deviation level) for the ratio $\epsilon'/\epsilon = (3.3 \pm 1.1) \times 10^{-3}$. Fermilab experiment E731 [10] which collected over 300K $K_L^0 \to 2\pi^0$ decays is expected to report a result for ϵ'/ϵ later this year with comparable or greater precision. Whether a non-zero value of ϵ'/ϵ is confirmed or (especially) if an inconsistency appears future higher precision experiments with perhaps 10^8 $K_L^0 \to 2\pi^0$ events will be required for the next generation of experiments studying the origin of CP violation. This would allow the statistical precision to approach the 10^{-4} level, representing an order of magnitude improvement and presenting a severe challenge to the standard model. Of course, knowledge of systematic uncertainties must also be improved commensurately. Hence, the kaon factory has a definite role to play by permitting the creation of extremely clean beams (e.g. by charge exchange) while maintaining sufficient intensity or, perhaps, by employing entirely new techniques. New experiments at an upgraded Fermilab Main Injector have also been suggested [11].

One promising approach to the CP violation problem that may merit kaon factory intensities for future work is the production of pure, tagged K^0 and $\overline{K}^0$ states in $p\bar{p}$ annihilation. This method which is being pursued for the first time at LEAR [12], employs the reactions $p\bar{p} \to K^0K^-\pi^+$ and $p\bar{p} \to \overline{K}^0K^+\pi^-$ (branching ratios $\sim 10^{-3}$) in which the $K^0(\overline{K}_0)$ state is tagged by $K^-\pi^+(K^+\pi^-)$. Time dependent CP violating asymmetries

$$A_f(t) = \frac{\Gamma(\overline{K}^0 \to f)(t) - \Gamma(K^0 \to f)(t)}{\Gamma(\overline{K}^0 \to f)(t) + \Gamma(K^0 \to f)(t)} ,$$

where f represents a final state such as $\pi^0\pi^0$, can be measured with estimated precision comparable to the current round of tests at CERN and Fermilab. These measurements of ϵ'/ϵ as well as the associated phases and other CP violating K_S and K_L decay modes will be done with quite a different set of systematic uncertainties than in the CERN and Fermilab experiments and, thereby, will represent an important consistency check.

A further use of antiprotons in the quest for information on CP (or T) violation may come to fruition at the kaon factory by studying asymmetries of hyperon polarizations in reactions such as $p\overline{p} \to \Lambda\overline{\Lambda}$ and $p\overline{p} \to \overline{\Xi}^+\Xi^-$ as discussed by Hamann [13]. Such reactions could provide a clean way to study CP violation outside the neutral kaon system since large hyperon polarizations (required by parity conservation to be transverse to the production plane) occur in the context of initial and final states with definite CP properties. Non-zero measurement of one of several possible observables would constitute definite evidence for $\Delta S = 1$ CP violation (since due to baryon number conservation there is no final state mixing). Predictions for the CP violating asymmetries are at the 10^{-4} level while current experiments on $p\overline{p} \to \Lambda\overline{\Lambda}$ [14] and $J/\psi \to \Lambda\overline{\Lambda}$ [15] are in the neighborhood of 10^{-2} and new experiments at LEAR may obtain an additional factor of ten in sensitivity (see ref. 13). To confront the standard model predictions one or two orders-of-magnitude more intensity will be necessary. Studies of cascade production $p\overline{p} \to \overline{\Xi}^+\Xi^-$ which are particularly attractive due to self-analyzing decay modes $\Xi \to \Lambda\pi$ (up-down asymmetry measured) and $\Lambda \to p\pi$ (Λ polarization measured) can occur with intense higher energy $\overline{p}$ beams at $\sqrt{s} > 2.64$ GeV/c.

Additional sources of CP or T violation could be revealed in rare kaon decay processes involving measurements of the transverse muon polarization in $K \to \pi\mu\nu$ ($K_{\mu 3}$) decays not expected in the standard model. The presence of nonzero muon polarization transverse to the decay plane is an indicator of T violation due to the T-odd product $\sigma_\mu \cdot (\overline{P}_\mu \times \overline{P}_\pi)$, where σ_μ is the muon polarization, and $\overline{P}_\mu$ and $\overline{P}_\pi$ are the muon and pion momentum vectors. In $K_{\mu 3}$ decay this effect might be due to the interference between the two form factors $f_+(\overline{P}_K + \overline{P}_\pi)$ and $f_-(\overline{P}_K - \overline{P}_\pi)$, since T-invariance requires f_+ and f_- to be relatively real. The results derived from measurement of the transverse μ polarization can be expressed in terms of Imξ, where Im$\xi \propto \langle \overline{\sigma}_\mu \cdot (\overline{P}_\mu \times \overline{P}_\pi) \rangle m_K$ and $\xi = f_-/f_+$. The results of a $K^+ \to \pi^0\mu^+\nu_\mu$ study [17] gave Im$\xi = -0.016 \pm 0.025$. Combining this with a $K_L^0 \to \pi^+\mu^-\nu_\mu$ experiment [16] (keeping in mind the possibility of complications due to electromagnetic final state interactions), the result is Im$\xi = -0.010 \pm 0.019$. Although, a null value for Imξ is consistent with the expectation based on the standard CKM model, the Weinberg Higgs model [18] of CP violation would predict Im$\xi \sim 10^{-3}$, an order of magnitude below the present limit. $K_{\mu 3}$ studies are ripe for a new generation of experiments using newer techniques and significant progress could be made even prior to the kaon factory era. Another semileptonic kaon decay, K_{e4}, has been examined theoretically by Castoldi, Frère and Kane [19] as a promising reaction to study for testing T-invariance.

The decay $K_L^0 \to \pi^0 e^+e^-$ is a rare example of a reaction which can proceed through both CP-conserving and CP-violating paths at potentially comparable rates. Since K_L^0 consists of the CP odd state K_2 with a small admixture of the CP even state K_1, decays proceeding through two virtual photons and through a single virtual photon are, respectively, possible. Various calculations indicate the CP-conserving and CP-violating amplitudes may be comparable and, furthermore, that the CP-violating components due to the mass matrix ($\Delta S = 2$) and the direct 2π amplitude ($\Delta S = 1$) may also be comparable [20]. Essential theoretical work is in progress to understand this reaction and new experiments have been mounted at BNL, FNAL and KEK (see ref. 2). Because the ranges of calculated values for the CP violating components and the CP conserving components (both due to mixing and direct contributions) are wide and overlap, there would be considerable difficulty in interpreting an observation of $K_L^0 \to \pi^0 e^+e^-$ based on the rate alone. A measurement of $K_S^0 \to \pi^0 e^+e^-$ (estimated to be at the 10^{-10} to 10^{-8} level [21]) would provide the most reliable input for determining the CP violating part of the $K_L^0 \to \pi^0 e^+e^-$ amplitude due to mixing (i.e., the K_1 component). There may be sufficient variation in Dalitz plots to enable one to distinguish the CP violating from the CP conserving components if adequate statistics were available (a formidable task in light of the small branching ratio expected). Sehgal [22] calculated the phase of the 2γ amplitude and the interference between the 1γ and 2γ contributions to arrive at another possible observable, a CP violating asymmetry between e^+ and e^- energies. Littenberg [2] has suggested measurement of the time dependence. In any case, since the branching ratio is expected to lie in the 10^{-12} to 10^{-11} region and significant statistics will be necessary to unravel the various contributions, $K_L^0 \to \pi^0 e^+e^-$ is certainly an important reaction for study at the kaon factory to help elucidate the mechanism of CP violation.

Further evidence of CP violation would be observation of longitudinal muon polarization in $K_L^0 \to \mu\mu$ decay. In the context of the standard model [23], the longitudinal polarization

$$P_L = \frac{N_L - N_R}{N_L + N_R} \sim 10^{-3} ,$$

where N_L (N_R) is the number of left- (right-) handed muons. However, much larger values can occur in Higgs models of CP violation. Here is an excellent example of a process which can be used to search for new effects such as alternate sources of CP violation while ultimately attempting to confront the standard model prediction. No experiments have yet been done and it appears that the intense beams at a kaon factory would be required to reach the 10^{-3} level.

Another interesting prospect for studying CP violation is the reaction $K_L^0 \to \pi^0\nu\bar{\nu}$. Although the branching ratio is expected to be in the 10^{-12} region making the experimental problems appear daunting, high motivation is provided by the indication that the only significant contribution comes from direct CP violation in the decay amplitude [24]. Aside from the small branching ratio and the obvious difficulty of interpreting the single π^0 signal, this process would avoid the ambiguities inherent in studying $K_L^0 \to \pi^0e^+e^-$ discussed above.

C. Lepton flavor violation

Searches for rare kaon decays not expected in the standard model could also contribute dramatic new information. Lepton flavor violating (LFV) interactions are strictly absent in the standard model with massless neutrinos because neither the intermediate vector bosons nor the Higgs particle have LFV couplings. However, in many extensions of the standard model LFV interactions appear naturally, leading to decays like $K_L^0 \to \mu e$ and $K^+ \to \pi^+\mu e$. Among these are models in which flavor violations are mediated by horizontal gauge bosons, additional neutral Higgs particles, vector or pseudoscalar leptoquarks and supersymmetric particles. The mass regions probed by rare kaon processes reach scales of order 100 TeV/c^2, which are inaccessible to direct experiments at any existing or planned high energy accelerator. Table 1 (from ref. 1) gives a sample of the mass regions probed by current experiments. Thus, although kaon decay experiments are generally performed at relatively low energies, their implications are relevant and complementary to studies done at the highest energy facilities.

Table 1: Mass bounds from different processes.

Process	Higgs scalars (GeV/c^2)	Pseudoscalar leptoquarks (TeV/c^2)	Vector leptoquarks (TeV/c^2)	Experimental value
$\frac{\Gamma(K_L^0 \to \mu\bar{e})}{\Gamma(K_L \to \text{all})}$	11	8	149	$<3 \times 10^{-10}$ [a]
$\frac{\Gamma(K_L^0 \to \mu\bar{\mu})}{\Gamma(K_L \to \text{all})}$	4.7	3.6	62	9×10^{-9} [b]
$\frac{\Gamma(K_L^0 \to e\bar{e})}{\Gamma(K_L \to \text{all})}$	8	2.6	108	$<1.2 \times 10^{-9}$ [c]
$\frac{\Gamma(K^+ \to \pi^+\mu e)}{\Gamma(K^+ \to \text{all})}$	1	0.5	5.6	$<1.8 \times 10^{-9}$ [d]
$\frac{\Gamma(\mu \to e\gamma)}{\Gamma(\mu \to \text{all})}$	0.3	–	–	$<4.9 \times 10^{-11}$ [e]
$\frac{\Gamma(\mu \to ee\bar{e})}{\Gamma(\mu \to \text{all})}$	2.6	–	–	$<1.0 \times 10^{-12}$ [f]
$\frac{\Gamma(\mu Z \to eA)}{\Gamma(\mu Z \to \nu Z')}$	22	22	118	$<4.6 \times 10^{-12}$ [g]
$\Delta m(K_L^0 - K_S^0)$	150	–	–	3.5×10^{-15} GeV [b]

[a]W.R. Molzon, Proc. Rare Decay Symp., Vancouver (1988).
[b]Particle data group, Phys. Lett. 170B, (1986) 1.
[c]E. Jastrzembski *et al.*, Phys. Rev. Lett. 20, (1988) 2300.
[d]M. Zeller, Proc. Rare Decay Symposium, Vancouver (1988).
[e]R.D. Bolton *et al.*, Phys. Rev. Lett. 56, (1986) 2461.
[f]U. Bellgardt, Nucl. Phys. B299, (1987) 1.
[g]S. Ahmad *et al.*, Phys. Rev. D38, (1988) 2102.

The ideas of quark and lepton substructure are motivated by the proliferation of fundamental particles and the lack of understanding provided by present theories. The existence of common substructure entities could conceivably lead to intergenerational rearrangement processes in which quarks and leptons are interchanged. Examples of reactions in which the net change of generation number is zero are $K_L^0 \to \mu e$, $K^+ \to \mu^+\nu_e$, and $K^+ \to \pi^+\mu^+e^-$. In some models these processes are mediated by exchange of heavy bosons which distinguish the generations.

The present round of rare decay experiments at BNL and KEK has already begun to produce significant new results as indicated in table 1. Because of the leverage that these experiments have in probing the standard model as well as in searching for new effects it can be expected that additional work taking advantage of new technology will be pursued at these facilities and that further improvements will be realized at the kaon factory. To chose an example, AGS experiment 791 [25] is now aimed at a sensitivity of approximately 10^{-11} for $K_L^0 \to \mu e$ decay. If the final result is null, then a strong motivation exists, as it does now, for pursuing this reaction since it is one of the most favored in models which extend the boundaries of the standard model. A positive result would create a whole industry of experiments to explore the new effect. It appears that with evolutionary improvements to the beam, chamber systems and triggering and data acquisition systems that a level of sensitivity of $< 10^{-13}$ could be achieved at the kaon factory using beams of one to two orders of magnitude higher intensity than are presently available. More drastic revisions to the approach, such as attempting to increase the acceptance by an order of magnitude are worthy of consideration as well. Here, experience at the meson factories is relevant. The two orders of magnitude in flux compared with "pre-factory" machines coupled with advances in technology have already led to five orders of magnitude improvements in sensitivity of rare muon decay experiments with major advances still underway.

D. Other rare processes

Other allowed rare kaon processes that might be accessible at the kaon factory could enable searches for exotic effects or new particles. The decays $K \to \pi e^+e^-$ and $K \to \pi\mu^+\mu^-$, involving both neutral and charged mesons, can be used to search for light scalar particles, e.g. Higgs particles which decay via $\mathrm{H} \to l^+l^-$. A new limit from BNL E787 [26] on the search for light Higgs particles has been obtained for $2\mathrm{m}_\mu < \mathrm{m_H} < 320$ MeV/c^2:

$$\frac{\Gamma(K^+ \to \pi^+\mathrm{H}\,,\ \mathrm{H} \to \mu^+\mu^-)}{\Gamma(K^+ \to \mathrm{all})} < 1.5 \times 10^{-7}\ (90\%\mathrm{C.L.})\ .$$

From the same data an upper limit on the branching ratio for $K^+ \to \pi^+\mu^+\mu^-$ was found to be $< 2.1 \times 10^{-7}$ (90% C.L.). Several experiments have been performed to search for heavy neutrinos in $M \to l\nu$ decays, where $M = \pi$ or K and $l = \mu$ or e. Heavy neutrinos ν_i are not prohibited by any model (although there are some cosmological and other indirect constraints on masses and lifetimes) and could be mixed with the dominantly coupled light neutrinos if the weak eigenstates ν_e, ν_μ, $\nu_\tau \ldots$ are distinct from the mass eigenstates.

Other rare kaon decays, such as $K_L^0 \to \gamma\gamma$, $K_S^0 \to \gamma\gamma$ and $K^+ \to \pi^+\gamma\gamma$ are of interest due to the unknown aspects of low energy QCD and as additional sources of information on CP violation. Long distance (i.e., mesonic) effects generally contribute significantly to the uncertainties of calculations of such processes.

3. CONCLUSION

The kaon factory will provide a new capability for high precision, high sensitivity particle physics experiments that is unique and complementary to many current and proposed efforts. The high intensity will be especially useful in confronting rare processes. Not only will higher statistical sensitivity be achievable but, in addition, one will be able to trade flux for purity making for potentially reduced systematic uncertainties and backgrounds. The sample of inviting possibilities discussed above is by no means meant to be complete or even representative of the program at the kaon factory which will be driven by future physics issues and priorities. What is more certain is that as new intense beams with hundredfold higher fluxes become available, they will be used in imaginative and productive ways to significantly aid in the advancement of physics.

References

[1] D.A. Bryman, Int. Jour. Mod. Phys. A 4, (1989) 79.

[2] L.S. Littenberg and J.S. Hagelin, to be published.

[3] J. Ellis and J.S. Hagelin, Nucl. Phys. B217, (1983) 189.

[4] D. Rein and L.M. Sehgal, Aachen preprint PITHA 89-03 (1989).

[5] Y. Asano *et al.*, Phys. Lett. 107B, (1981) 159.

[6] R. Shrock, *Proc. DPF Summer Study of Elementary Particle Physics and Future Facilities*, ed. R. Donaldson *et al.*, (Snowmass, 1982) p. 291.

[7] See Y. Kuno *et al.*, *Proc. Workshop on Rare Decays*, ed. J.-M. Poutissou *et al.*, (Vancouver, 1988), to be published; and D. Marlow *et al.*, Proc. of the 4th Family Workshop, (UCLA, 1989), to be published.

[8] M. Atiya *et al.*, IEEE Nucl. Sci. Symposium, Orlando, FL, Nov. 1988.

[9] H. Burkhardt *et al.*, Phys. Lett. 206B, (1988) 169.

[10] FNAL E731; M. Woods *et al.*, Phys. Rev. Lett. 60, (1988) 1695; L.K. Gibbons *et al.*, Phys. Rev. Lett. 61, (1988) 2661.

[11] B. Winstein, G.J. Bock and R. Coleman, FNAL EFI 89-01.

[12] L. Adiels *et al.*, Proposal for experiment PS195 CERN/PSCC/85-6, PSCC/P-82 (1985).

[13] See N. Hamann, *Proc. Workshop on Rare Decays*, *ibid.*

[14] P.D. Barnes *et al.*, Phys. Lett. 189B, (1987) 249.

[15] M.H. Tixier *et al.*, Phys. Lett. 212B, (1988) 523.

[16] S.R. Blatt *et al.*, Phys. Rev. D27, (1983) 1056.

[17] W.M. Morse *et al.*, Phys. Rev. D21, (1980) 1750.

[18] S. Weinberg, Phys. Rev. Lett. 37, (1976) 651.

[19] P. Castoldi, J.-M. Frère and G.L. Kane, Univ. Michigan preprint UM-TA-88-26.

[20] See C.O. Dib, I. Dunietz and F.S. Gilman, to be published, SLAC-PUB-4762, and J. Flynn and L. Randall, LBL preprint UCB-PTH-88-21 and references therein.

[21] M.K. Gaillard and B.W. Lee, Phys. Rev. D10, (1974) 897.

[22] L.M. Sehgal, Phys. Rev. D38, (1988) 808.

[23] P. Herczeg, Phys. Rev. D27, (1983) 1512; L. Wolfenstein, *Selected Topics in Electroweak Interaction*, ed. J.M. Cameron *et al.*, (World Scientific, Singapore, 1978), p. 236.

[24] L. Littenberg, to be published.

[25] R.D. Cousins *et al.*, AGS experiment 791.

[26] M. Selen, Ph.D. dissertation, Princeton University (1980), unpublished.

Z. Phys. C – Particles and Fields 46, S73–79 (1990)

Zeitschrift für Physik C Particles and Fields

Problems in neutrino physics

F. Scheck

Institut für Physik, Johannes-Gutenberg-Universität, Postfach 3980, D-6500 Mainz, Fed. Rep. of Germany

1 Introduction

The three "known" neutrinos ν_e, ν_μ, and ν_τ, as well as their postulated heavy companions, have been the subject of intense experimental and theoretical research over the last decade. And yet, all the basic questions that we raised when the first, controversial, results on the mass of electron neutrino from triton beta decay were announced and when the new generation of experiments on lepton family changing processes started, remain essentially unanswered, in spite of the impressive effort that went into neutrino research [1]. So we may just repeat some of these questions and then see what we know about them and what remains to be done. The most important question is, of course,

(i) Are neutrinos massive?

The present experimental limits on the masses of ν_e, ν_μ, ν_τ, obtained after many years of hard and skillful experimental work are

$m(\nu_e) < 18$ eV (from ^{3}H beta decay) [2]

$m(\nu_\mu) < 250$ keV (from $\pi \rightarrow \mu\nu$ decay) [3]

$m(\nu_\tau) < 35$ MeV (from $\tau \rightarrow 5\pi^\pm \nu_\tau$ decay) [4].

These numbers, however impressive experimental achievements they represent, must be compared with typical expectations within any of the local gauge theories built to incorporate and to extend the minimal standard model. For instance, grand unified theories based on the structure group SO(10) would yield in a most optimistic version, 10^{-5} eV for the ν_e, 10^{-2} eV for the ν_μ and only little more than that for ν_τ.

In fact, it is rather natural to have nonvanishing neutrino masses, within the context of any of the "more unifying" theories that have been proposed to "explain" the standard model. We certainly do not know of any general principle which would impose strict masslessness for electrically neutral fermions. It is clear that the first question is open and that there remains a large window of ignorance between the expected values and the limits quoted above. Question (i) is then followed by many more, all of which are related in some way or other to the mass sector of neutral leptons. They include

(ii) Why are the masses ν_e, ν_μ, ν_τ so small and where are their heavy partners?

Indeed, practically all realistic models of neutral lepton masses endow the light neutrinos with heavy partners (see below) which should show up, some way or other, in weak interaction processes.

(iii) Are neutrinos Majorana particles?

As I will show below, it is natural that at least some neutral leptons are Majorana (i.e. self-conjugate) rather than Dirac (i.e. charge-carrying) particles. Majorana fermions in some sense, are more fundamental than Dirac fermions.

(iv) Do the mass eigenstates mix in the weak interactions? If so, what is the pattern of neutrino mixing?

Once neutrinos are massive and their masses are not degenerate, it is likely that the states which couple to weak interaction vertices are not identical with the eigenstates of the mass matrix. There are, however, two pecularities as compared to the better known case of quarks: Firstly, the mixing of quarks is defined as a physical, observable phenomenon because their strong and electro-weak interactions obey different selection rules. Neutrinos have no other interaction than weak (and gravitational) interactions. So what defines the reference system with respect to which the weak interaction eigenstates are rotated? Secondly, neutrinos being eigenstates of helicity (at least to a very good approximation), the mixings of right-handed and of left-handed states are largely independent and can, in fact, be quite different.

(v) Do neutrino oscillations occur? And how are these oscillations modified in the presence of matter?

(vi) Does neutrinoless double β-decay $(\beta\beta)_{o\nu}$ occur?

(vii) What can we expect for the decay modes and rates of massive neutrinos? Are the rates sufficiently large to circumvent the bound on neutrino masses from cosmology?

Massiveness of neutrinos and state mixing in the neutral lepton sector unavoidably raises the question of existence and nature of additively conserved quantum numbers of neutrinos and their charged partners. In fact, it is very likely that lepton family changing processes will become predictable once the leptonic mass sector is understood. This is the more attractive possibility, at least, than to invoke some new very high energy scale, otherwise untestable, if $\mu\rightarrow e\gamma$ were indeed found. Therefore, to our list we should add a few more questions such as

(viii) Is total lepton number L conserved? Are the individual family numbers L_f conserved, f standing for e, μ, τ?

(ix) Are processes with $\Delta L_f \neq 0$ more likely to occur at higher energies?

The present state of experimental knowledge is easy to summarize: There is no direct evidence for any massive neutrino. No oscillations have been observed, no admixture of heavy neutrinos (or, for that matter, their supersymmetric partners) in the dominant light neutrino states has been found (e.g. in $\pi\rightarrow e\nu$, $\pi\rightarrow\mu\nu$, $K\rightarrow e\nu$, etc.). The limits on squares of mass differences and/or on mixing matrix elements have been improving rather impressively over the last years but there remain windows which are not explored because they are not accessible for kinematic reasons [5]. No neutrinoless double β-decay was positively observed but the resulting limits on the ν_e (Majorana) mass are somewhat weak because of possible destructive interference of different mass eigenstates [6]. Finally, as yet no family number changing process was observed, the limits on branching ratios such as $\Gamma(\mu\rightarrow e\gamma)/\Gamma(\mu\rightarrow all)$ having attained the level of 10^{-12}.

All in all, this state of affairs is disappointing but not discouraging: Disappointing because the hope was that as the precision of experiments had improved so impressively, there would be at least one sector where signals of new physics beyond the standard model would show up in an unambiguous way. It is not discouraging, though, because the characteristic scales of that new physics may be such that its effects at intermediate energies may be just around the corner.

2 Majorana particles are more fundamental than Dirac particles

Ever since the possibility of neutrinos being massive was considered the question was raised whether they are Majorana (i.e. self-conjugate) particles or Dirac particles like the electron (i.e. carrying some additively conserved quantum number). In this section I will substantiate and illustrate the following statements:

- Majorana particles are the natural building blocks in a theory of spin-$\frac{1}{2}$ particles in interaction with the weak gauge bosons and the Higgs particles,
- Diagonalization of the neutral lepton's mass matrix normally leads to Majorana eigenstates. Dirac states appear only when two or more mass eigenvalues are degenerate,
- Only when some global U(1)-symmetry is imposed on the theory, corresponding to an additively conserved quantum number such as L, L_f, etc., the mass eigenstates (originally Majorana states) are linearly combined to Dirac states.

As is well-known fermion fields carry four degrees of freedom, two for the spin, e.g. helicity ±1, and two from particle-antiparticle conjugation - unless the field Ψ and its charge conjugate Ψ_c are proportional to one another. Indeed, if $\Psi_c=\eta\Psi$, with η a phase, then only two degrees of freedom remain. The question is how these degrees of freedom are realized. The standard picture is that $\nu_f(-)$, the neutrino of family f with helicity -1 carries $L_f=+1$, while $\bar{\nu}_f(+)$, the antineutrino with helicity +1, carries $L_f=-1$. These states are the ones which couple to $W^{\pm}$- and Z^0-bosons, all vertices involving the neutrino field Ψ_{ν_f} being coupled to them via the combination (v-a) of vector and axial vector currents. Neutrinos $\nu_f(+)$ with positive helicity and antineutrinos $\bar{\nu}_f(-)$ with negative helicity do no couple to anything. So, with this assignment, neutrinos seem to be Dirac particles: ν_f and $\bar{\nu}_f$ carry opposite lepton (family) numbers.

A closer look shows that this conclusion is not justified and that the assignment is not unique. First of all, as long as neutrinos are massless (or, for that matter, have degenerate masses) there is no distinction between Dirac and Majorana particles. Any pair of mass-degenerate fields of one kind can be linearly combined to a field of the other kind. Further confusion stems from the fact that weak interactions seem to couple exclusively to the left-handed combination (v-a) so that one cannot distinguish between the aforementioned helicity selection rule and the conservation of lepton family number L_f.

We note in passing that the helicity of ν_μ in the decay $\pi^+\to\mu^+\nu_\mu$ is now known to be -1 within less than 4 per mil, $|h(\nu_\mu)|>0.9968$ (90% C.L.). This follows from the Berkeley-Northwestern-TRIUMF measurement of the asymmetry in $\mu^+\to e^+$ ($\nu\bar{\nu}$) decay near the kinematic end point [7] and from angular momentum conservation, as was pointed out by Fetscher [8]. Near the end point the differential decay probability has the form

$$d\Gamma = \Gamma_o \; (1 - P_\mu \frac{\xi\delta}{\rho} \cos\Theta) \;,$$

where P_μ is the polarization of μ^+ from $^+$-decay, Θ is the angle between the muon spin and the positron momentum, while ξ, δ and ρ are the well-known asymmetry and decay parameters of muon decay. Obviously, neither the combination $P_\mu\xi\delta/\rho$ nor P_μ alone can exceed 1. As the former was measured to be larger than 0.9968, one can conclude that P_μ itself must lie between this value and 1, and therefore, by angular momentum conservation, that the helicity of ν_μ is indeed very close or equal to its maximal value.

Let $\nu^i(+)$ and $\nu^i(-)$ be Majorana neutrinos with helicity ±1. Then, under CPT, $\nu^i(+)\underset{CPT}{\leftrightarrow}\nu^i(-)$. If the states $\nu^1(\pm)$ and $\nu^2(\pm)$ have the same mass then

$$\nu_D(\pm) := \nu^1(\pm) + i\,\nu^2(\pm)$$

is a Dirac state, and $\nu^1(\pm)-i\nu^2(\pm) = (\nu_D(\pm))_c$ is its charge conjugate. Under the effect of CPT, we have $\nu_D(+)\underset{CPT}{\leftrightarrow}(\nu_D(-))_c\equiv\bar{\nu}_D(-)$ (possibly up to phases/signs). Whatever the mechanism is which is responsible for creating the neutral lepton masses, Lorentz invariance alone allows for terms of the kind

$$(\bar{\Psi}\Psi),\; (\bar{\Psi}_c\Psi),\; \text{and}\; (\bar{\Psi}\Psi_c) \;.$$

The first of these conserves any U(1)-quantum number such as L_f, attached to the field Ψ, the other two terms change that quantum number by 2 units. Expressing the 4-spinor Ψ in terms of right- and left-handed two-component fields φ and χ, respectively, (these are the well-known Weyl-Van-der-Waerden spinors),

$$\Psi = \begin{pmatrix} \varphi \\ \psi \end{pmatrix}$$

there are two classes of mass terms [9], viz. the Dirac terms

$$\mathcal{L}^{D}_{mass} = m_D(\chi^*,\varphi) - m_D^*(\varphi^*,\chi), \qquad (1)$$

and the Majorana terms

$$\mathcal{L}^{M}_{mass} = \frac{1}{2}\left[m_1(\varphi,\varphi) + m_1^*(\varphi,\varphi)^* - m_2(\chi,\chi) - m_2^*(\chi,\chi)^*\right] \qquad (2)$$

(I tend to avoid the conventional notation $m_D\bar{\Psi}_L{}^+\Psi_R + m_D^*\bar{\Psi}_R{}^+\Psi_L$ for the former, $m_1\Psi_R{}^+\sigma^2\Psi_R + m_2\Psi_L{}^+\sigma^2\Psi_L$+h.c. for the latter, because of the possible confusion due to the noncommutativity of the bar on $^-$ and the index left or right, $(\overline{\Psi_L}) = (\bar{\Psi})_R$ etc.) The quantities m_1, m_2, m_D are complex parameters with dimension of a mass. Both forms (1) and (2) obey Lorentz invariance and hermiticity. The bilinear operator $(\cdot,\cdot)$ is meant to be a true scalar under Lorentz transformations and under symmetry group G which is imposed on the theory. If G is nontrivial, Φ and χ are multi-component fields corresponding to a representation of G. As this is not essential for our discussion we assume that this does not occur, i.e. that there is just one such two-component field Φ and similarly one such χ field. We then define two Majorana fields

$$\Phi^{(1)} := \begin{pmatrix} \chi^* \\ \chi \end{pmatrix} \quad \begin{pmatrix} \varphi^{(1)} \\ -\varphi^{(1)*} \end{pmatrix} ;$$

$$\Phi^{(2)} := \begin{pmatrix} \varphi \\ -\varphi^* \end{pmatrix} \quad \begin{pmatrix} \varphi^{(2)} \\ -\varphi^{(2)*} \end{pmatrix} \qquad (3)$$

and use the fields $\varphi^{(1)} \equiv \chi^*$ and $\varphi^{(2)} \equiv \Phi$ as the basis of states for diagonalizing the mass sector. (This is a simple notational trick. The details are worked out in Ref.9.) The complete mass Lagrangian, sum of eqs.(1) and (2) is then

$$\mathcal{L}_{mass} = -\frac{1}{2}\sum_{i,j=1}^{2} m_{ij}\,(\varphi^{(i)},\varphi^{(j)}) + \text{h.c.} \qquad (4)$$

where the matrix $\{m_{ij}\}$ is given by

$$\{m_{ij}\} = \begin{pmatrix} m_1 & m_D \\ m_D & m_2 \end{pmatrix} \qquad (5)$$

Diagonalization of this matrix is straightforward [9,10,11]. Setting

$$m_{1/2} = \mu_{1/2}\, e^{i\varphi_{1/2}}, \quad m_D = \mu_D\, e^{i\varphi_D} \qquad (6)$$

with μ_1, μ_2, μ_D real and positive, the eigenvalues λ_1 and λ_2 are found to be functions of the real parameters μ_1, μ_2, μ_D and $\Phi := \varphi_D - \frac{1}{2}(\varphi_1 + \varphi_2)$. The corresponding, orthogonal, eigenstates are obviously Majorana states, not Dirac. They can be combined to Dirac states only if $\lambda_1 = \lambda_2$. This, however, is accidental - unless an extra U(1) symmetry is imposed on $\mathcal{L}_{mass}$. The "accident" $\lambda_1 = \lambda_2$ occurs if either (i) $\mu_D = 0$ and $\mu_1 = \mu_2$, or (ii) $\mu_D \neq 0$, $\mu_1 = \mu_2$ and $\Phi \equiv \varphi_D - \frac{1}{2}(\varphi_1 + \varphi_2) = (2n+1)\frac{\pi}{2}$. One concludes therefore: As long as no U(1)-symmetry is imposed on the theory (i.e. no additively conserved charge-like quantum number is assumed from the start) the mass eigenstates of the neutral lepton sector are Majorana rather than Dirac states. In this sense Majorana fields are the natural building blocks in a theory of neutral leptons. Dirac particles appear only if there is a "conspiracy" which wants masses to be degenerate or if extra symmetry is imposed on the theory.

3 Models for the neutral mass sector, the see-saw mechanism

The structure (5) of the mass matrix, or, more generally, the form

$$\begin{pmatrix} (M_1) & (M_D) \\ (M_D) & (M_2) \end{pmatrix} \qquad (7)$$

where (M_i), (M_D) denote Majorana and Dirac submatrices, respectively, is typical for many extensions of the standard model such as left-right symmetric models based on the structure group $SU(2)_L \times SU(2)_R \times U(1)$, grand unified theories, or supersymmetric extensions thereof. The technical details of specific models sometimes complicate the picture even though the basic mechanism is very similar. In order to exhibit the salient features in a transparent manner we restrict the discussion to the simple case (5) of two basic states.

Fermion masses are thought to arise from Yukawa couplings to scalars (the Higgs particles) which develop nonvanishing vacuum expectation values. It is natural to assume that the mass parameter μ_D, eq. (6), has magnitude typical for charged leptons, i.e. $\mu_D \simeq m_f$, where m_f is the mass of the charged lepton in a given family $f = e, \mu, \tau$. This term can be generated by the usual Higgs doublet of the standard model. In contract to this, the left-type Majorana term μ_1 requires (at least) one new triplet of Higgs particles, if the mass Lagrangian is to obey the global symmetry of the theory. Such a triplet shows up in the parameter $\rho = M_W^2/M_Z^2 \cos^2\Theta_W = ((v_2)^2 + 2(v_3)^2/((v_2)^2 + (v_3)^2)$, where v_2 and v_3 are the vacuum expectation values of the neutral member of the doublet and triplet, respectively. The experimental value of ρ being very close to 1, v_3 can at most be a hundreth of v_2. Thus, unless the Yukawa coupling constants of the two terms are wildly different, one expects μ_1 to be very small as compared to μ_D: $\mu_1 \lesssim \frac{m_f}{100} \ll \mu_D$.

The origin of the right-handed Majorana term μ_2 is to be found outside the standard model. Indeed, extensions of the minimal model attribute this term to a new scale much larger than the typical scales of the standard model. Hence one expects μ_2 to be much larger than μ_D.

With the ordering $\mu_1 \ll \mu_D \ll \mu_2$ the eigenvalues of the mass matrix (5) are, approximately,

$$\lambda_1 \simeq \left|\mu_1 - \frac{\mu_D^2}{\mu_2} \cos 2\Phi\right|, \quad \lambda_2 \simeq \mu_2 \qquad (8)$$

corresponding to two Majorana mass eigenstates. Thus, there is always a light neutral lepton whose mass is controlled by the assumed Higgs triplet and/or by the ratio of the charged lepton mass ($m_f \simeq \mu_D$) over the new scale and the Dirac-Majorana interference phase $\Phi = \varphi_D - \frac{1}{2}(\varphi_1 + \varphi_2)$, as well as a heavy neutral partner whose mass reflects the scale(s) at which new physics comes in. There are then three possibilities:

(i) If $\mu_1/\mu_D \ll \mu_D/\mu_2$, the light neutrino masses follow the ones of their charged partners

$$m(\nu_e) : m(\nu_\mu) : m(\nu_\tau) = m_e^\alpha : m_\mu^\alpha : m^\alpha \qquad (9)$$

where the power α is 1 or 2, depending on whether μ_2 scales like m_f or is independent of the family. This is, in essence, the seesaw mechanism [12], a common feature of many models [1].

(ii) If $\mu_1/\mu_D > \mu_D/\mu_2$, the light particle (usual neutrino of one family) is somewhere in the interval $m_f^2/\mu_2 \lesssim m(\nu) \lesssim \frac{m_f}{100}$. Its mass vanishes only if $\cos 2\Phi$ vanishes and if there is no triplet spontaneous symmetry breaker [13].

(iii) The ratios μ_D/μ_2 and μ_1/μ_D, individually small as compared to 1, could be of similar magnitude, $\mu_D/\mu_2 = \beta\, \mu_1/\mu_D$ with equal or close to 1. In this case the light neutrino would have the mass

$$m(\nu) = \mu_1 \left|1 - \beta \cos 2\Phi\right| \qquad (10)$$

A curious special case is met when β and Φ take the values $\beta = 1$, $\Phi = 0$. Here $m(\nu)$ vanishes exactly, even though the mass matrix is nontrivial.

As we emphasized above, the more general case which is encountered in realistic extensions of the standard model exhibits a pattern very similar to our simplified case study provided the physical origin of the four blocks in the matrix (7) is as described above: (M_D) is similar to the charged lepton mass matrix, the left-type Majorana matrix (M_1) is due to additional Higgses whose vacuum expectation value is small in order not to disturb the ρ-value, while (M_2) contains the scales of the "new physics".

An example is provided by a grand unified theory based on the structure group SO(10) (and supersymmetric extensions thereof) [14]. This model has several attractive features: it is free of anomalies, provides a natural multiplet for the fermions of the standard model. It probably predicts the correct value for the Weinberg angle and manages to escape (marginally) the experimental constraint from the stability of the proton. Within a SO(10) GUT it is rather natural to have nonvanishing neutrino masses. Unfortunately they depend on the pattern and the details of spontaneous symmetry breaking and their values are fairly model independent. The scales of the symmetry breaking vacuum expectation values are set by the constraints provided by the absence of proton decay and by the correct extrapolation of the Weinberg angle to low energies. The best one can obtain in this framework [13]

$$\begin{aligned} &\text{for } \nu_e : 1.5 \times 10^{-6\pm1} \text{eV}, \\ &\text{for } \nu_\mu : 4.5 \times 10^{-3\pm1} \text{eV} \qquad (11) \\ &\text{for } \nu_\tau : \gtrsim 0.01 \text{ eV} . \end{aligned}$$

4 How can we learn more about neutrinos?

There are essentially three domains which look promising in trying to clarify the mysteries of neutrino physics.

(i) Lepton family changing processes

Experimental search for processes with $\Delta L_f \neq 0$ will continue by improving the present upper limits on $\mu \to e\gamma$, $\mu \to 3e$, $K_L^0 \to \mu e$, etc. by one to two orders of magnitude, in a new round of experiments at the meson factories, at BNL and possibly at the kaon factories. Here we wish to mention a process related to these, where substantial progress is to be expected soon: muonium-antimuonium conversion, $\mu^+ e^- \to \mu^- e^+$. If M denotes the initial muonium atom $\mu^+ e^-$, $\bar{M}$ its antipartner, the transition is assumed to be mediated by a Lagrangian $\mathcal{L}_{eff}$ of an effective local contact interaction and a typical coupling $G_{M\bar{M}}$ (to be compared to the Fermi constant G_F). The transition matrix element is then estimated as follows,

$$\begin{aligned} \langle \bar{M} | \mathcal{L}_{eff} | M \rangle &\simeq G_{M\bar{M}} \Psi_M(0) \Psi_{\bar{M}}(0) = \\ &= 1 \times 10^{-12} G_{M\bar{M}}/G_F \end{aligned} \qquad (12)$$

where Ψ_M and $\Psi_{\bar{M}}$ denote the orbital wave functions of M and $\bar{M}$, respectively.

In trying to estimate what one expects for the ratio $G_{M\bar{M}}/G_F$ on the basis of simple lepton number conserving, but individual family number violating interactions one finds that exchange of a doubly charged Higgs H^{++} in a left-right-symmetric model may give the largest contribution, typically

$$G_{M\bar{M}} \simeq 10^{-2} G_F \ (H^{++} \text{ exchange}).$$

Box diagrams involving exchange of neutrinos would contribute much less than that [15],

$$G_{M\bar{M}} \simeq 10^{-5} G_F \ (\text{Majorana neutrino exchange}).$$

The present experimental limit $G_{M\bar{M}} < 0.7\, G_F$ [16] is still well above these estimates. A new experiment proposed at SIN/PSI [17] will make use of the SINDRUM detector for identification of the Michel electrons from the μ^- in the final $\bar{M}$-state and will thereby be able to increase the sensitivity to $G_{M\bar{M}}/G_F$ by 3 to 4 orders of magnitude. If this is indeed achieved one will enter the range of realistic expectations for M to $\bar{M}$ conversion for the first time in a long series of experiments. This is an exciting prospect and there might be surprises waiting for us.

(ii) Neutrino oscillations

One will undoubtedly continue the search for neutrino oscillations of various kinds. Here, the kaon factories can play an important role by providing new intense neutrino sources. For instance, the EHF study group has proposed a $\nu_\mu \to \nu_e$ oscillation experiment, making use of the ICARUS detector in the Gran Sasso tunnel, which would be sensitive to mass ranges of the order of $\Delta m^2 \sim 10^{-4} (\text{eV})^2$, [18].

Unless positive evidence for oscillations is found somewhere, progress in this field will be slow. Because of the lack of firm predictions there are very many open windows where to search. Also, most analyses are made on the assumption of two states which mix and oscillate. As soon as three or more mass eigenstates are assumed to mix, the picture becomes more complicated and the information extracted from experimental limits is weakened.

(iii) Neutrinoless double β-decay $(\beta\beta)_{o\nu}$

One of the most promising keys to the mystery of neutrino properties is $(\beta\beta)_{o\nu}$ decay, double β-decay of nuclei without emission of neutrinos [19]. It is so important because it can occur only if a) the neutrino(s) which is (are) exchanged does (do) not carry any additively conserved quantum number, i.e. if it is a Majorana particle, and b) it is massive, or more precisely: the exchanged neutrinos are not degenerate in mass [20]. Therefore, if $(\beta\beta)_{o\nu}$ decay were found we would know at once that some neutrino states are massive and are Majorana states.

As to what these masses are, things will be more difficult to analyse. The transition rate, apart from nuclear physics effects which by now seem well under control [21], depends on an effective mass parameter

$$m_{eff} = \sum_n e^{i\alpha_n} |U_{en}|^2 m_n \qquad (13)$$

as well as on effective coupling terms of left- and right-handed currents $\langle\eta\rangle = \eta \sum_n U_{en} V_{en}$, $\langle\lambda\rangle = \lambda \sum U_{en} V_{en}$. Here n denote neutrino mass eigenstates, U_{en} and V_{en} mixing matrices of left and right fields, respectively, while η and λ are the coupling parameters of leptonic right times hadronic left currents and leptonic right times hadronic right currents, respectively. Thus double beta decay provides limits on right-handed weak interactions, though in a model dependent way. The corresponding limit on m_{eff} is found to be of the order of 2eV from the experimental bound for neutrinoless double β decay of ^{76}Ge

$$T^{o\nu}_{1/2}(^{76}Ge) > 4.7 \times 10^{23} y.$$

This, however, is only a lower limit on the neutrino masses m_n, because of the phases which appear in eq.(13) and which may give rise to destructive interferences in the sum over neutrino mass states. The phase factor $e^{i\alpha_n}$ is nothing but the ratio U_{en}/U^*_{en} and need not to be 1, so that individually large terms in the sum over n may cancel.

5 Conclusions

On the theoretical side we need new input and a fresh attempt at understanding the nature of the neutrinos. In spite of the impressive experimental progress we still know very little about ν_e and ν_μ, and even less about ν_τ. Most promising are the continuing investigation of oscillations, and of neutrinoless double β-decay. The most interesting forthcoming step in understanding leptonic quantum numbers may come from muonium-antimonium conversion where an experiment at the level of $G_{M\bar{M}}/G_F \sim 10^{-4}$ seems possible. Clearly, kaon factories such as KAON will provide several high quality beams which will allow to push lepton physics to new frontiers.

References

1. Good introductions and reviews include J.W.F.Valle, Lectures at Autumn School on Physics beyond the Standard model, Lisbon 1988, Nucl.Phys.Suppl. (to be published); F. Boehm and P. Vogel, Physics of Massive Neutrinos, Cambridge University Press, Cambridge 1987; B.Kayser, F.Gibrat-Debu and F.Perrier, The Physics of Massive Neutrinos, World Scientific, Singapore, 1988
2. M.Fritschi et al., Phys.Lett. B173(1986)485
3. R. Abela et al., Phys.Lett. B146(1984)431
4. H.Albrecht et al., Phys.Lett. B202(1988)104
5. The situation concerning neutrino oscillations and heavy neutrino admixtures is summarized in the Review of Particle Properties,Phys.Lett. B204(1988)1
6. see e.g. B.Kayser, Proc. of Seventh and Eigth Moriond Workshop on New and Exotic Phenomena, 1987, 1988
7. A.Jodidio et al., Phys.Rev. D34(1986)1967, E: Phys.Rev. D37(1988)237
8. W.Fetscher, Phys.Lett. B140(1984)117
9. F.Scheck: Leptons, Hadrons and Nuclei; North Holland, Amsterdam 1983
10. J.Schechter and J.W.F.Valle, Phys.Rev. D25(1982)774
11. H.Harari and Y.Nir, Nucl.Phys. B292(1987)251
12. M.Gell-Mann, P.Ramond and R.Slansky, in Supergravity, D.Freedman (editor), North Holland 1987
13. M.Depner, Diploma thesis, Mainz 1989, and M.Depner et al., to be published
14. See e.g. G.G.Ross, Grand Unified Theories, Benjamin/Cummings Publ.Co. (1984)
15. A.Halprin, Phys.Rev.Lett. 48(1982)1313
16. Yale-Heidelberg group, private communication and to be published
17. Yale-Heidelberg-ETH Zurich group, proposal at SIN/PSI 1988
18. P.Pistilli, in Proposal for a European Hadron Facility, J.F.Crawford (ed.), Mainz 1987
19. For reviews and references see e.g.: K.Muto and V.Klapdor, preprint MPIH- 1988-V28, Heidelberg; M.Doi, T.Kotani and E.Takasugi, Progr.Theor. Phys. Suppl. No.85 (1985)
20. B.Kayser, S.Petkov and S.P.Rosen, to be published
21. J.Suhonen, T.Taigel and A.Faessler, Nucl.Phys. A486(1988)91

Z. Phys. C – Particles and Fields 46, S81–87 (1990)

Zeitschrift für Physik C Particles and Fields

$\bar{P}P$ annihilation at rest: annihilation dynamics and meson spectroscopy

E. Klempt

Institut für Physik, Johannes-Gutenberg-Universität, Postfach 3980, D-6500 Mainz, Fed. Rep. of Germany

Abstract Proton-antiproton annihilations provide an excellent tool to study the dynamics of quarks and gluons in the confinement region. Annihilations at rest into two mesons show selection rules which can be related to symmetries between SU(3) invariant amplitudes. In the annihilation process not only conventional ($q\bar{q}$) mesons are produced but also states which do not fit into the meson nonets. It is shown how knowledge of the atomic state of the $\bar{p}p$ system at annihilation can be used to restrict the possible quantum numbers of meson resonances.

---

1 Introduction

Meson spectroscopy and the study of hadronic reactions will certainly be still an active field of research when KAON will come into operation. Kaon induced reactions and mesons produced by kaons scattered off nucleons will then likely be the most easily accessible physics; nevertheless it may be interesting to base expectations for KAON on recent progress on antiproton induced reactions and on the spectroscopy of mesons produces in $\bar{p}p$ annihilation.

The interest in meson spectroscopy and in hadronic reactions is based on the same question: is there an effective quark-quark interaction in the confinement regime? At small distances or large momentum transfers the quark-quark interaction can be well described by perturbative quantum chromodynamics. At large distances - large compared to the Compton wave length of the pion - hadronic reactions can be described by boson exchange processes. Yet we do not have a theory of hadronic interactions describing quark-quark interactions at distances which are typical for kaon induced reactions or antiproton-proton annihilation. The large number of models for the long-range quark-quark interaction, lattice gauge theories, QCD sum rules, 1/Nc expansion, flux tube models, bag models, "hybrid" models, ..., demonstrates the interest in finding a theory which is based on the successes of QCD but which is applicable in the confinement region.

Defining an effective interaction of quark-quark interactions is - also - an experimental question: can we isolate processes in which the effective quark-quark interaction manifests itself with sufficient clarity to allow to identify the underlying quark-quark interaction? Can we determine the dynamics of quarks and gluons in the confinement region?

There are two strategies which may be followed: the commonly accepted road is that of meson spectroscopy. Mesons are searched for, and there is considerable effort to identify mesons not fitting into SU(3) nonets. They may then be glueballs or hybrids, states in which the color degrees of freedom of QCD manifest themselves in new forms of hadronic matter; or they may be multiquark states, $qq\bar{q}\bar{q}$ states of baryonia which were searched for with great enthusiasm but which failed to be observed as narrow states.

The other strategy is to study the systematics of hadronic reactions. I will show that selection rules exist which seem to govern $\bar{p}p$ annihilation. And there is the exciting possibility that these selection rules do reflect the symmetry of the underlying quark-antiquark interaction.

A large fraction of results on $\bar{p}p$ annihilation still stems from the analysis of bubble chamber data [1]. In the first phase of LEAR one experiment - the ASTERIX experiment - investigated $\bar{p}p$ annihilation at rest with an electronic 2π detector [2]. Antiprotons from LEAR are stopped in a gaseous H_2 target; $\bar{p}p$ atoms are then formed which annihilate - with approximately equal probabilities - from S states or P states (with angular momentum l=0 or 1 between proton an antiproton). If the atomic cascade is observed in coincidence then events can be selected which are associated with the emission of a nD→2P transition X-ray. In these events annihilation occurred mostly from atomic P states.

In liquid H_2, annihilation proceeds dominantly via S states. The other experiments [3,4] at LEAR and at KEK - measuring single particle spectra ($\pi^{\pm},\pi^{0},\eta$) to determine two-body annihilations - used liquid H_2 targets.

2 Proton-Antiproton Annihilation

A strong selection rule was observed in $\bar{p}p$ annihilation into two strange mesons from S states of the $\bar{p}p$ system [5a]. Table 1 presents ratios of branching ratios for $\bar{p}p$ annihilation into two mesons. The data are grouped into subsets which are governed by the same SU(3) invariant couplings [6]. In annihilations into two strange mesons always one isospin component dominates. From the $^3S_1$ state of the $\bar{p}p$ atom, annihilation via the I=1 component of the $\bar{p}p$ wave function is strong while the branching ratio is small via the I=0 component. The opposite isospin selection rule may be present in annihilation from the $^1S_0$ initial states: Bettini et al. find a sizable contribution to $K^*\bar{K}$ + c.c. via the I=0 component of the $^1S_0$ $\bar{p}p$ wave function and a small contribution via I=1. This change of the isospin preference can be traced back to a change of the relative sign of the amplitude for $K^{*+}K^-$ and $K^{*0}\bar{K}^0$ production in a given final state [7]. The relative phase between production of two charged strange mesons and of two neutral strange mesons is always negative.

In the Quark Line Rule Model the annihilation processes can be described in annihilation and rearrangement diagrams. Annihilation into two strange mesons can then be described by four amplitudes $A_1^{\pm} + A_2^{\pm}$. The pattern of Table 1 then leads to the approximate symmetry $A_1^{\pm} + A_2^{\pm} \simeq 0$, $A_1^{\pm} \simeq A_1^{-} = A$. Similar arguments can be made for $\bar{p}p$ annihilations into $K\bar{K}$.

There are further indications that $\bar{p}p$ annihilation into two mesons can be described at the level of quark dynamics. Table 1 also presents ratios of branching ratios together with SU(3) ampli-

Table I: COMPARISION OF $\bar{p}p$ ANNIHILATION BRANCHING RATIOS WITH IDENTICAL FLAVOR STRUCTURE

RATIO			COMMENT	REF.
$\frac{(K^*K)_{even;I=0}}{(K^*K)_{even;I=1}}$	=	0.073 ± 0.043	1,2	a
$\frac{(K^*K)_{even;I=0}}{(K^*K)_{even;I=1}}$	=	0.033 ± 0.017	1,2	b
$\frac{(K\bar{K})_{even;I=1\ or\ 0}}{(K\bar{K})_{even;I=0\ or\ 1}}$	=	$0.08^{+0.16}_{-0.05}$	2,3	c
average	$\ll$	1 ; assumed to be zero		
QLR amplitude		$(A_1^+ + A_2^+)/A_2^+$		
$\frac{(K^*K)_{odd;I=1}}{(K^*K)_{odd;I=0}}$	=	0.79 ± 0.34	1,2	a
$\frac{(K^*K)_{odd;I=1}}{(K^*K)_{odd;I=0}}$	=	0.23 ± 0.07	1,2	b
$\frac{(K\bar{K})_{odd;I=0}}{(K\bar{K})_{odd;I=1}}$	=	$0.0046 \pm .0023$	2,3	c
average	$\ll$	1 ; assumed to be zero		
QLR amplitude		$(A_1^- + A_2^-)/A_2^-$		
$\frac{1}{3}\,\frac{\pi\eta}{\eta\eta}$	=	0.49 ± 0.20	4,5,6	d
$\frac{2}{3}\,\frac{\pi\omega}{\eta\omega}$	=	0.65 ± 0.21	4,5	e,f
$\frac{2}{3}\,\frac{\pi\omega}{\eta\omega}$	=	0.31 ± 0.23	4,5	d,g
$\frac{\rho\eta}{\omega\eta}$	=	$0.51^{+0.20}_{-0.09}$		d
$\frac{\rho\eta}{\omega\eta}$	=	1.4 ± 0.2	(omitted)	f
$\frac{1}{2}\,\frac{\rho\omega}{\omega\omega}$	=	0.81 ± 0.35	6	h,i
$\frac{\rho f_2}{\omega f_2}$	=	0.53 ± 0.15		h
average	=	0.525 ± 0.083		
QLR amplitude		$(2A_2^+ + R_2^+ + R_3^+)/(2R_1^+ + 2R_2^+)$		
$\frac{4}{9}\,\frac{\pi^0\pi^0}{\eta\eta}$	=	0.93 ± 0.36	4,5	d
$\frac{\rho^0\pi^0}{\omega\eta}$	=	0.87 ± 0.15	4,5	d,j
average	=	0.88 ± 0.14		
QLR amplitude		$2R_1^+/(2R_1^+ + 2R_2^+)$		
But: $\frac{\rho^0\rho^0}{\omega\omega}$	$\leq$	0.2		h,i
$\frac{(\pi^+\pi^-)_{odd}}{(\pi^+\pi^-)_{even}}$	=	0.130 ± 0.026	2,7	c,f
$2\,\frac{(\rho^+\pi^-)_{odd}}{(\rho^+\pi^-)_{even}}$	=	0.061 ± 0.026	2,8	h
$\frac{1}{3}\,\frac{(a_2^+\pi^-)_{odd}}{(a_2^+\pi^-)_{even}}$	=	0.71 ± 0.039	2,9	a,b,e,f
average	=	0.091 ± 0.017		
QLR amplitude		$(2A_2^- + R_2^-)/2R_1^-$		

1 The analysis of ref (b) supersedes that of (a): the same data were used, a minor error corrected and data on $\bar{p}n$ included.

2 Even and odd amplitudes are defined by the product of C parities of the initial state and of the two meson nonets.

3 The ratio $(T_0/T_1)^2$ is presented with $T_0^2 + T_1^2$ = b.r. (K^+K^-) + b.r. $(K^0\bar{K}^0)$ and $2T_0T_1$ = b.r. (K^+K^-) - b.r. $(K^0\bar{K}^0)$

4 Corrected for phase space by multiplication with the (inverse) ratio of decay momenta

5 Corrected for the $s\bar{s}$ component of the η wave function by multiplication of branching ratios involving one η by a factor of 3/2 (mixing angle - 19.5^0).

6 Branching ratios into two identical particles are multiplied by 2.

7 $\pi^+\pi^-$ odd refers to annihilations from the $^3S_1$ state, $\pi^+\pi^-$ even from $^3P_{0,2}$ states. The angular momentum barrier is different in the two decay modes therefore both ratios are normalized to the branching ratio for $\bar{p}p \to K^+K^- + K^0\bar{K}^0$

8 $(\rho^+\pi^-)$ odd refers to annihilation from the $^1S_0$ state, $(\rho^+\pi^-)$ even from $^3S_1$ states. The latter decay has two helicity amplitudes the former one. The ratio is corrected for this fact.

9 $(a_2^+\pi^-)$ odd refers to annihilation from the $^3S_1$ state, $(a_2^+\pi^-)$ even from the $^1S_0$ state. The latter decay mode has one helicity amplitude, the former three. The branching ratios are determined in four analyses on different final states, averaged and the error increased by a scaling factor of 1.5.

tudes derived from the quark line rule model. SU(3) breaking is taken into account by dividing the measured branching ratios by the decay momentum (= two-body phase space). In addition to the four A amplitudes there are four R amplitudes describing rearrangement. The data are grouped according to different SU(3) couplings. Obviously, the ratios do not depend on spin or orbital angular momentum excitations of the mesons but only on their quark content. This justifies a posteriori the use of quarks to describe p̄p annihilation at rest.

The ratios of Table 1 are approximately compatible with the values A=0.10(1); $R_1^+=R_3^+=R=0.23(2)$; $R_2^+=R_1^-=0$. These were derived by a fit to the data using A and R as only free parameters [8]. However, this analysis would prefer a slightly different solution.

Indeed, in the choice of R amplitudes there is some arbitrariness; the A amplitude could be different for production of $u\bar{u}$ and $d\bar{d}$ pairs or of $s\bar{s}$ pairs. Then R_2^+ and R_3^+ could both be small. Only the symmetry $A_1^\pm + A_2^\pm \approx 0$ must hold in this picture. This symmetry is strikingly evident in the experimental data, and it deserves an explanation. The fact that the two amplitudes $A_1^\pm$ and $A_2^\pm$ have the same size can be related to the absence of "diquark effects" in p̄p annihilation at rest [7]. The origin of the opposite phase of the two amplitudes is, however, completely unknown.

Finally, we note that the symmetry seems to be broken in p̄p annihilations into strange mesons from P states of the p̄p atom. The branching ratios for p̄p annihilation into K^+K^- and $K^0\bar{K}^0$ from P states [5c] differ by a factor of (3.3±1.0). Possibly, other effects become important at larger distances, and the quark line rule analysis fails to give "simple" results.

3 Meson Spectroscopy

Proton-antiproton annihilation is a rich source for the production of mesons. This is true for annihilation from S states and for annihilation from P states. In this contribution I would like to show how the information on the initial state of the p̄p atom can be used to determine the spin of a resonance (or to support the J^{PC} assignment).

Figure 1 shows the (KKπ) invariant mass recoiling against a $\pi^+\pi^-$ pair in p̄p annihilation at rest. The data of Fig. 1a are bubble chamber data [9], those of Fig. 2b stem from the ASTERIX experiment [10]. For this data set a special X-ray trigger was used enhancing the fraction

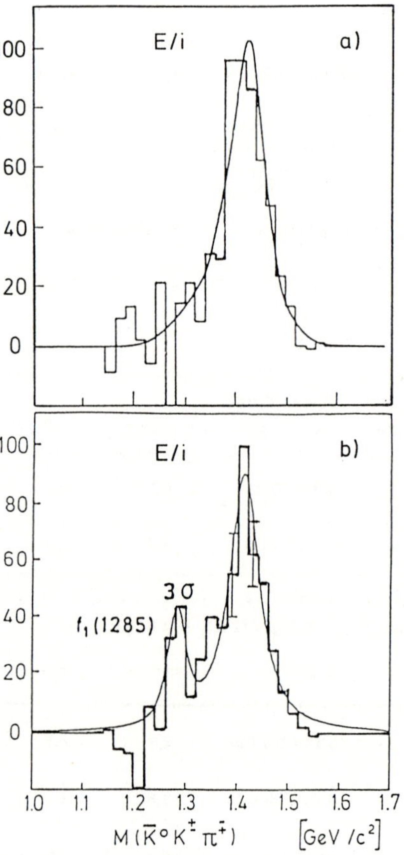

Fig.1. The $K\bar{K}\pi$ invariant mass spectrum produced in p̄p annihilation at rest in the reaction $\bar{p}p\to\pi^+\pi^-(K\bar{K}\pi)$. a) liquid H_2; b) gaseous H_2 at NPT

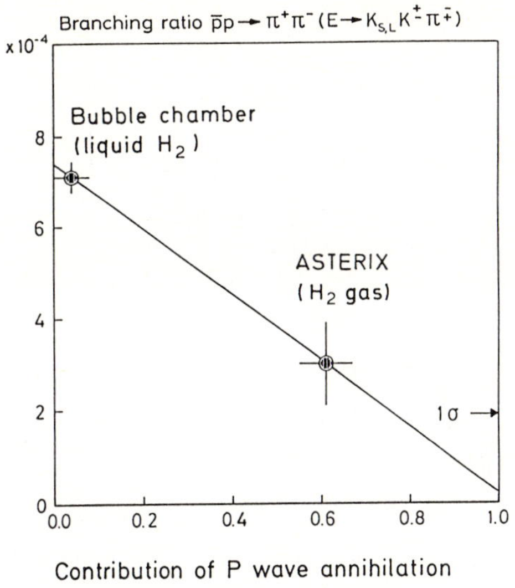

Fig.2. Branching ratio for $\bar{p}p \rightarrow \pi^+\pi^- E$ as a function of the contribution of P wave annihilation

of P wave annihilation. The two spectra look very similar apart from the appearance of the $f_1(1285)$ meson in the P wave data. The branching ratio for the production of E mesons decreases, however, with the fraction of P wave annihilation increasing (Fig. 2).

This pattern is easily understood if the assumption is made (which can be supported experimentally) that the two pions recoiling against the E have relative zero and that the angular momentum between dipion and E vanishes. Then the quantum numbers J^{PC} of a heavy meson produced from P states can only be 2^{++}, 1^{++}, 1^{+-} or 0^{++} while a particle produced from S states must be 0^{-+} or 1^{--}. Indeed, the f_1 (1285) with quantum numbers 1^{++} is only produced from P states. The E meson, produced only from S states, can therefore not be 1^{++} but only 0^{-+}. These are the quantum numbers which were originally determined [9] after the discovery [11] of the E in $\bar{p}p$ annihilation at rest in 1963! The resonance is therefore likely identical with the resonance [12] observed in radative J/ψ decay. The most recent analysis [12] suggests that there are two resonances: one resonance at 1420 MeV also observed in $\bar{p}p$ annihilation at rest and in other hadronic reactions, and a resonance at about 1500 MeV. This would imply that three pseudoscalar resonances may exist: $\eta(1280)$, $\eta(1420)$, $\eta(1500)$; only two of them can, of course, be accommodated in the nonet of pseudoscalar radial excitations. Hence the third particle must be an object of different type, possibly a glueball.

A second example in which knowledge on the initial state can be used to support a spin-parity assignment is a 2^{++} resonance observed in $\bar{p}p \rightarrow \pi^+\pi^-\pi^0$ at LEAR [13] and - very likely - previously [14] in $\bar{p}p \rightarrow \pi^0\pi^0\pi^0$ and in $\bar{p}n \rightarrow \pi^+\pi^-\pi^-$. The resonance is of special interest since it does not seem to fit into the conventional $q\bar{q}$ meson nonets while an interpretation of a $qq\bar{q}\bar{q}$ or $N\bar{N}$ state seems very plausible.

Figure 3 shows the squared $\pi^+\pi^-$ invariant mass spectrum for bubble chamber data (a), for data where antiprotons are stopped in gas (b), and for data tagged as P wave annihilation (c). With increasing P wave contribution two resonances show up which are not observable in Fig. 3a: one resonance is the well known $f_2(1270)$ the other peak was called AX (1565). From the absence of its charged partners isospin zero can be deduced. Hence the possible spin-parities are 0^{++} or 2^{++}. The resonance is more clearly visible if mass instead of mass squared is plotted (Fig. 4). A phase shift analysis showed a phase motion in the 2^{++} wave. But also the production mechanism can be used to favour this J^{PC} assignment: the reaction $\bar{p}p \rightarrow \pi^0 f_2(1270)$ can proceed via the 0^{-+} state of the $\bar{p}p$ atom with two units of angular momentum, or from the 1^{++} or 2^{++} initial states with one unit of angular momentum. Obviously, the former production mechanism is suppressed. A similar suppression is expected if the AX(1565) is 2^{++}.

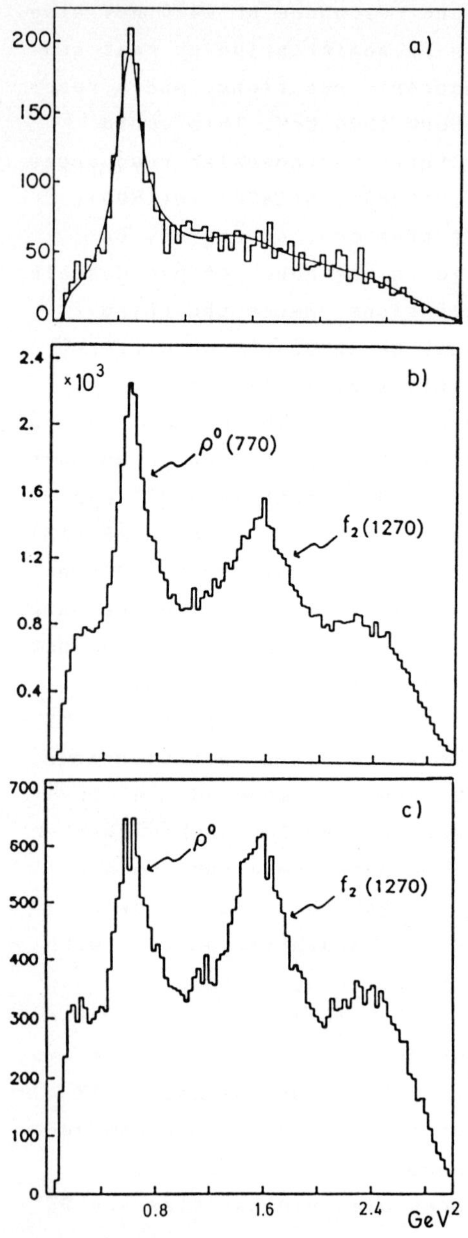

Fig.3. The $\pi^+\pi^-$ invariant mass square from $\bar{p}p \to \pi^+\pi^-\pi^0$ annihilations. a) liquid H_2 ("S-wave") b) gaseous H_2 ("S- and P-wave") c) with trigger on L-X-rays (P-wave")

However, if the AX(1565) is 0^{++}, then it could be produced from the 0^{-+} state of the $\bar{p}p$ atom with zero angular momentum. In this case, a large AX(1565) production should be expected in $\bar{p}p$ annihilation in a bubble chamber. This is obviously not the case, and $J^{PC}=2^{++}$ is deduced from

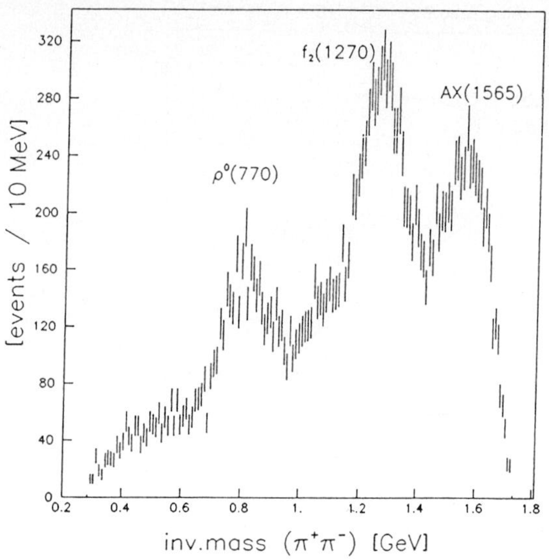

Fig.4. The $\pi^+\pi^-$ invariant mass spectrum (as Fig.3c, but with a linear mass scale)

the production mechanism without using the result of the phase shift analysis.

4 Conclusion

I have tried to convince you that $\bar{p}p$ annihilation offers a very good chance to test models of the quark-quark interaction in the confinement region. There is strong evidence that the quark degrees of freedom become manifest in $\bar{p}p$ annihilation, and there is even the chance that the selection rules observed in $\bar{p}p$ annihilation reflect the symmetry of the quark-antiquark gluon vertex.

The identification of unconventional (non-$q\bar{q}$) mesons and the determination of their mass, width, decay modes and quantum numbers should also help to test models of quark-quark interactions. In $\bar{p}p$ annihilations a resonance, the E/i, is produced which is one of the prime glueball candidates. The AX(1565)- on the other hand - is likely a multiquark state. These two resonances show the potential of $\bar{p}p$ annihilations to search for new states which are produced when three quarks and three antiquarks merge, annihilate, rearrange, interact to form the outgoing meson resonances.

This work has been partly funded by the German Federal Minister for Research and Technology (BMFT) under the contract number 06 MZ 223.

References

1. R. Armenteros, B. French: High energy physics.Burkop, E.H.S. (ed.),Vol. IV, p. 284, London:Academic Press 1969; see also ref 2 for up-dated results
2. R. Armenteros et al., "Study of p̄p Interaction at Rest in a H_2 Gas Target at Lear", PS171, CERN
3. G. Backenstoss et al., "Investigation on Baryonium and Other Rare p̄p Annihilation Modes", PS182, CERN
4. A. Angelopoilos et al., "Search for Bound NN̄ States", PS183, CERN

5a. B. Conforto et al.:Nucl.Phys. B3 469 (1967)
b. A. Bettini et al.: Nuovo Cimento A63, 1199 (1969)
c. C. Amsler, in: pp̄ Interactions and Fundamental Symmetries, ed.by K. Kleinknecht and E. Klempt, North Holland Amsterdam 1989 p. 213
d. L. Adiels et al.:Z.Phys. C42, 49 (1989)
e. M. Chiba et al.:Phys.Rev. D38, 2021 (1988)
f. M. Chiba et al.:Phys.Rev. D39, 3207 (1989)
g. S. Carius: Ph.D. thesis, Stockholm (1986)
h. see Refs. 1,2
i. M. Bloch et al.:Nucl.Phys. B23, 221, (1970)
j. B. May et al.: submitted to Z.Phys. C
k. P. Espigat et al.:Nucl.Phys. B36, 93 (1972)
l. J. Diaz et al.:Nucl.Phys. B16, 239 (1970)

6. U. Hartmann, E. Klempt, J. Körner: Phys.Lett. B155, 163 (1985)
7. E. Klempt: Z.Phys. A331, 211 (1988)
8. U. Hartmann, E. Klempt, J. Körner: Z.Phys.A - Atomic Nuclei 331, 217 (1988)
9. P. Baillon et al.:Il Nuovo Cimento 50A, 393 (1967)
10. K.D. Duch et al.: submitted to Z.Phys.C
11. R. Armenteros et al.:Proc.Sienna Int. Conf.Elem.Part., Vol.I 287 (1963)
12. see L. Köpke and N. Wermes: Phys.Rep. 174, 67 (1989)
13. B. May et al.:Phys.Lett. B225,450(1989)
14. L. Gray et al.:Phys.Rev.D27, 307 (1983)

Z. Phys. C – Particles and Fields 46, S89–92 (1990)

Zeitschrift für Physik C Particles and Fields

Exotic mesons

F. E. Close

Oak Ridge National Laboratory*, Oak Ridge, TN 37831, USA and
University of Tennessee, Knoxville, TN 37996, USA

Abstract

I discuss the status of exotic mesons – states that do not fit into the simple $q\bar{q}$ model.

Significant in proving the validity of the quark model in the late 1960's were the two following phenomena:

(i) Mesons occur in nonets with $| Q | \leq 1$ and $|$ strangeness $| \leq 1$. There were many fruitless searches for "exotic states of the first kind" such as mesons with charge or strangeness 2.

(ii) Mesons have spin parity and charge conjugation, J^{PC}, that are consistent with them being made from (nonrelativistic) spin-1/2 quark and antiquark. Thus one cannot form 0^{--}, nor the sequence 0^{+-}, 1^{-+}, 2^{+-}, etc. There were fruitless searches for such "exotics of the second kind".

Then along came the notion of color, which is the source of the force between quarks. This naturally explained nature's preference for $q\bar{q}$ and qqq colorless systems. However, it also appears to predict multiquark states such as $q^2\bar{q}^2$ with charge and strangeness belonging to the category "exotics of the first kind". If the color forces are not color ionic but instead more akin to covalent (analogous to the familiar electromagnetic case), then these states will consist of two color singlet mesons bound in a "molecule". Weinstein and Isgur [1] argue that $K\bar{K}$ form 0^{++} bound states just below the KK threshold, consistent with the known $f_0(975)$ and $a_0(980)$. It is possible that exotics of the first kind can be avoided or are so broad that they do not show up [2]. It is an open question as to whether other such "molecules" exist.

If we go beyond mere color to the full QCD theory, then we have colored gluons, presumably mutually attracting to form glueballs by the same color forces that are familiar to us from quark systems. It is possible to form "oddball" states such as 0^{--}, but models and lattice simulations for QCD suggest that they are significantly above 2 GeV. Thus exotics of the second kind may be avoided (temporarily).

Isoscalar flavorless glueballs are not the easiest objects to identify. Thus in the late '70's and early '80's a flurry of interest developed in the possible existence of hybrid states containing both quarks and gluons as dynamical constituents. In models the lightest supermultiplet contains a 1^{-+} nonet, which could be around 1.5 GeV in mass according to some QCD sum rules [3] and bag model phenomenology [4].

So now we are looking for exotics of the second kind. Had such a state been found in 1968, it would have undermined the naïve quark model. If one is found tomorrow, it will be greeted with enthusiasm as "proof" for the quark-gluon theory of hadronic particles. This interesting turnaround is the stuff of theses on the philosophy of science. It is relevant since GAMS claim to have identified such a state [5] in $\pi^- p \to (\eta^0 \pi^0)n$. If true, this is the most important discovery in hadron physics in many years. There have been criticisms raised on the analysis that led to the claim for this state [6]. These criticisms remain to be answered. If they are answered satisfactorily, then we have a clear signpost for the existence and mass scale of exotic states. If not, then we go on searching.

These are exotics which models predict should exist and are on the menu for experiment. With the exception

---

*Operated by Martin Marietta Energy Systems, Inc. under Contract DE-AC05-84OR21400 with the U.S. Department of Energy

of the above 1^{-+} state, there are no other clear candidates around for "second-kind exotics".

If, by exotic, we mean states that do not readily fit into the simple quark model, then we find that there are other candidates.

Some of these have allowed quantum numbers but are unusual in some other property. Examples include:

(i) A 1^{--} state decaying [7] into $\phi\pi$. The ϕ contains $s\bar{s}$ and the π has I=1; thus the initial state seems to be "$s\bar{s}$ with I=1", and hence *prima facie* a $q^2\bar{q}^2$ state [8]. This has only been seen in one experiment and needs confirmation. Lipkin and I noted [9] that the near-degeneracy in mass of this 1^{--} and the GAMS 1^{-+} state [5] would be natural if they are the G-parity eigenstates of 10–10* $q^2\bar{q}^2$ multiplets. However, this would imply that a whole set of (unseen) states should exist.
A detailed study on the empirical limits on such states, given what (little) we know on the above 1^{--}, 1^{-+} examples, is required. One consequence is that if 1^{-+} is this $q^2\bar{q}^2$, its decay to $\eta\pi$ dominates over $\eta'\pi$. However, if it is $q\bar{q}g$, the $\eta'\pi$ mode should dominate [9]. Thus, it should be possible to determine its constitution.

(ii) Metastable states such as $\xi(2230)$ that show up in "new" processes – notably $\psi \to \gamma K\bar{K}$. This narrow state has led to a variety of fanciful interpretations such as (qs)–$(\bar{q}\,\bar{s})$ baryonium, its narrow width being a consequence of the high angular momentum barrier between the qs and the $\bar{q}\,\bar{s}$.
However, it now appears that this state could be a conventional $4^{++}(s\bar{s})$, the partner of the $f_4(2030)n\bar{n}$. LASS [10] see a state in this mass region and determine its spin-parity from partial-wave analysis. Its mass fits in well with the expectations in quark potential models [11] and gives us confidence in these models. If they can describe the states up at 2–3 GeV and are known to work well down at 1–1.5 GeV, then it is very likely that they are a reliable guide in the intermediate region. We can use these model spectra to discriminate other "exotic" candidates such as the proliferating 0^{-+} around 1.4 GeV or the $f_2(1720)$, "θ", whose production and decays seem to be somewhat unusual, or the three 2^+ states claimed in $\phi\phi$ above 2 GeV. More of all this later.

(iii) "Utterly exotic" – such as the U(3.1) states [12]. They are heavy and narrow, which is already exotic, and occur in exotic charges. The simplest explanation is that this state is a statistical fluctuation.

(iv) "Phano-exotics" such as $1^{-+}(1405)$ [5]. This has J^{PC} inaccessible to (nonrelativistic) $q\bar{q}$. It could be $q^2\bar{q}^2$, as already mentioned, or the first hybrid $q\bar{q}g$.

I will now look at these states in turn. I will not review "predicted" exotics, such as $H(uuddss)$ or C-S exotics $cu\bar{s}\bar{d}$ $c\bar{s}uud$, which have not yet been sighted.

$\xi(2230)$: A would-be exotic that isn't

a) What is it? This state is seen in $\psi \to \gamma K^+K^-$ and $\gamma K^0_S K^0_S$. Thus it has $J^{PC} = 0^{++}, 2^{++}, 4^{++} \ldots$.

b) Why "exotic"? It is rather narrow ($\Gamma < 50$ MeV) and isolated, and produced in $\psi \to \gamma X$ – the theorists' favored glueball channel.

c) Theory. Glueball, $s\bar{s}g$ hybrid and $(qs)_{3^*}$ – $(\bar{q}\,\bar{s})_3$ baryonium have been suggested.

d) Problem. One could not *a priori* eliminate the glueball (the ratio of $\eta\eta/K\bar{K}$ branching fractions could be a test – see Ref. 13). The baryon would lead one to expect a whole genre, none of which is seen. (The dog that didn't bark in the night was significant for Sherlock Holmes; analogously the absence of a whole plethora of states is not to be overlooked.)

e) News. LASS see a signal [10] that is consistent with the $\xi(2230)$ in the process $K^-p \to (K\bar{K})\Lambda$ which has $J^{PC} = 4^{++}$. As such, it may be the $f_4(s\bar{s})$ partner of $f_4(n\bar{n})$, or a mixture of them, in any event a canonical quark model state whose mass is in excellent agreement with expectations in the quark model.

f) Moral. The nonrelativistic quark model is in good shape through 2.2 GeV, and we may, with increased confidence, use it to identify other misfits.

$\hat{\rho}(1480)$ – or $c(\phi\pi)$ – a (not so far-out?) exotic

a) What is it? Lepton-F see this [7] in $\pi^-p \to \hat{\rho}p \to (\phi\pi^0)p$. Its mass is 1480 MeV and $\Gamma = 130\pm60$ MeV.

b) Why? The decay products are $\phi(s\bar{s})$ and π. Thus $\hat{\rho}$ appears to have I=1 and hidden strangeness, and as such is *prima facie* exotic. A further datum supporting this is that

$$B(\hat{\rho} \to \phi\pi) \geq \frac{1}{2}B(\hat{\rho} \to \omega\pi)\ .$$

c) Theory. Lipkin and I suggested long ago that 1^{--} exotics of this sort could occur in this mass region [8]. However, there may be a very nonexotic interpretation of this state. Achasov and Kozhevikov point out [14] that OZI violation, as in $\phi \to \rho\pi$, may be misleading here. The 2^3S_1 ρ' can decay to $\phi\pi$ via *real* intermediate states:

$$\rho' \to \bar{K}K^* \to \bar{K}(K\pi) \to (\bar{K}K)\pi \to \phi\pi\ .$$

In a specific model they calculate that the B.R. could be 0.5%.

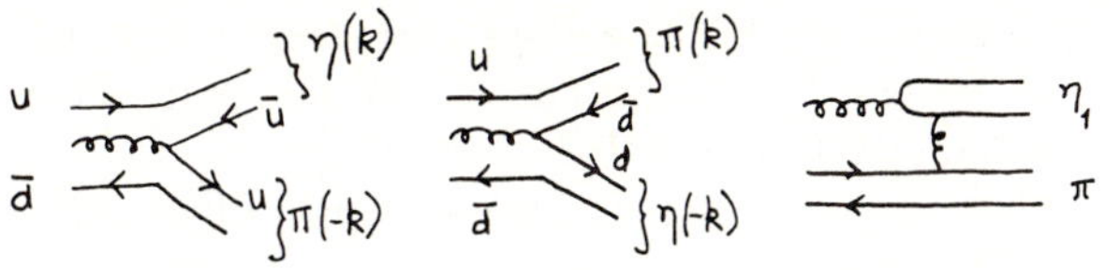

Figure 1

d) Problem. In another recent paper Achasov has pointed out [15] a problem with the decays of $\rho'(1600)$, namely that in theory

$$B(\rho' \to \omega\pi) >> B(\rho' \to K^*K) .$$

However, empirically $B(\rho' \to K^*K) = 9\pm2\%$ whereas no sign of $\omega\pi$ exists. This is a problem in its own right, but in addition, it probably undermines the attempt to identify the $\hat{\rho}(\phi\pi)$ as ρ'. Note that $B(\hat{\rho} \to \omega\pi) < 2B(\hat{\rho} \to \phi\pi)$ according to Lepton-F. This limit on $\omega\pi$ is inconsistent with the observed K^*K rate of $9\pm2\%$ for the $\rho'(1600)$. The identification of $\hat{\rho}$ with ρ' is in trouble.

e) Postscript. The $\rho'(1600)$ is now identified as two states: $\rho(1480)$ and $\rho(1600)$, the 1^3D_1 and 2^3S_1 $q\overline{q}$ states. This might complicate the above arguments. These criticisms of the $\hat{\rho} \equiv \rho'$ have been independently discussed, and in more detail than here, in Ref. 16.

U(3.1): A far-out exotic

a) What is it? This is produced in the Σ^-Be at CERN and in nA at BIS–2. It is metastable and decays into $\Lambda\overline{p}$ and pions [12].

b) Why? Because of its high mass (3.1 GeV), metastability and, most specifically, charge modes. The $U^+ \to \Lambda\overline{p}\pi^+\pi^+$ requires exotic composition: the Λ hints at a strange quark being present, and so to get the charge, a minimum composition is $sud\overline{d}\,\overline{d}$. There still remains the problem of why so narrow and so alone.

c) Theory. $J^P = 4^-$ M-baryonium has been invoked. My worries here parallel those outlined in the case of $\xi(2230)$.

One proposal is that this is a bound state of Λ, $\overline{p}$ and exotic meson M^{++}. This can be tested in that it predicts existence of $U^{---}(\Lambda\overline{p}\pi^-\pi^-)$. But it requires $B\overline{B}$ binding, an exotic meson M and, moreover, binding of M to $B\overline{B}$. This explanation is as exotic as the U(3.1); however, I can offer no more than hope that future data will show it to be a statistical fluctuation or will reveal other states in sufficient numbers to guide us. A single isolated narrow state of this kind would be most awkward to explain.

$H(1405)1^{-+}$: A canonical phano-exotic

a) What is it? This is seen by GAMS [5] in $\pi^-p \to (\eta\pi^0)n$.

b) Why? $J^{PC} = 1^{-+}$ cannot be accommodated in the nonrelativistic $q\overline{q}$ model.

c) Theory. Either $q^2\overline{q}^2$ in L=1 or a hybrid $q\overline{q}g$. Bag models [4] and QCD sum rules [3] had predicted that exotic 1^{-+} states could exist in the 1.4 GeV mass region, hence the excitement over this claimed signal.

d) Future tests. There have been criticisms of the data analysis which need to be answered; as this state is potentially the most significant discovery in meson spectroscopy in recent years, the importance of answering any criticisms cannot be overstressed.

The $\eta\pi$ decay has been seen. Lipkin and I noted that a hybrid would have [9]

$$B(\eta'\pi) >> B(\eta\pi) .$$

The essential feature is that a selection rule [9] prevents decay by figure 1a,b; only the OZI-violating topology [figure 1c] can contribute, and here the gluon produces η_1, which is more η' than η.

Some remarks about gluonic hadrons

If the glueball mass scale is above 1.5–2 GeV, then it is probable that some hybrid states will be lighter than glueballs. In naïve models this is clear; a glueball consisting of at least two gluons whereas a hybrid requires only a single gluon to be excited. In string models, where hybrids correspond to excitation of the string, hybrids and glueballs are predicted at around 2 GeV. We may indeed hope, therefore, that hybrid states may be accessible to a facility such as KAON.

We may imagine a day when several hybrid and glueball states have shown up among the many other resonances; we will then have the task of deciding which is which. Many of the states will probably be mixtures of $q\overline{q}$ (or qqq) and gluonic states and overpopulation of multiplets would be a clue. The ordering of the various J^{PC} may help us. One has for low states

$$\begin{aligned} m(0^{++} \simeq 2^{++}) &: \; q\overline{q},\; q^2\overline{q}^2,\; gg,\; gq\overline{q} \\ 0^{++} \text{ isolated} &: \; KK \text{ molecule} \\ 0^{-+} &: \; gg,\; gq\overline{q};\; (q\overline{q})^* . \end{aligned}$$

The $2S(q\overline{q})$ multiplet of 0^{-+} begins to be complete and any further 0^{-+} with masses below 1.7 GeV would call for degrees of freedom beyond quarks.

The photon has been, and will continue to be, a definitive probe of the internal structure of hadrons. Magnetic moments reveal the spin-flavour correlations within baryons, and photoproduction of hybrid baryons is subject to rather distinctive selection rules [17].

In the meson sector the leptonic widths

$$\Gamma(\rho, \omega, \phi, \psi) \to e^+e^-$$

are in the ratio 9:1:2:8, attesting to the quark flavour content within those mesons. The $M1$ transitions $V \to P\gamma$ probe the flavour content of the pseudoscalar mesons.

If a state X is produced in $\psi \to \gamma X$, then the subsequent decays $X \to \gamma V$ ($V = \rho, \omega, \phi$) can tag the flavour content of X. If $X = n\bar{n}$ then $X \not\to \gamma\phi$; conversely if $X = s\bar{s}$ then $X \not\to \gamma\rho, \gamma\omega$. If X is a glueball then there will be a single peak, common to $\gamma\rho$, $\gamma\omega$, $\gamma\phi$ in relative strengths 9:1:2 (but whose overall strength is model dependent). For $n\bar{n}$ or $n\bar{n}g$ with I=0 one expects $\gamma\rho : \gamma\omega = 9 : 1$ and some 300 MeV higher in mass will be a second peak seen in $\gamma\phi$ with relative strength twice that of the lower mass $\gamma\omega$ peak. If the states X are orthogonal mixtures of $n\bar{n}$ and $s\bar{s}$, the relative radiative strengths will help determine the mixing angle. These tests may prove useful in disentangling the nature of $\eta(1440)$ where a $\gamma\rho$ signal has been seen.

References

[1] J. Weinstein and N. Isgur, Phys. Rev. D 27, (1983) 588.

[2] R.L. Jaffe, Phys. Rev. D15, (1987) 281.

[3] I. Balitsky *et al.*, Z. Phys. C33, (1986) 265;
J. Govaerts *et al.*. Nucl. Phys. B284, (1987) 674;
J. Lattore *et al.*, Z. Phys C34, (1987) 347.

[4] T. Barnes and F. Close, Phys. Lett. 116B, (1982) 365;
T. Barnes, F.E. Close, and F. deViron, Nucl. Phys. B224, (1983) 241;
M. Chanowitz and S. Sharpe, Nucl. Phys. B222, (1983) 211.

[5] GAMS collaboration, M. Boutemeur, Procs., 22 Rencontre de Moriond (Editions Frontières, Paris, 1987);
F. Binon, in Ref. 12.

[6] W. Dunwoodie, private communication;
S. Tuan, R. Dalitz, and T. Ferbel, University of Hawaii report, August 1988.

[7] Lepton-F collaboration, S.I. Bityukov *et al.*, Phys. Lett. 188B, (1987) 383;
V. Kubarovski, L. Landsberg, and V. Obraztsov, Serpukhov report 88–63 (1988).

[8] F.E. Close and H.J. Lipkin, Phys. Rev. Lett. 41, (1978) 1263.

[9] F.E. Close and H.J. Lipkin, Phys. Lett. 196B, (1988) 245.

[10] LASS collaboration, SLAC–PUB–4652.

[11] D.P. Stanley and D. Robson, Phys. Rev. D21, (1980) 3180;
S. Godfrey and N. Isgur, Phys. Rev. D32, (1985) 189.

[12] V. Smith, Proc. Brookhaven Workshop on Glueballs, Hybrids and Exotics, August 1988.

[13] F.E. Close, p. 851 *et seq.* in Rep. Prog. Phys. 51, (1988) 833 (1988) and Proc. of APS meeting, Storrs, Connecticut, August 1988.

[14] N. Achasov and A. Kozhenikov, Phys. Lett. 207B, (1988) 199.

[15] N. Achasov, Phys. Lett. 209B, (1988) 373.

[16] L. Landsberg, Ref. 12, and V. Kubarovski *et al.*, Ref. 7.

[17] T. Barnes and F.E. Close, Phys. Lett. 128B, (1983) 277.

Z. Phys. C – Particles and Fields 46, S93–103 (1990)

Zeitschrift für Physik C Particles and Fields

Mesons, hybrids, and glueballs: the search for new forms of hadronic matter

S. Godfrey

Guelph-Waterloo Program for Graduate Work in Physics, Department of Physics, University of Guelph, Guelph, Canada N1G 2W1

Abstract

I will review some of the outstanding issues in meson spectroscopy. The most important qualitative issue is, whether hadrons with explicit gluonic degrees of freedom exist. To answer this question requires a much better understanding of conventional $q\bar{q}$ mesons. I therefore begin by examining the status of conventional meson spectroscopy and how the situation can be improved. I then survey the expected properties of hybrids to give guidance to experimental searches. I briefly mention glueballs, since they are likely to be more difficult to find. The final section discusses multiquark systems, as they are likely to be important in the mass region under study and will have to be understood better.

1. INTRODUCTION

It is almost twenty years since the birth of Quantum Chromodynamics [1],the theory of the strong interactions, and it is not yet clear what the physical states of the theory are. Since it is possible that it may be decades before we have a thorough theoretical understanding of QCD, we must rely heavily on the insights we can gain from experiment and QCD based models. To a large extent our understanding of hadron structure is based on the constituent quark model in which mesons are made of a quark and antiquark and baryons are made of three quarks [2,3]. However, an important consequence of QCD was the expectation that *exotic* hadrons beyond the naive quark model should also exist; hybrids, glueballs, and $q\bar{q}q\bar{q}$ states. The problem is that none of these exotics have been unambiguously identified. What has happened to them? In this overview I examine how these states might be found.

As an operational definition I will refer to states predicted by the constituent quark model as *conventional* hadrons and those lying outside the quark model as *exotic* hadrons. I emphasize, however, that there is nothing fundamental about the quark model and the physical states of the theory should be based on the gauge invariant operators one can construct in QCD and will in general include gluonic excitations [4]. Nevertheless, states predicted by the quark model are the only ones that have been unambiguously identified. As a result, a major preoccupation of hadron spectroscopists is the attempt to discover the new types of hadronic matter — the ***hybrids*** which have constituent quarks and an excited glue degree of freedom and ***glueballs*** which have no valence quark content whatsoever [5]. A serious impediment to the discovery of such states is the sad shape of hadron spectroscopy. In particular, none of the light meson spectra is well mapped out.It is therefore of utmost importance to sort out light meson spectroscopy so that we have a template against which to compare exotic candidates.

Despite the "deplorable" shape of conventional meson spectroscopy, cleaning it up will likely elicit nothing more than a big yawn from the physics community at large. What is needed is the discovery of something dramatic — the new forms of hadronic matter, and the primary purpose of the next generation of hadron experiments should be to discover glueballs, hybrids, and other exotic hadrons. The discovery of such states is important to understand *soft* QCD.

This overview is divided into three sections: In section 2, I will discuss issues in conventional meson spectroscopy and how progress might be made. Section 3 describes hadrons with excited gluonic degrees of freedom. Hybrid properties are described in some detail to demonstrate the contrasts in different models of hadron structure, and because they look to be the best hope for finding gluonic excitations in hadrons. This is followed with a brief description of glueballs. For completeness, in section 4, I comment briefly on multiquark states, although they will receive a more complete discussion in F. Close's contribution. I conclude the overview with some general comments on the path to progress in hadron spectroscopy.

2. CONVENTIONAL MESONS

The Quark Model is over 25 years old! It is a useful tool for understanding hadron spectroscopy, but we still don't understand why it works. It is time to fill in some of the missing states so that we can either verify the model, find out where it needs to be refined, or possibly show it is wrong. In our quest for *exotic* hadrons we should not forget that *conventional* $q\bar{q}$ mesons can also tell us much about the nature of confinement. For instance, we have little more than a qualitative understanding of spin orbit splittings in light hadrons which has led to the misidentification of some hadronic states. Finally, if we are to make progress in finding *exotic* hadrons we must have a much better understanding of the *conventional* hadron background, the template against which exotics will be studied.

2.1 Quark Model vs Experiment

In the constituent quark model conventional mesons are bound states of a spin $\frac{1}{2}$ quark and a spin $\frac{1}{2}$ antiquark, bound by a phenomenological potential which has some basis in QCD. The quark and antiquark spins combine to give total spin 0 or 1 which is coupled to the orbital angular momentum L. This leads to meson parity and charge conjugation given by

$$P = (-1)^{L+1} \qquad (1)$$

$$C = (-1)^{L+S} \qquad (2)$$

resulting in the meson states of table 1. Thus, a meson with $J^{PC} = 1^{-+}$ would be forbidden in the constituent quark model as would a doubly charged meson m^{++}.

Since we cannot at this point calculate hadron properties from first principles, we must rely on QCD motivated models, which exist in the literature, to help interpret experimental resonances. In what follows, for purposes of illustration, I will compare experimental data to the results of a particular model with which I am most familiar and which I believe is probably the most comprehensive calculation of meson properties [6,7]. Figure 1 compares the quark model predictions for the strange meson spectrum to the experimentally observed states [8,9]. A striking feature of the spectrum is its regularity. There is good agreement for the masses of the leading orbital excitations between the quark model and experiment, supporting the picture of linear confinement. Although the first p-wave multiplet is more or less complete, we have very little information on higher orbitally excited multiplets and radially excited multiplets. Information on both types of excitations are important to

Table 1: The quantum numbers of the conventional $q\bar{q}$ mesons.

		J^{PC}
L=0	S=0	0^{-+}
	S=1	1^{--}
L=1	S=0	1^{+-}
	S=1	$\{0^{++}, 1^{++}, 2^{++}\}$
L=2	S=0	2^{-+}
	S=1	$\{1^{--}, 2^{--}, 3^{--}\}$
⋮	⋮	⋮

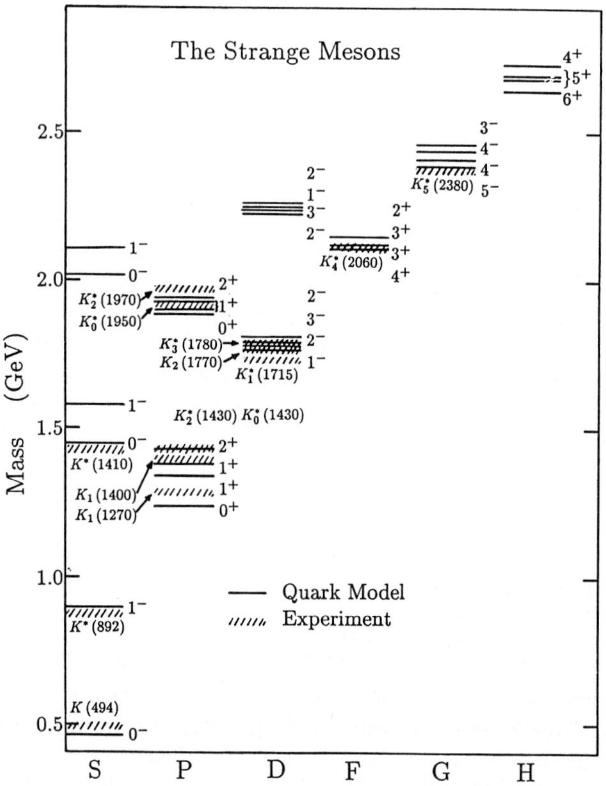

Figure 1: The strange meson spectrum. The quark model predictions are taken from Godfrey and Isgur [6], and the experimental values from Aston *et al.* [8,9].

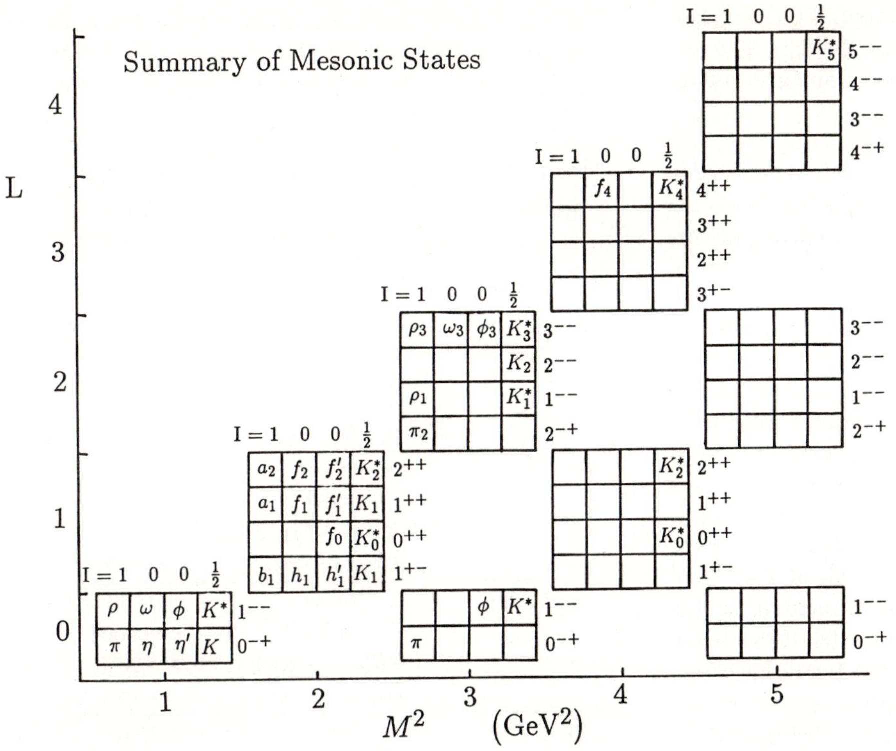

Figure 2: Summary of mesonic states with light quarks.

understand confinement. For example, in orbitally excited multiplets there are two contributions to the spin orbit interaction. The first comes from the short range Lorentz vector one-gluon-exchange and the second comes from the long range linear confinement piece which is supposedly a Lorentz scalar type interaction. The two contributions have opposite signs, so at some point the ordering of multiplets should invert with respect to the ordering of the lower L mesons. For instance the p-wave mesons have the ordering $M(^3P_2) > M(^3P_1) > M(^3P_0)$, which can be compared to the quark model predictions for $M(^3G_3) > M(^3G_4) > M(^3G_5)$. Knowing where this inversion occurs would be useful for understanding confinement [10].

For the case of the radially excited mesons even less is known experimentally for the light quark mesons. Without completing at least some of these multiplets we can hardly say we understand the meson spectrum.

One can do a systematic study of all light mesons with similarly discouraging results. Rather than list all of them in detail, we summarize the status of light meson spectroscopy in figure 2. One sees that the strange meson spectrum is fairly representative of the entire light meson spectrum. Starting with the p-wave multiplet, which is the multiplet most filled, we find that even this is far from being well understood. The scalar mesons (0^{++}) are in a state of confusion due to the possible interpretation of the $a_0(980)$ and $f_0(980)$ as $q\bar{q}q\bar{q}$ states [11–13]. In the 1^{++} sector the $f_1(1420)$ (formerly the E(1420)) has long been considered to be the $^3P_1$ $s\bar{s}$ meson. Recently the LASS group discovered another state [9] which appears to be a more likely candidate, the $f_1(1530)$. The isovector member of the multiplet, the a_1, has also been subject to debate because of its apparently different measured mass in τ decay [14]. Finally, the $^1P_1$ $s\bar{s}$ state (h_1') has yet to be confirmed. Turning to the higher mass multiplets, both radially excited and orbitally excited, we find that they are even less understood. Some of my assignments (or lack of assignments) in figure 2 are subject to debate, but this lack of a consensus underlines the fact that far too little is known about light meson spectroscopy.

2.2 Some Puzzles in Mesons

In addition to the obvious searches for the missing mesons, there are numerous puzzles in meson spectroscopy which may be hints of new types of hadronic matter. In what follows, I will mention several of the current puzzles. It is quite possible that the examples I discuss below will be understood by the time the KAON factory becomes operational, but I have little doubt that there will be new, more interesting ones to be solved.

$\eta(1440)$ (formerly the $\iota(1440)$): Because this pseudoscalar state is seen in the gluon rich J/ψ radiative decay the ι is considered to be a glueball candidate [5]. The impediment to coming to a definitive conclusion about this state is that two isoscalar radially excited pseudoscalars (2^1S_0) are expected to lie in this mass region, so until we obtain a more complete understanding of the 2^1S_0 nonet, the issue will remain cloudy. Another problem is the confusion involving the misidentification of the $E(1420)$. The bottom line is that we need a general cleaning up of the environment in this mass region.

$f_2(1750)$ (formerly the $\theta(1750)$): This state is seen in J/ψ radiative decay to $K\overline{K}$, so it is also a glueball candidate [5]. Although a number of 2^{++} states, both radially excited p-waves and orbitally excited f-waves, are expected, none seem to fit with the θ. Also disconcerting is the fact that it was not seen in $K\overline{K}$ by the LASS group. Are we seeing a glueball?

$f_1(1420)$ (formerly the E(1420)): There has been considerable confusion in the 1400 MeV mass region for the 1^{++} and 0^{-+} sectors for several years. The one state that was thought to be understood was the axial vector E(1420). Recently, however, the LASS group established the existence of another axial vector meson at about 1530 MeV which appears to be a much stronger candidate for the 1^{++} member of the p-wave $s\bar{s}$ multiplet [9]. If this is the case, what is the E(1420), and does this puzzle have anything to do with the the $\iota(1440)$ puzzle? My guess is that the E(1420) is a $K^*\overline{K}$ molecule as is the $\iota(1440)$. Both lie just above $K^*\overline{K}$ threshold and perhaps they are weakly bound molecules analogous to the explanation of the 0^{++} $\delta(980)$ and $S^*(980)$ mesons.

Besides the puzzles I have mentioned above, there are numerous other ones. For instance, what is the explanation of the g_T 2^{++} tensor mesons seen at Brookhaven in $\pi p \to K\overline{K}$ [15]. Are they glueballs, or are they conventional mesons, or do they have a totally different explanation? Above, I mentioned that the 0^{++} $\delta(980)$ and $S^*(980)$ mesons are thought to be $K\overline{K}$ molecules. If this is indeed the case, where are the $^3P_0$ $q\bar{q}$ states? I would expect (or at least hope) that some of these puzzles will be solved by the time the KAON factory becomes operational, but, as I said at the beginning of this section, there will no doubt be new puzzles to take their place.

Table 2: Quark Model predictions for the properties of the missing L=2 mesons. The masses and widths are given in MeV. The masses come from ref. 6 and the widths from ref. 7.

Meson State	Property	Prediction
$\eta_2(1^1D_2)$	Mass	1680
	width	~ 400
	$BR(\eta_2 \to a_2\pi)$	$\sim 70\%$
	$BR(\eta_2 \to \rho\rho)$	$\sim 10\%$
	$BR(\eta_2 \to K^*\overline{K} + c.c.)$	$\sim 10\%$
η_2' (1^1D_2)	Mass	1890
	width	~ 150
	$BR(\eta_2' \to K^*\overline{K} + c.c.)$	$\sim 100\%$
K_2 (1^1D_2)	Mass	1780
	width	~ 300
	$BR(K_2 \to K^*f(1280))$	$\sim 30\%$
	$BR(K_2 \to \rho K)$	$\sim 20\%$
ω_1 (1^3D_1)	Mass	1660
	width	~ 600
	$BR(\omega_1 \to B\pi)$	$\sim 70\%$
	$BR(\omega_1 \to \rho\pi)$	$\sim 15\%$
K_2 (1^3D_2)	Mass	1810
	width	~ 300
	$BR(K_2 \to K^*(1420)\pi)$	$\sim 50\%$
	$BR(K_2 \to K^*\pi)$	$\sim 30\%$
ρ_2 (1^3D_2)	Mass	1700
	width	~ 500
	$BR(\rho_2 \to [a_2\pi]_S)$	$\sim 55\%$
	$BR(\rho_2 \to \omega\pi)$	$\sim 12\%$
	$BR(\rho_2 \to \rho\rho)$	$\sim 12\%$
ω_2 (1^3D_2)	Mass	1700
	width	~ 250
	$BR(\omega_2 \to \rho\pi)$	$\sim 60\%$
	$BR(\omega_2 \to K^*\overline{K})$	$\sim 20\%$
ϕ_2 (1^3D_2)	Mass	1910
	width	~ 250
	$BR(\phi_2 \to K^*\overline{K} + c.c.)$	$\sim 55\%$
	$BR(\phi_2 \to \phi\eta)$	$\sim 25\%$

2.3 Hunting Missing States

Given our unsatisfactory knowledge of light meson spectroscopy, how do we go about finding the missing states and solve some of the puzzles? One can see from figure 2 that as we go to higher mass, the number of states multiply rather rapidly so that in general we will have to find the missing states in a large background of other states. It is therefore highly unlikely that we will have any success in unravelling the spectroscopy by bump hunting. Rather, we will need high statistics experiments to perform partial wave analysis to filter by J^{PC} quantum numbers. To assist us in this process a guide to the expected properties will be useful. To this end we turn to quark model predictions for the expected masses and decay modes [6,7]. As an example of what such a search would entail we examine the missing states of the L=2 meson multiplet whose quark model predictions are listed in table 2.

Starting with the $\eta_2(1^1D_2)$, we expect it to be almost degenerate in mass with its non-strange isoscalar partner, the π_2 (formerly the $A_3(1680)$). From table 2 we see that it is expected to be rather broad and it decays predominantly through the a_2 isobar, which in turn decays to $\rho\pi$. The final state is expected to have 4π's making it rather complicated to reconstruct the original η_2 resonance. The $\rho_2(1^3D_2)$ will also decay dominantly to a 4π final state. The ω_2 decays to the simpler $\rho\pi$ final state with a moderate width, but since it has a similar mass as the $\pi_2(1680)$, which also decays to $\rho\pi$, it is possible that it is masked by the π_2. The $s\bar{s}$ states, the $\eta_2'(1^1D_2)$ and the $\phi_2(1^3D_2)$, are both relatively narrow and one would expect that they would have been observed. In fact, the LASS group has recently reported seeing them in $K\overline{K}\pi$ [8]. The likely reason that they have been so difficult to find is that they are produced rather weakly. The strange mesons, the $K_2(1^1D_2)$ and the $K_2(1^3D_2)$, lie at around 1800 MeV. They decay to some relatively simple channels and are predicted to have a moderate width. It is possible that they have been sighted as the $L(1770)$'s. The final state in the table is the broad 1^{--} ω_1. Structure has been seen in this mass region, but the experimental situation is likely confused due to the nearby broad 2^3S_1 1^{--} state, which is expected to lie at around 1450 MeV and overlaps and interferes with the 1^3D_1 state.

One can perform a similar analysis of other multiplets. In general, however, given how complicated the meson spectrum is, it appears that a good starting place would be the strange mesons. The reason for this is that in the strange meson sector we don't have the additional problem of deciding whether new states are glueballs or conventional mesons, and we don't have the additional complication of mixing between isoscalar states due to gluon annihilation. Following this a detailed survey of ϕ states would be useful, since they form a bridge between the heavy quarkonia ($c\bar{c}$ and $b\bar{b}$) and the light quark mesons and would help us understand the nature of the confinement potential.

2.4 Future Directions

From the preceding sections we conclude that we need a far better understanding of meson spectroscopy before we can say that we understand it. The first step is find some of the missing states. It will be important to fill in both the orbitally excited multiplets and the radially excited multiplets. These missing states will lie in a large background of other states. To find them we will need unprecedented statistics along with a partial wave analysis to filter the J^{PC} quantum numbers, as results from the LASS spectrometer group have shown us. In figure 3a we see the invariant mass distribution of the $\overline{K}^0\pi^+\pi^-$ system [8]. The K_2^* and the K_3^* show up clearly but little else. However, the partial wave analysis (PWA) results shown in figure 3b show structure in many more channels [8]. This sort of analysis is only possible with high statistics. The LASS group recorded about 100 million events to tape, and they estimate that in order to do better a next generation experiment will need at least a factor of 10 more events, preferably a factor of 50. Statistics alone will not be enough. A next generation detector should have good uniform 4π acceptance so that holes in the detector do not distort the moments of the PWA. Another improvement would be good neutral detection so that π's and η's etc. are not missed.

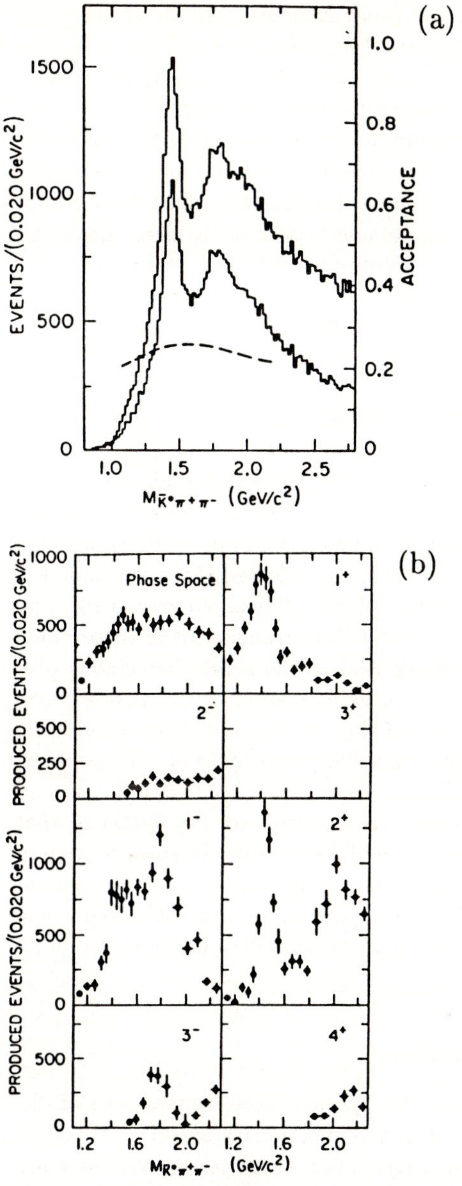

Figure 3: (a) The $\overline{K}^0\pi^+\pi^-$ invariant mass distribution. The outer histogram is for the entire data sample whereas the inner histogram is for events with $t' \leq 0.3$ $(\mathrm{GeV/c})^2$.
(b) The spin-parity wave sums. The intensities of the summed partial waves with the same J^P are plotted as a function of mass (from ref. 8).

We will also need to develop new experimental and theoretical techniques to study broad resonances. From the experimental side it is clear that it will not be easy to identify a broad resonance in a background of other broad resonances as well as in the presence of new production thresholds. From the theoretical side most quark model calculations have treated mesons in the valence quark limit without considering the influence of coupling to production and decay channels. We are at the point in our understanding that these effects can no longer be ignored as they can make significant changes to the observed hadron masses and decay properties. In any case, it will be necessary to understand how to obtain observed cross sections starting with the underlying spectrum when decay channel coupling is taken into account along with final state interactions of the decay products.

Conventional hadron spectroscopy may not be the sexiest physics but it is good, solid bread and butter physics that needs to be done.

3. GLUONIC EXCITATIONS [5]

The complication in QCD which makes it so difficult to solve is the presence of boson-boson interactions required by gauge invariance, which in perturbation theory gives rise to rather complicated three and four boson couplings. Although it is difficult to extract physical properties from the QCD Lagrangian, it is these gluon self couplings which lead to the belief that gluons play a dual role in QCD; as mediators of the strong force, like in the conventional $q\overline{q}$ mesons and qqq baryons, and as constituents in glueballs and hybrids. The problem at present is that it is not even clear what the correct degrees of freedom are (relevant to soft QCD) so that the predictions of the different models must be viewed with caution. Nevertheless the discovery of glueballs or hybrids would be strong confirmation of QCD and would be very important for understanding QCD in the "soft" region.

3.1 Hybrid Mesons

In searching for hybrids there are two ways of distinguishing them from conventional states. One approach is to look for an excess of observed states over the number predicted by the quark model. The drawback to this method is that it depends on a good understanding of hadron spectroscopy which, as we have seen, we do not yet have and so cannot reliably rule out a state as a conventional meson. The other approach is to search for quantum numbers which are not consistent with quark model predictions.

The first step in the second approach is to enumerate the hybrid J^{PC} quantum numbers. To do this in a model independent manner obeying gauge invariance we form gauge invariant operators [16,17] from a colour octet $q\overline{q}$ operator, $\overline{\psi}\Gamma\frac{\lambda}{2}\psi$, and a gluon field strength, $F_{0i} \approx \vec{E}^a$ or $F_{ij} \approx \vec{B}^a$, where Γ is a Dirac tensor and λ_a are the generators of SU(3) colour:*

$$\vartheta = (\overline{\psi}\Gamma\frac{\lambda_a}{2}\psi) \otimes (\vec{E}^a \text{ or } \vec{B}^a) \qquad (3)$$

This composite operator is known as an *interpolating field* and is equally applicable to the various approaches ranging from the bag model to the flux tube model, in addition to more rigorous lattice gauge theory calculations. It shows that a state with its quantum numbers can be created from the vacuum with this operator. For example, the interpolating field for the 1^{-+} exotic state is given by $(\overline{\psi}\vec{\gamma}\psi) \times \vec{B}$ and for the 2^{+-} by $(\overline{\psi}\vec{\gamma}\gamma_5\psi) \otimes \vec{B}$. The quantum numbers of the low lying hybrids are given by:

2^{++}	2^{-+}	$\underline{2^{+-}}$	2^{--}
1^{++}	$\underline{1^{-+}}$	1^{+-}	1^{--}
0^{++}	0^{-+}	$\underline{0^{+-}}$	$\underline{0^{--}}$

Higher J operators can also be constructed but they are presumably higher in mass and more difficult to produce. The underlined $q\overline{q}g$ states have exotic quantum numbers not present in the constituent quark model. If these exotic state are sufficiently low in mass and do not have exceedingly large widths, they could provide the ***smoking gun*** evidence for hybrids which we are seeking. Their discovery would unambiguously signal hadron spectroscopy beyond the quark model.

3.1.1 Hybrid Masses

The Bag Model [19–23]

In the bag model [22] hadrons are viewed as a region of space enclosing a fixed number of quarks and gluons with the model made Lorentz invariant by the addition of a surface pressure term, B_0, to the Lagrangian density. Inside the bag the quark fields, ψ, obey the free Dirac equation:

$$\begin{aligned} (\not{p} - m)\psi &= 0 \qquad \text{inside S} \\ \psi &= 0 \qquad \text{outside S} \end{aligned} \qquad (4)$$

along with the boundary conditions that 1) there is no colour current through the bag surface, and 2) pressure balance determines the bag surface, i.e., on the equilibrium bag surface the external pressure term B_0 exactly balances the Zitterbewegung pressure due to the quarks and gluons held within the bag. The lowest energy solutions have quarks in $1S_{1/2}$, $1P_{1/2}$, $1P_{3/2}$ eigenmodes. Gluons in the bag obey the free Helmholtz equation:

$$\begin{aligned} (\nabla^2 + \omega^2)\vec{A}^a &= 0 \qquad \text{inside S} \\ \vec{A}^a &= 0 \qquad \text{outside S} \end{aligned} \qquad (5)$$

subject to the same boundary conditions. The solutions are the transverse electric (TE) and transverse magnetic (TM) cavity resonator modes:

*One has to be careful when constructing interpolating fields as they can be misleading if subject to constraints. For example, the time component of the electromagnetic current $\overline{\psi}\gamma_0\psi$ has $J^{PC} = 0^{+-}$ quantum numbers, but because the current is conserved, the operator has zero coupling to any massive state [18].

$$\Longrightarrow \left\{ \begin{array}{ll} TE & J^{PC} = 1^{+-} \\ TM & J^{PC} = 1^{--} \end{array} \right. \quad \text{(lowest modes).} \qquad (6)$$

In the zeroth order bag model the mass of a hadron is simply the sum of the quark and gluon constituent energies and the energy of the bag itself. To go beyond the zeroth order bag model involves including contributions from gluon exchange [19,20].

To make hybrids we combine a colour octet gluon with a $q\bar{q}$ pair in a colour octet to obtain a colour singlet:

$$(q\bar{q})_8 \times g_8 = (q\bar{q}g)_1 + \ldots . \qquad (7)$$

Thus the lowest $q\bar{q}g$ hybrid meson multiplets are constructed from a colour octet $q\bar{q}$ with $J^{PC} = 0^{-+}$ or 1^{--}, each in the $J^P = (1/2)^+$ mode, and a gluon in the lightest TE mode with $J^{PC} = 1^{+-}$ resulting in the following lowest lying hybrids:

$$2^{-+},\ 1^{-+},\ 1^{--},\ 0^{-+}.$$

The SU(3) flavour quantum numbers of a hybrid are those of the component $q\bar{q}$ pair so that hybrid mesons span the familiar SU(3) flavour nonets. However, the I=0 and I=1 states are not degenerate because in the isoscalar hybrids, the relative ease of internal annihilation of the $q\bar{q}$ pair, which is already in a colour octet, shifts the mass.

Three groups have completed major calculations of the hybrid spectrum in the bag model including spin dependent forces due to gluon exchange (which I have not discussed) [19–21]. Some representative results are shown in figure 4 along with constituent quark model predictions for conventional mesons. The different groups are in reasonable agreement for the splittings but differ on the multiplet mass. The most important differences arise from the choice of parameters and the treatment of spin independent effects; Chanowitz and Sharpe have included self energies and find that the overall scale is sensitive to these undetermined values.

The 0^{-+}, 2^{-+}, and 1^{--} hybrid nonets are near in mass to $q\bar{q}$ states with the same quantum numbers which can result in considerable mixing. This can only confuse the situation when determining if a state is a hybrid or conventional meson. Thus, the discovery of such states would be difficult to be convincing, because they are also candidates for conventional states.

The Flux Tube Model

The flux tube model of hadrons [24,25] is based on the strong coupling limit of QCD [26] with its parameters fixed from the familiar meson and baryon sectors. The significant difference between the flux tube approach and the bag model approach is that the eigenstates of the strong coupling limit of (lattice) QCD consist of, not quarks and gluons as in the bag model, but quarks on lattice sites connected by arbitrary paths of flux links, or

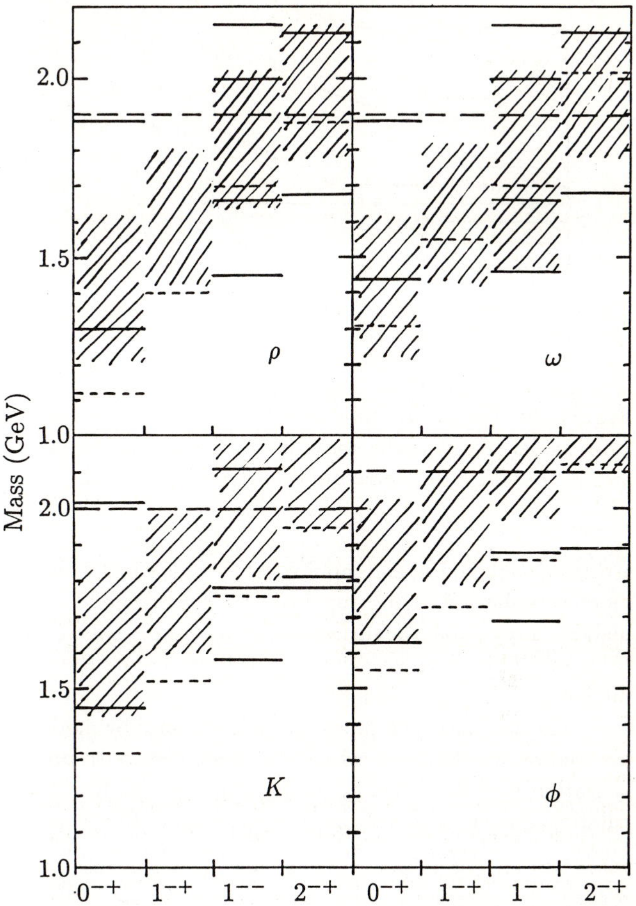

Figure 4: Hybrid mass predictions. The short dashed lines are the bag model predictions of Barnes Close and deViron [19]. The shaded region are the bag model predictions of Chanowitz and Sharpe for a range of values of the quark and gluon self energies [20]. The long dashed lines are the flux tube model predictons of Isgur and Paton [24]. The solid lines are the conventional $q\bar{q}$ predictions of the relativized quark model [6].

in the absence of quarks, of arbitrary closed loops of flux. It is assumed that the flux tube picture survives departures from the strong coupling limit, or in other words, that the flux tubes do indeed form a reasonable set of basis states, and that the adiabatic treatment of the flux tubes in the presence of quark motion is reasonable.

In this picture the string states define adiabatic quark potentials analogous to the nuclear potentials in molecular physics where adiabatic surfaces are defined for the nuclear motion based on the faster moving electronic potentials. We should then expect a tower of quark states built on each string adiabatic surface, figure 5. In the flux tube model conventional hadrons correspond to gluonic fields in the ground states. There are two types of hybrids in this model, vibrational hybrids which correspond to excitations of the quantum string into higher

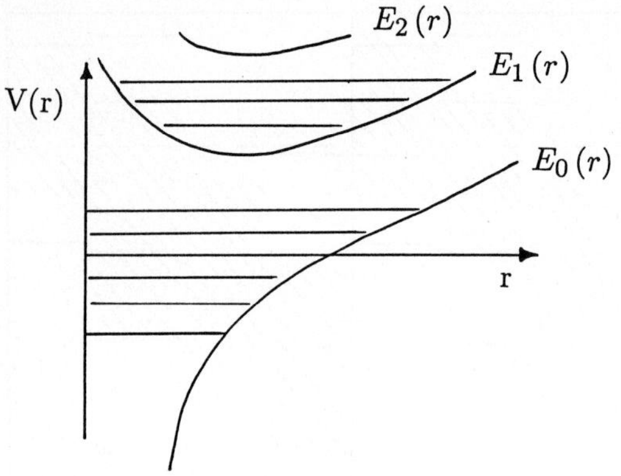

Figure 5: The quark adiabatic potentials of the flux tube model. $E_0(r)$ is the conventional hadron potential and $E_1(r)$ and $E_2(r)$ correspond to hybrid potentials.

string normal modes, and topological hybrids which have more complicated string topologies and correspond to higher energy adiabatic surfaces. The latter are expected to be much higher in energy so we will not discuss them further.

The adiabatic potentials are characterized by mode occupation numbers, n^i_m where the index i refers to polarization and the index m to the string mode. In calculating masses it is assumed that low lying flux tube modes can be treated in a nonrelativistic small oscillation approximation. The first excited state is doubly degenerate with phonons of transverse vibration with $\sigma = \pm 1$ angular momentum about the $q\bar{q}$ axis, and, when combined with spin, we get the 8 nearly degenerate nonets of hybrid mesons:

$$J^{PC} = 0^{\pm\mp} \quad 1^{\pm\mp} \quad 2^{\pm\mp} \quad 1^{\pm\pm} \tag{8}$$

with masses approximately 1.9 ± 0.1 GeV. Among these states are three J^{PC} exotic nonets with nine neutral members having $J^{PC} = 2^{+-}, 1^{-+}$, and 0^{+-}. These results should be contrasted to the bag model where there is no such degeneracy because the TM mode is much higher in mass than the TE mode. These results are compared to bag model results in figure 4.

3.1.2 Hybrid Decays

In the previous section we came to the conclusion that the most promising approach for finding hybrids is to look for ones with exotic quantum numbers. Even so, there are numerous states to consider so, for the sake of brevity, we start with 1^{-+} exotics as an example. Possible decays are given by:

$$\begin{aligned}
\rho_g &\to [\underline{\pi\eta}, \underline{\pi\eta'}, \pi\rho, K^*\overline{K}, \eta\rho, \ldots]_P \\
&\to [\pi B, \pi D, \eta A_1, KQ_{1,2} \ldots]_S \\
\\
\omega_g &\to [K^*K, \pi\pi(1300), \underline{\eta\eta'}, \ldots]_P \\
&\to [A_1\pi, \overline{K}Q_{1,2}, \ldots]_S \\
\\
K^*_g &\to [\pi K, \eta K, \phi K, \eta' K, \ldots]_P \\
&\to [\pi Q_{1,2}, KA_1, KB, \ldots]_S \\
\\
\phi_g &\to [K\overline{K}(1400), KK^*, \underline{\eta\eta'}, \ldots]_P \\
&\to [\overline{K}Q_2]_S \\
&\to [\overline{K}Q_1]_D
\end{aligned} \tag{9}$$

The underlined decays to two distinct pseudoscalars in a relative P-wave is a unique signature of the 1^{-+} state.

Given this long list of decays we turn to the various models for guidance to which modes are likely to be dominant. One would naively expect S-wave mesons in the $\pi\eta$ or $\pi\rho$ channels to be dominant due to the large available phase space. However, a common feature of the various models is the selection rule that the gluonic excitation cannot usually transfer its angular momentum to the final state meson pairs as relative angular momentum, but must instead appear as internal orbital angular momentum of the $q\bar{q}$ pairs.[†] This eliminates $\pi\eta$, $\pi\rho$, $\pi\eta'$, and $K^*\overline{K}$. The selection rule suppresses the decay channels which would likely be large and may make hybrids stable enough to appear as conventional resonances while at the same time explaining why hybrids with exotic J^{PC} have yet to be seen; they do not couple strongly to simple final states.

In particular, in the bag model the dominant decays occur when the valence gluon forms a colour octet $J^{PC} = 1^{+-}$ $q\bar{q}$ pair in which either q or $\bar{q}$ is in a P-wave mode, figure 6. The bag then contains two $q\bar{q}$ colour octets which after rearrangement fall apart into $q\bar{q}$ singlets, one in an S-wave ground state with $J^{PC} = 0^{-+}, 1^{--}$ and the other in an L=1 $J^{PC} = (0,1,2)^{++}, 1^{+-}$, i.e. $J^{PC} = 1^{-+} \to \pi D$ or ηA_1.

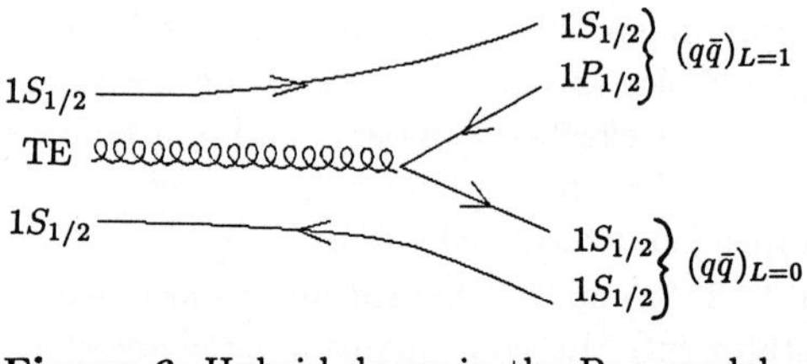

Figure 6: Hybrid decay in the Bag model.

The flux tube model also predicts that the low lying hybrids will decay preferentially to final states with one ground state S-wave and one excited P-wave meson, $b_1(1235)\pi$, $a_2(1320)\pi$, $K^*_2(1420)\overline{K}$, $\pi(1300)\pi, \ldots$, rather than two ground state mesons like $\pi\pi$, $\rho\pi$, $K\overline{K}$. The reason for this is that the relative coordinates of the two final state mesons are parallel to the initial meson axis and so cannot absorb the unit of string angular momentum about the initial meson axis, figure 7. This selection rule is broken for final states with different spatial wavefunctions [27].

[†]I note, however, that this selection rule is not absolute.

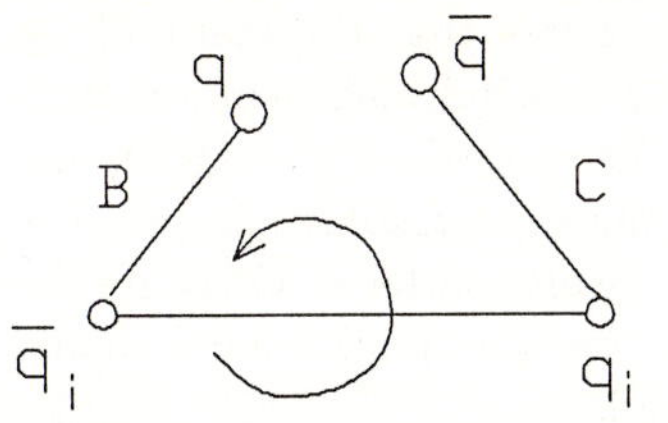

Figure 7: Hybrid decay, $H \to BC$ in the flux tube model.

Table 3: The dominant decays of the low-lying exotic hybrid mesons (from ref. 25).

Hybrid State	[decay mode]$_{L \text{ of decay}}$	Part. Width [MeV]
$\tilde{\rho}_2^{+-}(1900)$	$\to [\pi a_2]_P$	450
	$\to [\pi a_1]_P$	100
	$\to [\pi h_1]_P$	150
$\tilde{\omega}_2^{+-}(1900)$	$\to [\pi b_1]_P$	500
$\tilde{\phi}_2^{+-}(2100)$	$\to [K\overline{K}_2^* + c.c.]_P$	250
	$\to [\overline{K}Q_2 + c.c.]_P$	200
$\tilde{\omega}_1^{-+}(1900)$	$\to [\pi a_1]_{S,D}$	100, 70
	$\to [\pi\pi(1300)]_P$	100
	$\to [\overline{K}Q_2 + c.c.]_S$	100
$\tilde{\rho}_1^{-+}(1900)$	$\to [\pi b_1]_{S,D}$	100, 30
	$\to [\pi f_1]_{S,D}$	30, 20
$\tilde{\phi}_1^{-+}(2100)$	$\to [\overline{K}Q_1 + c.c.]_D$	80
	$\to [\overline{K}Q_2 + c.c.]_S$	250
	$\to [\overline{K}K(1400) + c.c.]_P$	30
$\tilde{\rho}_0^{+-}(1900)$	$\to [\pi a_1]_P$	800
	$\to [\pi h_1]_P$	100
	$\to [\pi\pi(1300)]_S$	900
$\tilde{\omega}_0^{+-}(1900)$	$\to [\pi b_1]_P$	250
$\tilde{\phi}_0^{+-}(2100)$	$\to [\overline{K}Q_1 + c.c.]_P$	800
	$\to [\overline{K}Q_2 + c.c.]_P$	50
	$\to [\overline{K}K(1400) + c.c.]_S$	800

I consider in further detail the flux tube model predictions which are listed in table 3. The flux tube model predicts that the $\tilde{\rho}_2^{+-}$, $\tilde{\rho}_0^{+-}$, and $\tilde{\phi}_0^{+-}$ are probably too broad to appear as resonances. The $\tilde{\omega}_1^{-+}$ decays mainly to $[a_1\pi]_S$ and $[\pi(1300)\pi]_P$ with $\Gamma \sim 100$ MeV which would make it difficult to reconstruct the original hybrid given the broad widths of the final state mesons. Similar problems also make the $\tilde{\phi}_1^{-+}$ difficult to find. According to the flux tube model the best bets for finding hybrids are: $\tilde{\rho}_1^{-+}$, $\tilde{\omega}_2^{+-}$, $\tilde{\omega}_0^{+-}$, and $\tilde{\phi}_2^{+-}$.

What we conclude from all this is that the favoured final states all contain broad p-wave mesons. To reconstruct the original resonance an isobar analysis will be essential, and to do this we will again need unprecedented statistics to pull a signal from the background.

3.2 Pure Glue States

In addition to hybrids, it is also expected that hadrons will exist with no valence quark content at all. The predictions for glueball masses vary considerably from calculation to calculation. I will optimistically use the flux tube model predictions as a guide because of their agreement with lattice calculations [28,29]. These are shown in figure 8. The flux tube model predicts that the lowest glueball is a 0^{++} at 1.5 GeV with all other states above 2 GeV and the lowest J^{PC} exotic at around 2.5 GeV.‡ Because of the uncertainty in the scalar meson sector, it seems likely that it would be very difficult to distinguish a scalar meson from the poorly understood conventional mesons. In fact, one can generalize that it will be difficult to unambiguously determine that any state with conventional quantum numbers is a glueball due to the dense background of conventional mesons. The best bet will then be to find glueballs with exotic quantum numbers. Unfortunately these states are all expected to have mass greater than 2.5 GeV and so will be difficult to find. Glueballs, therefore, do not seem to be the best place to start our search for gluonic hadrons.

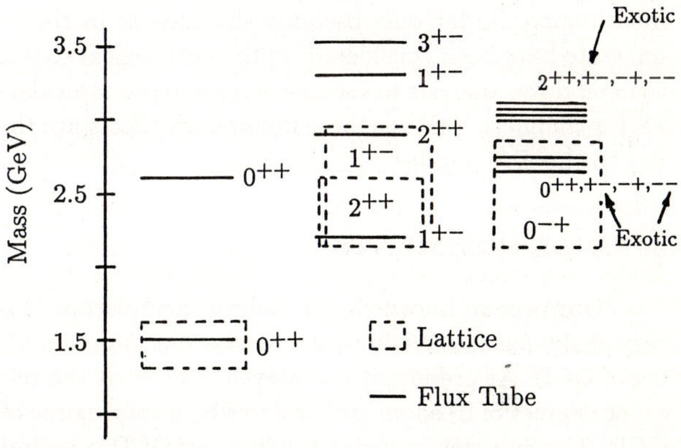

Figure 8: The low lying glueball mass spectrum. The flux tube results come from ref. 24 and the lattice results from ref. 29.

4. MULTIQUARK STATES

Multiquarks are discussed in detail in the contribution of F. Close [30]. Here, I briefly discuss their relevance to meson spectroscopy. Upon considering $qq\bar{q}\bar{q}$ systems we find that the colour couplings are not unique as they are in mesons and baryons, and whether or not multiquark states exist is a dynamical question. It is possible that multiquark states exist as bound states [4], but it is also possible that $qq\bar{q}\bar{q}$ configurations lead to meson-meson potentials [13]. Both must be taken into account when attempting to unravel the meson spectrum.

‡I note that the bag model predictions are considerably lower.

A study of the J^{PC} sector of the $qq\overline{q}\overline{q}$ system found that weakly bound $K\overline{K}$ "molecules" exist in the isospin zero and one sectors in analogy to the deuteron. It was suggested that these two bound states be identified with the $f_0(975)$ and $a_0(980)$ (the S^* and δ). The meson-meson potentials which come from this picture, when used with a coupled channel Schrödinger equation, reproduce the observed phase shifts for the δ and S^* in $\pi\pi$ scattering [13]. The $K\overline{K}$ molecules are the exception, however, as the model predicts that in general the $qq\overline{q}\overline{q}$ ground states are two unbound mesons.

There is evidence that meson-meson potentials have to be considered in other processes as well. In the reaction $\gamma\gamma \to \pi^+\pi^-$ the meson-meson potentials are needed along with $q\overline{q}$ resonances to reproduce the $\gamma\gamma \to \pi^+\pi^-$ cross section data [31]. Enhancements in the production of low invariant mass $\pi\pi$ pairs have been observed in a number of processes; $\eta' \to \eta\pi\pi$, $\psi' \to J/\psi\pi\pi$, $\Upsilon(nS) \to \Upsilon(mS)\pi\pi$, and $\psi \to \omega\pi\pi$. Similar enhancements have also been seen in some $K\pi$ channels in $\overline{p}p \to K\overline{K}\pi$. The conclusion drawn from these examples is that final state interactions arising from meson-meson potentials will play a central role in understanding the 1 to 2 GeV mass region. So far only pseudoscalar mesons in the final state have been considered, so the next logical step is to extend the analysis to vector-vector and pseudoscalar-vector channels. Perhaps these multiquark effects are the key to the E/ι puzzles.

5. FINAL COMMENTS

Our present knowledge of hadron spectroscopy is a very shaky foundation on which to base our understanding of QCD. At present, it is not even clear what the relevant degrees of freedom are for describing this regime of QCD. The first step in understanding *soft* QCD is to find the missing conventional $q\overline{q}$ states. Unless we understand conventional hadrons better, it will be very difficult to make progress in finding evidence for the gluonic degree of freedom in the hadron spectrum which should be the outstanding issue for the next generation of experiments.

Although it is conceivable that hybrid states with non-exotic quantum numbers could be identified as being excess states beyond those predicted by the quark model, given the very broad range of predictions for hybrid masses, I very much doubt that this is the most fruitful approach. It is more likely that, to be successful, a hybrid search should focus on the exotic properties of hybrids which would offer unambiguous evidence of new physics. The most noncontroversial such characteristic is that all models agree that one of the lowest hybrids will have exotic J^{PC} quantum numbers 1^{-+} with mass about 1.6 ± 0.3 GeV. The presence of a resonance signal in this channel would be strong evidence for the discovery of a hybrid, so it seems sensible that this be the place to begin any experimental search.

One should also appreciate that the study of $q\overline{q}$, hybrids, glueballs, and $qq\overline{q}\overline{q}$ is an indivisible subject since they are all governed by the same theory — QCD, and require the same experiments. In fact, some take the view that the only objects that really matter in QCD are the gauge invariant operators and the quark model should not be taken seriously at all.

To really unravel the meson spectrum in the 1 to 3 GeV mass region and make progress in hadron spectroscopy will take unprecedented statistics. It is important that we study many properties of hadrons in many different channels. The data will come from many different machines, not only the KAON factory. For instance, J/ψ radiative decays studied at e^+e^- colliders look at different aspects of hadron spectroscopy, as does two gamma physics properties studied at higher energy e^+e^- colliders or in $p\overline{p}$ annihilation. It will be necessary to do isobar analysis with broad intermediate states. To succeed we will need high intensity machines like the KAON factory and use more advanced detectors collecting a wide range of data. Then with hard work and probably some luck we may finally understand the strong interactions. Finally, we should not discount the possibility that genuine surprises could still emerge from careful experimental study.

ACKNOWLEDGEMENTS

This work was supported by the Natural Science and Engineering Research Council of Canada.

References

[1] M. Fritzch, M. Gell-Mann, and H. Leutwyler, Phys. Lett. 47B, (1971) 365 ; D.J. Gross and F. Wilczek, Phys. Rev. D8, (1973) 3497 ; S. Weinberg, Phys. Rev. Lett. 31, (1973) 494 .

[2] G. Zweig, CERN prepr. 8419/TH.412; 8182/Th.401 (1964; unpublished); M. Gell-Mann, Phys. Lett. 8, (1964) 214.

[3] For an overview of the current status of the quark model: S. Godfrey, *The Constituent Quark Model: A Status Report* Nuovo Cimento A (1989, in press).

[4] R.L. Jaffe, K. Johnson, and Z. Ryzak, Ann. of Phys. 168, (1986) 344.

[5] For other reviews of the subject see for example: T. Barnes, *The Exotic Atoms of QCD: Glueballs, Hybrids, and Baryonia*, Proc. of the School of Physics of Exotic Atoms, Erice Sicily, 1984; *The Bag Model and Hybrid Mesons*, Proc. of the SIN School of Strong Interactions, Zuoz, Switzerland, SIN report (1985); M. Chanowitz, *Meson Spectroscopy Viewed from J/ψ Decay: Gluonic States at BEBC*, LBL report LBL-24355 to be published in the Proc. of the Charm

Physics Symposium, eds. M.-H. Ye and T. Huang (Gordon and Breach, New York, 1987); F. Close, *Gluonic Hadrons*, Rep. Prog. Phys. 51, (1988) 833; S. Godfrey, *An Overview of Hybrid Meson Phenomenology*, Proc. of the BNL Workshop on Glueballs, Hybrids, and Exotic Hadrons, ed. S.U. Chung, Upton New York, 1988 (AIP, New York, 1989); N. Isgur, *Hadron Spectroscopy: An Overview with Strings Attached*, *ibid.*

[6] S. Godfrey and N. Isgur, Phys. Rev. D32, (1985) 189, and D34, (1986) 899; S. Godfrey, Phys. Rev. D31, (1985) 2375 and references therein.

[7] R. Kokoski and N. Isgur, Phys. Rev. D35, (1987) 907.

[8] D. Aston *et. al.*, Nucl. Phys. B292, (1987) 693; D. Aston *et. al.*, SLAC report SLAC-PUB-4202, appears in the Proc. of the SLAC Summer Institute on Particle Physics, Stanford California, 1986; D. Aston *et. al.* Phys. Lett. 180B, (1986) 308; D. Aston *et. al.* Phys. Lett. 201B, (1988) 169; D. Aston *et. al. Proc. of the BNL Workshop on Glueballs, Hybrids, and Exotic Hadrons* ed. S.U. Chung, Upton New York, 1988 (AIP, New York).

[9] D. Aston *et al.* Phys. Lett. B201, (1988) 573.

[10] S. Godfrey, Phys. Lett. 162B, (1985) 367.

[11] R.L. Jaffe, Phys. Rev. D15, (1977) 267,281; D17, (1978) 1444.

[12] J. Weinstein and N. Isgur, Phys. Rev. Lett. 48, (1982) 659; Phys. Rev. D27, (1983) 588.

[13] J. Weinstein and N. Isgur, University of Toronto report, UTPT-89-03 (1989; unpublished).

[14] See e.g. M. G. Bowler, Phys. Lett. B182, (1986) 400; N. Isgur, C. Morningstar, and C. Reader, Phys. Rev. D39, (1987) 1357.

[15] A. Etkin *et al.* Phys. Lett. B201, (1988) 568.

[16] T. Barnes, *The Bag Model and Hybrid Mesons*, Proc. of the SIN School of Strong Interactions, Zuoz Switzerland, SIN report (1985).

[17] For a more thorough discussion see R.L. Jaffe, K. Johnson, and Z. Ryzak, ref. 4.

[18] I thank T. Barnes and R. Jaffe for explaining this to me.

[19] T. Barnes, F.E. Close, and F. deViron, Nucl. Phys. B224, (1983) 241.

[20] M. Chanowitz and S.R. Sharpe, Nucl. Phys. B222, (1983) 211.

[21] M. Flensberg, C. Peterson, and L. Sköld, Z. Phys. C22, (1984) 293.

[22] For earlier work see P. Hasenfratz, R.R. Horgan, J. Kuti, and J.M. Richard, Phys. Lett. 95B, (1980) 299; F. de Viron and J. Weyers, Nucl. Phys. B195, (1981) 391.

[23] A. Chodos *et al.* , Phys. Rev. D9, (1974) 3471; T. de Grand *et al.* Phys. Rev. D12, (1975) 2060.

[24] N. Isgur and J. Paton, Phys. Lett. 124B, (1983) 247; Phys. Rev. D31, (1985) 2910.

[25] N. Isgur, R. Kokoski, and J. Paton, Phys. Rev. Lett. 54, (1985) 869.

[26] See for instance J. Kogut, Rev. Mod. Phys. 51, (1985) 659; 55, 775 (1983); M. Creutz, *Quarks, Gluons, and Lattices*, Cambridge University Press (Cambridge, 1983).

[27] The selection rule was originally pointed out by the Orsay group in the context of constituent gluon models.

[28] G. Schierholz, *Status of Lattice Glueball Mass Calculations*, Proc. of the BNL Workshop on Glueballs, Hybrids, and Exotic Hadrons, ed. S.U. Chung, Upton New York, 1988 (AIP, New York, 1989).

[29] C. Michael and M. Teper, Univ. of Oxford preprint (1988).

[30] F. Close, these proceedings.

[31] T. Barnes, K. Dooley, N. Isgur, Phys. Lett. 183B, (1987) 210.

Z. Phys. C – Particles and Fields 46, S105–110 (1990)

Zeitschrift für Physik C Particles and Fields

Some aspects of strange baryon decays*

F. Myhrer**

Center for Theoretical Physics, Laboratory for Nuclear Science and Department of Physics, Massachusetts Institute of Technology, Cambridge, MA 02139, USA

Abstract

The $SU(3)$ relations of the baryons' semileptonic decays and the baryons magnetic moments are compared in a model to illustrate how $SU(3)$ breaking effects will appear. The baryon magnetic moments break $SU(3)$ through the explicit quark mass-dependence whereas the g_A/g_V $SU(3)$ breaking-effects are not so apparent.

1. INTRODUCTION

The semi-leptonic decays of hyperons have been reviewed by Gaillard and Sauvage [1] where references to the literature can be found. In this discussion, based on work in collaboration with H. Høgaasen [2], I will concentrate on possible $SU(3)$ violations of the decays $B' \to Be\overline{\nu}$. From Lorentz covariance and parity invariance the matrix elements between spin-1/2 baryons for vector and axial vector currents are

$$\langle B'(p')|V_\mu(q)|B(p)\rangle \sim \overline{u}_{B'}(p')\left\{g_V(q^2)\gamma_\mu - i\frac{f_2(q^2)}{M+M'}\sigma_{\mu\nu}q^\nu + \frac{f_3(q^2)}{M+M'}q_\mu\right\}u_B(p) \tag{1a}$$

and

$$\langle B'(p')|A_\mu(q)|B(p)\rangle \sim \overline{u}_{B'}(p')\left\{g_A(q^2)\gamma_\mu - i\frac{g_2(q^2)}{M+M'}\sigma_{\mu\nu}q^\nu + \frac{g_3(q^2)}{M+M'}q_\mu\right\}\gamma_5 u_B(p) \tag{1b}$$

Here M and M' are the masses of the baryons B and B', and the four-momentum transfer is $q = p - p'$. Time-reversal invariance requires that all six Lorentz-scalar form factors are real functions. Since I will concentrate on the direct determination of the ratio $g_A(0)/g_V(0)$ for the hyperon decays, the Cabibbo angle dependence is not explicitly written in Eqs. (1). The two form factors f_2 and g_2 are the so-called second-class form factors ($f_2 = 0$ from CVC in the $SU(3)$ limit, $M = M'$). The induced scalar and pseudoscalar form factors f_3 and g_3 are neglected in this discussion since in the electronic decay rates the terms involving f_3 and g_3 are multiplied by $(m_e/M)^2 \lesssim 0.25 \cdot 10^{-6}$. The weak magnetism term, f_2, is related to the nucleon anomalous magnetic moment since it is assumed that the hadronic weak vector currents and electromagnetic currents are parts of an $SU(3)$ octet. At the quark level we only have first class currents (the currents are simply of the $V - A$ type). However, as we shall see, the baryon decays calculated in quark models can give $g_2(0) \neq 0$ for hyperon decays.

The effect of an $SU(3)$ breaking term in the Hamiltonian is subject to the Ademollo–Gatto theorem [3] which effectively says that to first order in the breaking parameter, the quark mass difference $\Delta m = m_s - m_u$, we get a difference in particle masses but not in the vector coupling constant g_V. Here I will analyze this in a model and keep comparing with calculations of the baryon magnetic moments where $SU(3)$ breaking are obvious due to baryon or quark mass differences.

2. THE SU(3) LIMIT

The following discussion will mainly be based on the MIT bag model [4] and its extension, the chiral bag

*This work is supported in part by funds provided by the U.S. Department of Energy (D.O.E.) under contract #DE-AC02-76ER03069, and by a grant by the National Science Foundation.

**On sabbatical leave from the University of South Carolina.

model [5] where a classical pseudoscalar field surrounds the quark core so that the axial vector current is continuous in space. In an additive quark model (the quarks are independent of each other apart from a "trivial" center of mass motion) the baryons' magnetic moments or the axial weak decays are described by the quark one-body operators $\Theta(i)$ as

$$\Theta = \left\langle B' \left| \sum_{i=1}^{3} \Theta(i) \right| B \right\rangle \tag{2}$$

where for the quark magnetic moment operator

$$\Theta(i) = Q(i)\sigma_z(i)\mu^Q(i) \tag{3}$$

or for the quark "axial charge" operator

$$\Theta(i) = \tau(i)\sigma_z(i)g_A^Q(i) \ . \tag{4}$$

Here $Q(i)$ and $\sigma(i)$ are the charge and the spin of the ith quark, respectively. The operator $\tau(i)$ changes quark $d \to u$ ($\Delta S = 0$) or $s \to u$ ($\Delta S = 1$). The jth quark has the Dirac wave function

$$\Psi(j) = \frac{1}{\sqrt{4\pi}} \begin{pmatrix} F(kr) \\ iG(kr)\sigma(j)\cdot\hat{\mathbf{r}} \end{pmatrix} u(j) \tag{5}$$

where the Pauli-spinor $u = \binom{1}{0}$ or $\binom{0}{1}$ and the quark momentum k is determined by the MIT confinement condition $F = -G$ for $r = R$. Then the confined quark magnetic moment of Eq. (3) is

$$\mu^Q = -\frac{2}{3}\int_0^R dr\, r^3 F\cdot G \tag{6}$$

and another bag integral gives the "axial charge" of Eq. (4) as

$$B \equiv g_A^Q = \int_0^R dr\, r^2 \left(F^2 - \frac{1}{3}G^2\right) \tag{7}$$

where the normalization $\int_0^R dr\, r^2(F^2+G^2) = 1$ is used. In the $SU(3)$ limit using massless u, d and s quarks ($m =$ 0 MeV) the MIT model gives

$$\mu_0^Q = \frac{1}{2\omega_0}\frac{4x_0-3}{6(x_0-1)} \simeq \frac{0.826}{2\omega_0} \sim 1.92\text{n.m.} \tag{8}$$

for $R = 1$ fm, where the confined quark energy is $\omega_0 = k = x_0/R$ and $x_0 = 2.043$ [4].

The explicit dependence of μ^Q on the quark energy ω and indirectly its mass ($\omega R = [x^2 + (mR)^2]^{1/2}$) is indicated in Eq. (8). From the expression for the bag integral B or the "axial charge," Eq. (4), it is clear that B does not depend explicitly on the quark energy ω.

$$B_0 = \frac{x_0}{3(x_0-1)} \simeq 0.65 \tag{9}$$

As will be discussed, B of Eq. (7), will depend on the quark masses [6], m, and for broken $SU(3)$ will increase with increasing differences Δm between the quark masses [4]. In fact, the static bag model predicts that B increases linearly with Δm for Δm small [7]. For increasing value of the quark mass ($m_u = m_d = m_s \simeq 125$ MeV for $R = 1$ fm) the value of B increases to $B \simeq 0.72$. For very heavy quarks, the value of $B = 1$ as is seen from Eq. (7) since $G \to 0$ as $m \to \infty$. For *equal mass quarks* and using the usual $SU(6)$ relations among the baryon magnetic moments and the baryons g_A/g_V, we find for example ($R = 1$ fm and $m = 0$ MeV),

$$\mu(\text{proton}) = \frac{4}{3}\mu(u) - \frac{1}{3}\mu(d) = \mu^Q = 1.92\text{n.m.} \tag{10a}$$

$$\mu(\Lambda) = \mu(s) = -\frac{1}{3}\mu^Q = -\frac{1}{3}\mu\,(\text{proton}) \simeq -0.66 \text{ n.m.} \tag{10b}$$

and for the neutron decay ($m = 0$ MeV)

$$\frac{g_A}{g_V}(n \to p) = \frac{5}{3}B_0 \simeq 1.09 \ . \tag{10c}$$

Please, note how close the value of $\mu(\Lambda)$ in Eq. (10b) is to the experimental value of -0.614 n.m.

However, in all quark models we do have perturbative one-gluon-exchange (OGE) corrections to the three-quark system as illustrated in fig.1. For example, these corrections lead to the $\Sigma - \Lambda$ mass splitting [4,8]. These OGE corrections also introduce correlations among the quarks and give small but important corrections to the additive quark model results above. Here I will sketch the derivations; for the original derivations see Ushio and Konashi [9], Kobzarev *et al.* [10] and later Høgaasen and Myhrer [11]. The strength of this OGE correction is determined by the $N - \Delta$ mass difference in the static MIT bag. This OGE color magnetic interaction is written as

$$H' = \sum_{i\neq j} c_{ij}\lambda^a(i)\lambda^a(j)\sigma(i)\cdot\sigma(j) \tag{11}$$

where $\lambda^a(i)$ are the color $SU(3)$ operators ($a = 1,\ldots,8$) and c_{ij} depend on the strong coupling constant α_s, the values of the u, d and s quark masses and are spatial integrals over the quark wave functions. When the baryons are probed by a photon or a $W^\pm$ via the ith quark operator $\Theta(i)$, Eqs. (3) or (4), we find in the additive quark model using Eq. (2) the results in Eqs. (10). The color magnetic interaction in Eq. (11) gives a correction $\delta\Theta$ to Eq. (2) and Eqs. (10) which is illustrated in fig. 2 and which can be parametrized as

$$\delta\Theta = \left\langle B' \left| \sum_{i\neq j} b^{\Theta}(i,j)\,\{\Theta(i), \sigma(i)\cdot\sigma(j)\} \right| B \right\rangle \tag{12}$$

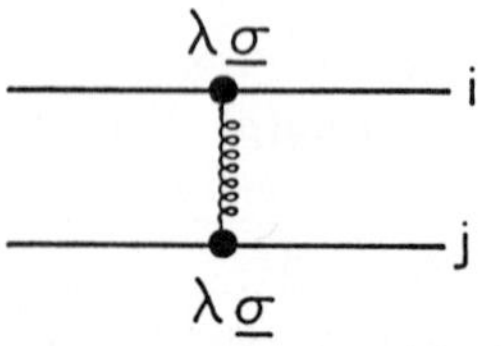

Figure 1: The one gluon exchange between the quarks i and j.

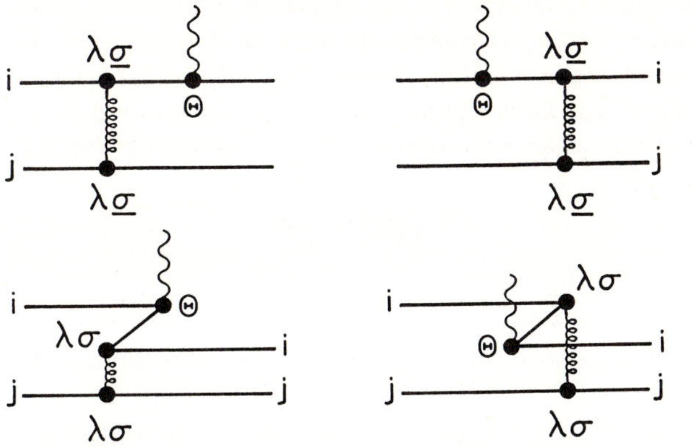

Figure 2: The color magnetic tree diagrams of the OGE correction to the additive quark model. The fermion propagator is calculated using the mode sum over confined quark states.

where the coefficients $b^\Theta(i,j)$ are spatial integrals of the quark wave functions. (For the proper operators in Eq. (12) see Høgaasen and Myhrer [11].)

In the $SU(3)$ limit, for massless quarks, the axial vector operator Eq. (4) gives the OGE correction

$$\delta g_A = G \sum_{i \neq j} \langle B' | \tau(i)\sigma_z(j) | B \rangle \quad (13a)$$

where

$$G \equiv -\frac{4}{3} b^A(i,j) \simeq 0.05 \quad (13b)$$

(when we use $\alpha_s = 2.2$ as determined from the $N-\Delta$ mass difference). A quark mass of $m \simeq 125$ MeV will change the value of G slightly. For $m \simeq 125$ MeV, $B = 0.72$ as stated, and using these two values we find the g_A/g_V ratios for the different baryon decays as given in table 1.

From table 1 we [2] find the values of the $SU(3)$ antisymmetric and symmetric coupling amplitudes F and D to be

Table 1: The ratio g_A/g_V in the $SU(3)$ limit from a model calculation compared to experiments. Here the quark masses $m_u = m_d = m_s = 125$ MeV and the bag radius $R = 1$ fm. The experimental numbers are from the Particle Data Group [14].

B'→ B	MIT bag + OGE	$SU(3)$ amplitudes	Exp.
$n \to p$	$\frac{5}{3}B + G = +1.25$	$F + D$	1.259
$\Sigma^- \to n$	$-\frac{1}{3}B - 2G = -0.34$	$F - D$	-0.36 ± 0.05
$\Lambda \to p$	$B = +0.72$	$F + \frac{1}{3}D$	0.696 ± 0.025
$\Xi^- \to \Lambda$	$\frac{1}{3}B - G = +0.19$	$F - \frac{1}{3}D$	0.25 ± 0.05

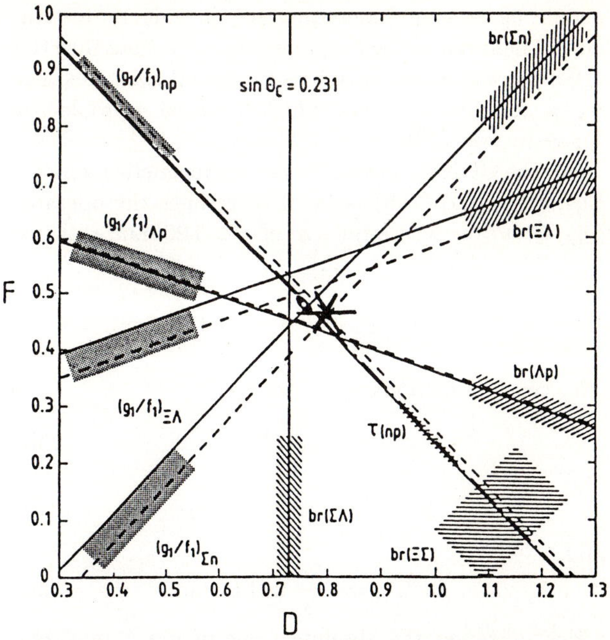

Figure 3: The D and F values in semileptonic decays. The figure is from the review of Gaillard and Sauvage [1]. Their optimal result for D and F is the black ellipse at the center. This includes asymmetry measurements as well as branching ratio (br) measurements. Our model result is marked by an asterisk.

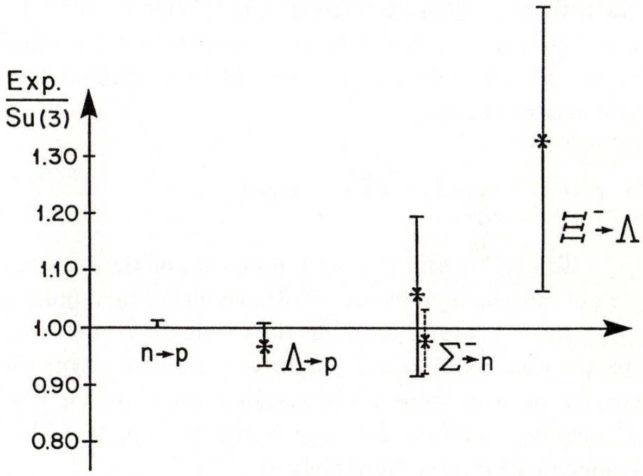

Figure 4: The ratios of the direct asymmetry measurements of g_A/g_V and our g_A/g_V $SU(3)$ model results for four semileptonic baryon decays are plotted. The errors are the experimental ones taken from the particle Data Group [14]. The recent Fermilab experiment [26] is the dashed line. For perfect $SU(3)$ symmetry all the ratios should be equal to one.

$$F = 0.455 \quad \text{and} \quad D = 0.795 \,. \quad (14)$$

This value is marked on the "standard plot", fig. 3, taken from Gaillard and Sauvage [1]. In fig. 4 we compare these model $SU(3)$ results with data to display explicitly which directly determined g_A/g_V (asymmetry measurements) should be measured more precisely in order to test the $SU(3)$ behavior of the coupling constants. Note that this OGE correction has the correct sign and size to explain g_A/g_V for the $\Sigma^- \to ne\overline{\nu}$ which has been a problem in quark model calculations [12].

The OGE correction (12) to the magnetic moments Eqs. (10a) and (10b) is equal to c times the operator $\sum_{i\neq j} Q(i)\sigma_z(j)$ where the size of $c \simeq 0.20$ n.m. [11] (for $\alpha_s = 2.2$ and massless quarks).

$$\delta\mu = c\sum_{i\neq j}\langle B'\,|Q(i)\sigma_z(j)|\,B\rangle \tag{15a}$$

This correction $\delta\mu$ gives, for example,

$$\delta\mu(\text{proton}) = \delta\mu(\Sigma^+) = 0 \tag{15b}$$

and

$$\delta\mu(\Lambda) = c/3 \sim 0.07\,\text{n.m.} \tag{15c}$$

and

$$\delta\mu(\Xi^-) = -\frac{2c}{3} \simeq 0.14\,\text{n.m.} \tag{15d}$$

which decreases the absolute value of the Λ magnetic moment. (The measured Λ magnetic moment includes a contribution of -0.04 n.m. from Σ^0, Λ wave function mixing [13].) This OGE correction is precisely the size needed to understand (in a model) why the magnetic moment (in units of n.m.) $\mu(\Xi^-) = -0.69 \pm 0.04 < \mu(\Lambda) = -0.613 \pm 0.004$ [14] as shown in ref. 11. Again, the OGE correction succeeds in correcting a difficult problem for the quark model. It should be remarked that the value of the Ξ^- magnetic moment should be confirmed by a new measurement.

3. CHIRAL BAG AND SU(3)

To explain the magnetic moments of the baryons, we not only have to break $SU(3)$ symmetry by requiring $m_s > m_u = m_d \simeq 10$ MeV but the pionic cloud surrounding the quark core is also important in order to get the correct nucleon values. The classical pseudoscalar field ϕ^k outside the bag is for large r (the mass $\mu_k \neq 0$ MeV is included for later convenience)

$$\Phi^k(r) = -\lambda^k \frac{g^k}{8\pi M}\sigma\cdot\nabla\left(\frac{e^{-\mu_k r}}{r}\right) \ . \tag{16}$$

The nucleon mass is M, and $k = 1,\ldots,8$ denote the flavor octet components (λ^k = Gell-Mann $SU(3)$ matrices). This field gives an axial current

$$A^k_\mu(r) \simeq -f_k\partial_\mu\Phi^k \tag{17}$$

outside the quark bag. We assume the radius R of the bag is large enough ($R \simeq 1$ fm) so we can treat this pseudoscalar field as a perturbative correction to the MIT bag. The quark axial current

$$A^k_\mu(r) = \overline{\Psi}(r)\gamma_5\gamma_\mu\lambda^k\Psi(r) \tag{18}$$

is discontinuous at the bag surface $r = R$. The requirement that A^k_μ is continuous for $r = R$ determines the pseudoscalar coupling strength $f_k g^k$ ($f_\pi = 93$ MeV). In the $SU(3)$ limit and for massless mesons ($\mu_k = 0$ MeV) the axial charge contributions from the pseudoscalar field is [15]

$$g^\pi_A(0) = \frac{1}{2}B \ . \tag{19}$$

In this chiral bag model it is easy to reproduce the measured baryon magnetic moments [14] (the kaon and η_8 contributions to the magnetic moments can be neglected due to their heavy masses [11,16]). The pion cloud contributes an isovector magnetic moment to p, n and Σ^+, Σ^- which together with the OGE correction (15a) discussed earlier give the correct isoscalar/isovector values, partly because this OGE correction does not contribute to either the proton or the Σ^+ magnetic moments (15b).

The g_A/g_V ratios for the different baryon decays are much more delicate. *In the $SU(3)$ limit* with all quark masses equal $m = 15$ MeV, we use the LAPP version [17] of the MIT bag model. In this model the quarks experience a central attractive potential to simulate that the quarks do not move freely in the bag. A potential that gives analytic solutions of the Dirac equation inside the bag is

$$V(r) = \beta\left(\frac{1}{R} - \frac{1}{r}\right) \quad ; \quad r \leq R \ . \tag{20}$$

A set of parameters that gives $g_A/g_V = 1.26$ for neutron decays is [18]: the bag pressure $B^{1/4} = 110$ MeV, the "zero point energy" $Z_0 = -2.5$, the quark-gluon coupling constant $\alpha_s = 2.4$ which, together with the pionic contribution, gives the correct $N-\Delta$ mass splitting, and $\beta = 0.93$ (with $m = 15$ MeV). With these parameters the proton mass, magnetic moment and r.m.s. radius are close to the measured numbers. Furthermore, with the above parameters the pion-nucleon coupling constant is calculated to be $g^2_{\pi N}/4\pi = 14.2$ which also is very reasonable. With these parameters we find the axial charge (7) $B = 0.528$ and $g^\pi_A = 0.33$ meaning the relation (19) is no longer correct when $\mu_\pi \neq 0$ MeV. The gluonic correction G of Eq. (13b) is still $\simeq 0.05$. This means that in the $SU(3)$ limit ($\mu_\pi = \mu_K = \mu_{\eta_8}$) we find that $g^\pi_A = g^K_A = g^{\eta_8}_A = 0.33$ and only $g_A(n \to p)/g_V = 1.26$ and $g_A/g_V(\Lambda \to p) = 0.73$ are changed in table 1. This model also gives a value for the integrated proton-spin structure function $g_1(x)$ [18] to be equal

$$\Gamma(p) = \int_0^1 dx\, g^p_1(x) = \frac{1}{6}\left\{g^Q_A + \frac{1}{2}\left(g^\pi_A + \frac{1}{5}g^{\eta_8}_A\right) - 2G\right\}$$

$$= 0.163. \tag{21}$$

Here we [18] have assumed the flavor-singlet η_0 is very heavy due to the $U(1)$ anomaly and its contribution to Eq. (21) is therefore neglected. To compare with the EMC experiment [19] we have to include the QCD radiative corrections, $(1 - \alpha_s/\pi)$ for flavor non-singlet, and estimate [18]

$$\Gamma(p) \times \text{QCD corrections} \simeq 0.145 \qquad (22)$$

This result is within the errors of the EMC-result [19] $0.114 \pm 0.012 \pm 0.026$. Here both the mesonic part [20] and the OGE-corrections [21], which is mainly a flavor $SU(3)$ singlet, contribute to this favorable result.

4. BROKEN SU(3) AND THE RATIO g_A/g_V

As already discussed, we have to break the $SU(3)$ symmetry and add the pionic cloud to reproduce the baryon magnetic moments. For semi-leptonic decays this means we cannot use the static bag model in calculating the weak form factors of Eqs. (1) since recoil corrections are important ($M' \neq M$). As shown [22–24], the quark contribution to the ratio g_A/g_V is no longer determined by the bag integral B of Eq. (7). They show that the recoil corrections introduce different bag-integrals of the quark wave functions. Lie-Svendsen and Høgaasen [22,23] have calculated the weak form factors of Eqs. (1) in two different Lorentz-frames using several bag models, and Kubodera *et al.* [24] have used the Cloudy Bag model. In table 2, taken from ref.[23] table 8, the calculated g_A/g_V values for several semi-leptonic decays are given (note that in ref. 22 the OGE corrections, G, was just added (see table 1) and the model parameters were not changed to refit the $n \to pe\bar{\nu}$ g_A/g_V value). The values of g_A/g_V in table 2 include differences in the quark masses $\Delta m \neq 0$, the meson masses $\mu_\pi \neq \mu_K$ and recoil corrections, $M \neq M'$ in Eqs. (1). As seen in table 2, apart from model calculations of the neutron decay which tend to give too small values for g_A/g_V, no large $SU(3)$ breaking is apparent even if $SU(3)$ is broken explicitly in this model [2].

A better quantity to measure to establish $SU(3)$ breaking in semi-leptonic decays is the g_2/g_A ratio which is process-dependent [22,23] and a non-zero value reflects $SU(3)$ breaking [12]. The form factor g_2 has been calculated in various bag models and the calculated sign of g_2 [22–25] is consistent with the recent Fermilab Σ^- decay experiment [26]. *A range* of "predicted" values for g_2/g_A for several bag models can be found in the work of Lie-Svendsen and Høgaasen [22,23].

In conclusion, we expect the standard $SU(3)$-flavor breaking mechanism $\Delta m = m_s - m_u \neq 0$ to give different g_A/g_V values for the different $\Delta S = 1$ and $\Delta S = 0$ processes but model calculations indicate small g_A/g_V $SU(3)$ deviations. A measurement of the ratio g_2/g_A will contribute to the model-understanding of the $SU(3)$ breaking. It is obvious from figs. 2 and 3 that no $SU(3)$ breaking has been established experimentally in semi-leptonic decays of hyperons.

Table 2: The ratio g_A/g_V for several semileptonic decays $B' \to Be\bar{\nu}$ calculated in the Breit frame for several bag-models: MIT bag, CBM = the Cloudy Bag Model with surface coupling, the Chiral Bag Model excluding the pion from the bag interior and the LAPP version of the bag (see the text). This table is taken from ref.[23] where details can be found.

PROCESS	MIT	CBM	CHIRAL	LAPP
$n \to p$	1.12	1.12	1.20	1.29
$\Lambda \to p$	0.73	0.73	0.78	0.71
$\Sigma^- \to n$	−0.31	−0.31	−0.33	−0.31
$\Xi^- \to \Lambda$	0.22	0.22	0.24	0.21
$\Xi^- \to \Sigma^0$	1.23	1.23	1.28	1.19

References

[1] J. M. Gaillard and G. Sauvage, Ann. Rev. Nucl. Part. Sci. 34, (1984) 351.

[2] H. Høgaasen and F. Myhrer, in preparation.

[3] M. Ademollo and R. Gatto, Phys. Rev. Lett. 13, (1964)264.

[4] T. DeGrand, R. L. Jaffe, K. Johnson and J. Kiskis, Phys. Rev. D12, (1975) 2060.

[5] A. W. Thomas, Adv. Nucl. Phys. 13, (1984) 1; F. Myhrer, in Quarks in Nuclei, W. Weise, ed. (World Scientific, Singapore, 1984), vol. 1, p. 325.

[6] E. Golowich, Phys. Rev. D12, (1975) 2108.

[7] Ø. Lie-Svendsen and H. Høgaasen, private communication.

[8] A. DeRujula, H. Georgi and S. L. Glashow, Phys. Rev. D12, (1975) 147.

[9] K. Ushio and H. Konashi, Phys. Lett. 135B, (1984) 468; K. Ushio, Phys. Lett. 158B, (1985) 71 and Z. Phys. C30, (1986) 115.

[10] I. Yu, Kobzarev et al., Sov. J. Nucl. Phys. 43, (1986) 803; M. I. Krivoruchenko, ibid. 40, (1984) 514, ibid. 45, (1986) 109; M. Krivoruchenko et al., ITEP-preprint #23–89 (1989).

[11] H. Høgaasen and F. Myhrer, Phys. Rev. D37, (1988) 1950.

[12] J. Donoghue and B. R. Holstein, Phys. Rev. D25, (1982)206.

[13] N. Isgur and G. Karl, Phys. Rev. D21, (1980) 3175; J. Franklin, D. B. Lichtenberg, W. Namgung and D. Carydas, ibid. 24, (1981) 2910.

[14] Particle Data Group, Phys. Lett. 204B, (1988) 1.

[15] R. L. Jaffe, in Point-Like Structures Inside and Outside Hadrons Proceedings of the 17th International School of Subnuclear Physics (Erice, A. Zichichi, ed. (Plenum Press, New York, 1982), p. 99.

[16] P. Gonzales, V. Vento and M. Rho, Nucl. Phys. A395, (1983) 446; F. Myhrer, Phys. Lett. 125B, (1983) 359.

[17] H. Høgaasen, J. M. Richard and P. Sorba, Phys. Lett. 119B, (1982) 272.

[18] H. Høgaasen and F. Myhrer, Phys. Lett. 214B, (1988) 123.

[19] The EMC Collaboration, J. Ashman et al., Phys. Lett. 206B, (1988) 364.

[20] A. W. Schreiber and A. W. Thomas, Phys. Lett. 215B, (1988) 141.

[21] F. Myhrer and A. W. Thomas, Phys. Rev. D38, (1988) 1633.

[22] J. O. Eeg and Ø. Lie-Svendsen, Zeit. Phys. C27, (1985) 119;

[23] Ø.Lie-Svendsen and H. Høgaasen, Zeit. Phys. C35, (1987) 239.

[24] K. Kubodera et al., Nucl. Phys. A439, (1985) 695; T. Yamaguchi et al., Sophia University preprint (1988).

[25] L. J. Carson, R. J. Oakes and C. R. Willcox, Phys. Rev. D33, (1986) 1356.

[26] S. Y. Hsueh et al., Phys. Rev. D38, (1988) 2056.

Z. Phys. C – Particles and Fields 46, S111–120 (1990)

Zeitschrift für Physik C Particles and Fields

Current status of E/f_1 (1420) and ι/η (1450)*

S.U. Chung

Brookhaven National Laboratory, Upton, NY 11973, USA

Abstract

The current status and future prospects are given of the $E/f_1(1420)$ and the $\eta(1430)$ region containing the $\iota/\eta(1450)$. These states are seen in the channels $K\overline{K}\pi$ and $\eta\pi\pi$.

1. Introduction

The BNL Workshop on Glueballs, Hybrids and Exotic Hadrons was held on August 29 - September 1, 1988 [1]. The main purpose here is to summarize and update several important issues raised at the workshop.

There are currently several outstanding problems regarding gluonic degrees of freedom in the light-quark sector. Among them are the $E/f_1(1420)$ with $J^{PC} = 1^{++}$ which may be a multiquark state and the $\iota/\eta(1450)$ with $J^{PC} = 0^{-+}$, a component of which may be a pseudoscalar glueball. A common thread exists between these two states; they both decay into the same final states—$K\overline{K}\pi$ and $\eta\pi\pi$. Because of this, the states found in these channels in the mass range between 1.3 to 1.6 GeV are sometimes referred to as the E/ι phenomenon.

This work parallels rather closely a more general review of the meson spectroscopy given by the author [2] in 1987, but with an emphasis on bringing up to date many of the materials covered in the review as well as commenting on the evolving current thoughts on the role gluonic and multiquark states play in the quarkonium family.

*This manuscript has been authored under contract number DE-AC02-76CH00016 with the U.S. Department of Energy. Accordingly, the U.S. Department retains a non-exclusive, royalty-free license to publish or reproduce the published form of this contribution, or allow others to do so, for U.S. Government purposes.

2. General Overview

In this section a general overview is given of the isoscalar sector of the light-quark meson spectroscopy.

Godfrey and Isgur [3] published in 1985 an ambitious paper in which a 'relativized' quarkonium model with QCD was used to describe all the $q\overline{q}$ mesons, from pions to upsilons, in a unified framework. Their model, referred to as the GI model in this review, gives a complete mass spectrum of the entire quarkonium family of meson states; their predictions therefore provide an excellent starting point for identifying glueballs (gg or ggg), hybrids ($q\overline{q} + g$) or multiquark states ($q\overline{q} + q\overline{q}$).

The light-quark quarkonia are composed of u, d or s quarks. In this review, the symbol n (for non-strangeness) is used as a generic term to stand for u and/or d. Thus, for isovector mesons, one should read

$$n\overline{n} = (u\overline{u} - d\overline{d})/\sqrt{2},$$

while, for isoscalar mesons, one has

$$n\overline{n} = (u\overline{u} + d\overline{d})/\sqrt{2}.$$

The GI model assumes that all the $n\overline{n}$ and $s\overline{s}$ states are decoupled or ideally mixed, except the η and the η' which are known to be nearly SU(3) octet and singlet states. In dealing with the masses of nonet family members which are ideally mixed, it is convenient to keep the following simple mnemonic in mind. Using the notation

$$M_N(q\overline{q}), \qquad N = 2I + 1$$

to denote the mass of a $q\overline{q}$ state with isospin I, one may write

$$\begin{aligned} M_3(n\overline{n}) &= M_1(n\overline{n}) \\ M_1(n\overline{n}) + M_1(s\overline{s}) &= 2M_2(n\overline{s}) = 2M_2(s\overline{n}) \end{aligned}$$

where the subscripts 3, 2 and 1 stand for isovector, isodoublet and isoscalar mesons, respectively. These relations follow from the premise that the mass difference within a multiplet arises solely from that between n and s quarks. Except for the ground-state π nonet, all the nonet members in the GI model satisfy roughly the above mass formulas, including the radial excitations.

In table 1 are given the 'established' isoscalar mesons along with the predictions of the GI model. The experimental data in the table come from the Review of Particle Properties by the Particle Data Group (April 1988) [4]. The listed particles are either from the Meson Summary Table or the Meson Full Listings, which are collectively referred to in this review as the PDG Tables.

The predicted masses in the GI model generally describe the experimental values to within 50 MeV, except the $h_1'(1380)$ whose mass deviates by 90 MeV. However, the $h_1'(1380)$ has been seen so far by a single group [5] and needs confirmation. It should be noted, on the other hand, that the masses of both the $h_1(1170)$ and the $h_1'(1380)$ are lower than those of the GI model, while the $D/f_1(1285)$ and the $D'/f_1'(1530)$ appear higher than predicted; this could indicate a need for further tuning in the $^1P_1$ and $^3P_1$ sector in this model.

The $\eta(1430)$ in table 1 is most likely composed of two separate resonances. The objects seen in hadroproductions are given here the name $\lambda/\eta(1420)$, and the name $\iota/\eta(1450)$ is reserved in this review for those seen in J/ψ radiative decays. A more detailed discussion is given in a later section.

The two 0^{++} states $S/f_0(975)$ and $\epsilon/f_0(1400)$ may in fact consist of four separate resonances as shown in table 1, according to an analysis by Au, Morgan and Pennington [6]. The GI model predicts two ground-state isocalars at masses 1090 and 1360 MeV. However, this region is generally thought to be populated by multiquark states [3,7,8].The ground-state quarkonia may then lie above 1090 MeV. Recently, the GAMS group reported two additional states near the $a_2(1320)$ in the $\pi\eta$ system (see M. Boutemeur [9]). This group detects an S-wave state at 1300 MeV; assuming ideal mixing the non-

Table 1: Isoscalar Mesons [a]

type [b]	Name	J^{PC}	Mass (MeV)	Width (MeV)	GI [c]	Δ [d]
1^1S_0	η	0^{-+}	548.8	1.05 keV	520	29
			± 0.6	± 0.15 keV		
1^1S_0	$\eta'(958)$	0^{-+}	957.50 ± 0.24	0.21 ± 0.02	960	[e]
2^1S_0	$\zeta/\eta(1280)$ [f]	0^{-+}	1279 ± 5	32 ± 10	1270[g]	9
	$\eta(1430)$ [h]	0^{-+}	1440 ± 20	60 ± 30		
	$\lambda/\eta(1420)$ [i]	0^{-+}	1421.9 ± 1.5	31–80		[j]
	$\iota/\eta(1450)$ [k]	0^{-+}	1451.9 ± 2.5	50–160		[j]
2^1S_0	$\eta'(2S)$	0^{-+}			1550[g]	
1^3P_0	$f_0(1P)$	0^{++}			1090	
1^3P_0	$S/f_0(975)$	0^{++}	976 ± 3	34 ± 6		
	$S_1/f_0(991)$	0^{++}	991	21		[l]
	$S_2/f_0(988)$	0^{++}	988	0		[l]
	$\epsilon/f_0(1400)$	0^{++}	1300–1470	118–460		[m]
	$\epsilon_1/f_0(900)$	0^{++}	900	350		[n]
	$\epsilon_2/f_0(1430)$	0^{++}	1430	200		[n]
1^3P_0	$f_0'(1P)$	0^{++}			1360	
	$G/f_0(1590)$	0^{++}	1587 ± 11	175 ± 19		
2^3P_0	$f_0(2P)$	0^{++}			1780	
2^3P_0	$f_0'(2P)$	0^{++}			1990	
1^3S_1	$\omega(783)$	1^{--}	782.0 ± 0.1	8.5 ± 0.1	780	[o]
1^3S_1	$\phi(1020)$	1^{--}	1019.41 ± 0.01	4.41 ± 0.05	1020	[o]
2^3S_1	$\omega(2S)$	1^{--}			1460	
1^3D_1	$\omega(1D)$	1^{--}			1660	
2^3S_1	$\phi(1680)$	1^{--}	1685	130		−5
			+75/−15	± 50	1690	
1^3D_1	$\phi(1D)$	1^{--}			1880	
1^1P_1	$h_1(1170)$	1^{+-}	1170 ± 40	335 ± 26	1220	−50
1^1P_1	$h_1'(1380)$	1^{+-}	1380 ± 20	80 ± 30	1470	−90[p]
2^1P_1	$h_1(2P)$	1^{+-}			1780	
2^1P_1	$h_1'(2P)$	1^{+-}			2010	
1^3P_1	$D/f_1(1285)$	1^{++}	1283 ± 5	25 ± 3	1240	43
	$E/f_1(1420)$	1^{++}	1422 ± 10	55 ± 3		
1^3P_1	$D'/f_1'(1530)$	1^{++}	1527 ± 5	106 ± 14	1480	47
2^3P_1	$f_1(2S)$	1^{++}			1820	
2^3P_1	$f_1'(2S)$	1^{++}			2030	
1^3P_2	$f_2(1270)$	2^{++}	1274 ± 5	185 ± 20	1280	−6
	$f_T/f_2(1430)$	2^{++}	1412–1480	30–150		[q]
1^3P_2	$f_2'(1525)$	2^{++}	1525 ± 5	76 ± 10	1530	−5
	$\theta/f_2(1720)$	2^{++}	1721 +2/−4	138 ± 11		
2^3P_2	$f_2(2P)$	2^{++}			1820	
2^3P_2	$f_2'(2P)$	2^{++}			2040	
1^3F_2	$f_2(1F)$	2^{++}			2050	
1^3F_2	$\xi/f_2(2220)$	2^{++}	2227 ± 6	21 +18/−14	2240	−13[r]
	$g_T/f_2(2010)$	2^{++}	2011 ± 70	202 ± 65		
	$g_{T'}/f_2(2300)$	2^{++}	2297 ± 28	149 ± 41		
	$g_{T''}/f_2(2340)$	2^{++}	2339 ± 55	319 ± 75		
1^1D_2	$\eta_2(1D)$	2^{-+}			1680	
1^1D_2	$\eta'_2(1D)$	2^{-+}			1890	
1^3D_2	$\omega_2(1D)$	2^{--}			1700	
1^3D_2	$\phi_2(1D)$	2^{--}			1910	
1^3D_3	$\omega_3(1670)$	3^{--}	1668 ± 5	166 ± 15	1680	−12
1^3D_3	$\phi_3(1850)$	3^{--}	1854 ± 9	110 +46/−33	1900	−47[s]
1^1F_3	$h_3(1F)$	3^{+-}			2030	
1^1F_3	$h_3'(1F)$	3^{+-}			2220	
1^3F_3	$f_3(1F)$	3^{++}			2050	
1^3F_3	$f_3'(1F)$	3^{++}			2230	
1^3F_4	$h/f_4(2050)$	4^{++}	2047 ± 11	204 ± 13	2010	37
1^3F_4	$f_4'(1F)$	4^{++}			2200	
1^1G_4	$\eta_4(1G)$	4^{-+}			2330	
1^1G_4	$\eta'_4(1G)$	4^{-+}			2510	
1^3G_4	$\omega_4(1G)$	4^{--}			2340	
1^3G_4	$\phi_4(1G)$	4^{--}			2520	
1^3G_5	$\omega_5(1G)$	5^{--}			2300	
1^3G_5	$\phi_5(1G)$	5^{--}			2470	

[a] The GI model assumes ideal mixing for all except for η and η'.
[b] Spectroscopic Notation: $n^{2S+1}L_J$ (n=radial quantum number).
[c] Predicted Mass (MeV) from the GI model.
[d] Δ= (Experimental Mass − GI Mass) in MeV.
[e] Assumed fixed at 960 MeV in the GI model
[f] The name ζ is often given to this state.
[g] Scheme P2 of the GI model (the two states are nearly ideally mixed).
[h] The $\eta(1430)$ is split into $\lambda/\eta(1420)$ and $\iota/\eta(1450)$.
[i] The name λ has been given here for those observed in hadroproductions.
[j] Masses and widths from the PDG Tables [4], p. 294.
[k] Observed in J/ψ decays.
[l] The $S(975)$ is split into the $S_1(991)$ and the $S_2(988)$, see [6].
[m] The mass and width from the PDG Tables [4], p. 292.
[n] The $\epsilon(1400)$ is split into the $\epsilon_1(900)$ and the $\epsilon_2(1430)$, see [6].
[o] Assumed fixed at 780 and 1020 MeV in the GI model.
[p] The mass and width from the PDG Tables [4], p. 292.
[q] The mass and width from the PDG Tables [4], p. 295.
[r] The mass and width from the PDG Tables [4], p. 314.
[s] The mass and width from the PDG Tables [4], p. 310.

strange isoscalar partner is expected at around the same mass.

The state $\xi/f_2(2220)$ is listed as a 2^{++} resonance in table 1, although its spin-parity assignment is uncertain at this time. Likewise, the $\phi_3(1850)$ is listed as a 3^{--} resonance in this table , even though its quantum numbers are not yet established.

One of the most important conclusions that can be drawn from examining table 1 is this: there appear to be extra states that seem to lie outside of the quarkonium picture of meson states. Table 1 shows that the first radial excitations of the pseudoscalars should occur at 1270 and 1550 MeV; the $\lambda/\eta(1420)$ and the $\iota/\eta(1450)$ cannot be fit in this picture and therefore appear as spurious states. Two ground-state isoscalar 1^{++} states occur at 1240 and 1480 MeV in the GI model and are filled in by $D/f_1(1285)$ and $D'/f_1'(1530)$ as seen in table 1, but the $E/f_1(1420)$ clearly appears as an extra state. Note that similar observations apply to the $G/f_0(1590)$, the $f_T/f_2(1430)$ and the $\theta/f_2(1720)$. These extra states point to a need to enlarge the quarkonium picture, by including the gluonic degrees of freedom and multiquark states.

The purpose of this review is to select two of these spurious states, namely the $E/f_1(1420)$ and the $\eta(1430)$ region [$\lambda/\eta(1420)$ and $\iota/\eta(1450)$], and discuss in some detail their current status and the future prospects.

3. $E/f_1(1420)$

The ground-state $J^{PC} = 1^{++}$ nonet contains the isovector $a_1(1270)$, the isodoublet Q_A a linear superposition of the $Q/K_1(1280)$ and $Q/K_1(1400)$, and the isoscalar $D/f_1(1285)$. According to the GI model (see table 1), the 1^{++} isoscalars should occur at 1.24 GeV for $n\bar{n}$ and 1.48 GeV for $s\bar{s}$. The $D/f_1(1285)$ with a (49 $\pm$ 6)% branching ratio into $\eta\pi\pi$ and only (11 $\pm$ 3)% into $K\bar{K}\pi$ is the dominantly $n\bar{n}$ isoscalar expected at 1.24 GeV. For its $s\bar{s}$ partner at 1.48 GeV, a natural choice now is the $D'/f_1'(1530)$ seen by two experiments [5,10], in the reaction $K^-p \to K\bar{K}\pi\Lambda$ at 4.2 and 11 GeV/c. It should be noted that the $E/f_1(1420)$ is not seen in the K^-p LASS data [5], indicating that it has little, if any, $s\bar{s}$ component (see fig. 1)

The best evidence that the $E/f_1(1420)$ is a spin-1 object comes from the study of photon-photon collisions at PEP (TPC/2γ and Mark II) and PETRA (JADE and CELLO) (see G. Gidal [11]). They have performed a comparative study of the reactions $\gamma\gamma \to K\bar{K}\pi$ and $\gamma\gamma^* \to K\bar{K}\pi$. The $K\bar{K}\pi$ spectra with tagged electrons (the photons off the mass shell) show the $E/f_1(1420)$ (see fig.2), whereas the untagged data (on-shell photons) do not show the $E/f_1(1420)$ at all (not shown). The implications are that the process $\gamma\gamma^* \to K\bar{K}\pi$ results in the $E/f_1(1420)$ formation, while the related process $\gamma\gamma \to K\bar{K}\pi$ does not; the only way to explain this is that the $E/f_1(1420)$ has spin-1, i.e. $J^{PC} = 1^{++}$ or 1^{-+}. Indeed, Chanowitz [12] suggested that the observed resonance might in fact be an exotic 1^{-+} state. It is seen in fig. 2 that the $D/f_1(1285)$ production is very much suppressed compared to that of the $E/f_1(1420)$, even though they both have spin-1. One may speculate that the $E/f_1(1420)$ is a multiquark state with enhanced coupling to photons due to its high quark charges, whereas the $D/f_1(1285)$ is a regular quarkonium state.

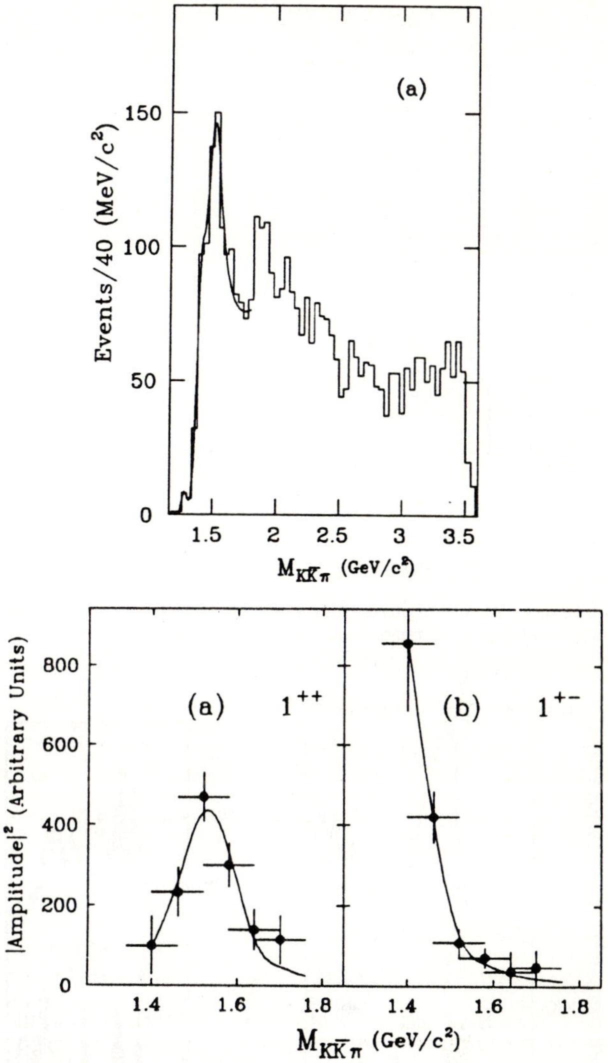

Figure 1: $K\bar{K}\pi$ mass distribution, and the 1^{++} and 1^{+-} waves, from LASS data [5].

The Mark III data [13] on J/ψ hadronic decays provide additional information on the $E/f_1(1420)$. Fig. 3 shows the $K\bar{K}\pi$ and $\eta\pi\pi$ spectra from J/ψ decays with the recoils identified as γ, ω and ϕ[14,15]. The $K\bar{K}\pi$ system recoiling off the ω shows the $E/f_1(1420)$ bump with mass (1442 $\pm$ 5 +10/−17) MeV and width (40 +17/−13 $\pm$ 5) MeV, and its spin-parity is consistent

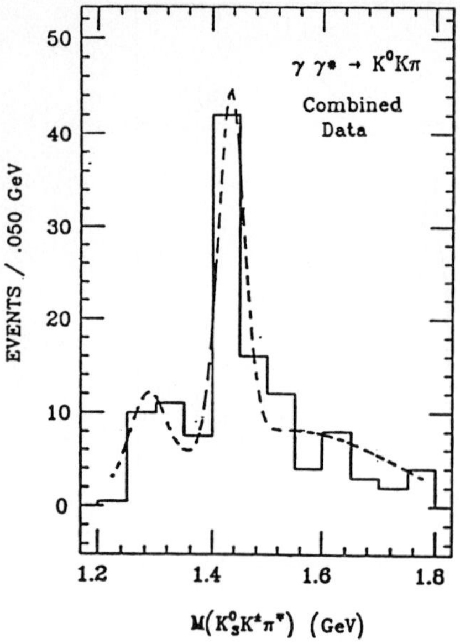

Figure 2: Combined $K\overline{K}\pi$ spectrum from $\gamma\gamma^*$ intial state (PEP and PETRA) [11].

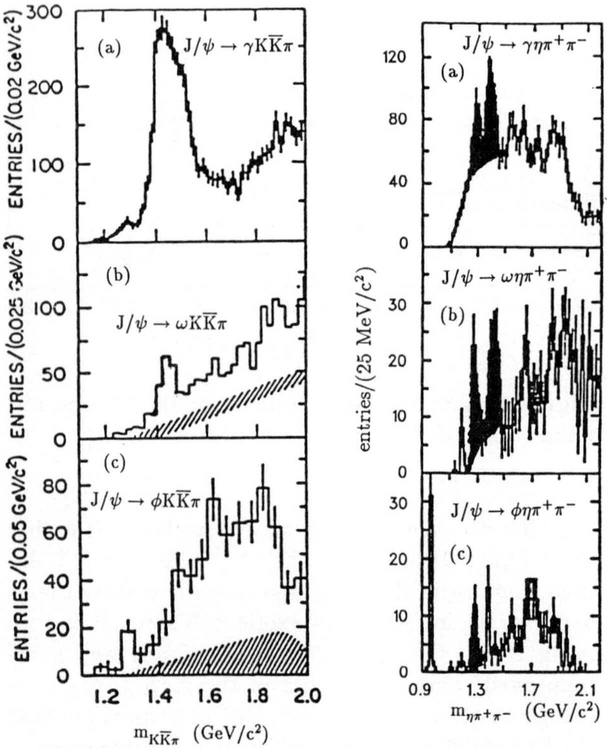

Figure 3: $K\overline{K}\pi$ and $\eta\pi\pi$ Mass Spectra from J/ψ decays (MarkIII) [14], with γ, ω and ϕ as recoiling particles.

with 1^+. The $K\overline{K}\pi$ system produced off the ϕ, on the other hand, shows no evidence of the $E/f_1(1420)$ bump but a small enhancement at the $D/f_1(1285)$ mass. These observations suggest that the $E/f_1(1420)$ has little $s\overline{s}$ component, consistent with the LASS result [5], and that the $D/f_1(1285)$ is not a pure $n\overline{n}$ state. It can be seen in the $\eta\pi\pi$ spectra that the $D/f_1(1285)$ is produced off both the ω and the ϕ. A note of caution is necessary here: it is frequently assumed that the state near 1300 MeV/c in the J/ψ radiative decays is the $D/f_1(1285)$. However, in the absence of a spin-parity analysis one must allow the possibility that the observed state is in fact the $\zeta/\eta(1280)$ and not the $D/f_1(1285)$ or perhaps a mixture of both.

Further information on the $E/f_1(1420)$ comes from the WA76 data at the CERN Ω Spectrometer. The authors have studied central production of the $K\overline{K}\pi$ system in the reaction $(\pi^+/p)p \to (\pi^+/p)(K\overline{K}\pi)p$ at 85 and 300 GeV/c [16]. From a Dalitz-plot analysis they conclude that the $D/f_1(1285)$ and the $E/f_1(1420)$ have both $1^{++}(K^*\overline{K})$ via S-wave (see fig.4). They have also studied $\eta\pi\pi$ in the same production mechanism [17]. They find again a $1^{++}[\delta/a_0(980)+\pi]$ at around 1420 MeV; they quote a branching ratio $B(E \to \eta\pi\pi)/B(E \to K\overline{K}\pi) = (6 \pm 4)\%$.

In any event, a 1^{++} E state should be considered established, in particular, in view of the $\gamma\gamma^*$ results. The $D/f_1(1285)$ and $D'/f_1'(1530)$ seem to fill up the 1^{++} isoscalar sector; then, what is the $E/f_1(1420)$? If the massless nature of gluons is preserved in a two-gluon bound system, then spin-1 is forbidden; therefore, the $E/f_1(1420)$ is unlikely to be a gluonium. Since its mass is relatively low, it is probably not a three-gluon bound state either. One may thus speculate that the $E/f_1(1420)$ is a multiquark state. High quark charges of a multiquark system can easily accommodate its relatively abundant production in the $\gamma\gamma^*$ channel. This may also account for its unusual production channels, namely $\gamma\gamma^*$, hadronic J/ψ decays and central production at SPS energies.

One could further argue that a multiquark state is not likely to be produced in conventional peripheral productions at lower energies, such as in the K^-p LASS data or the π^-p BNL data (see next section). Here, emphasis must be placed on the peripheral nature of the production mechanism; early bubble-chamber data with π^-p interactions at 4 GeV/c detected mostly the $E/f_1(1420)$ in the $K\overline{K}\pi$ system, but the analysis included high $-t$ events [18]. One suspects that the low $-t$ events did contain some $\eta(1430)$, but limited statistics (221 events) precluded its separation from the $E/f_1(1420)$ events.

4. $\eta(1430)$ Region

Experimentally, there exist three 0^{-+} states above η and η'. They are the $\lambda/\eta(1420)$ (given the name λ in this review for convenience), seen in hadronic π^-p and $\overline{p}p$ interactions, the $\zeta/\eta(1280)$, and the $\iota/\eta(1450)$ seen in J/ψ radiative decays. Neither the $\lambda/\eta(1420)$ nor the $\iota/\eta(1450)$ is shown to be rich in $s\overline{s}$ content, and both have masses far below the predicted 1.55 GeV to be the $s\overline{s}$ partner of the $\zeta/\eta(1280)$ (see table 1).

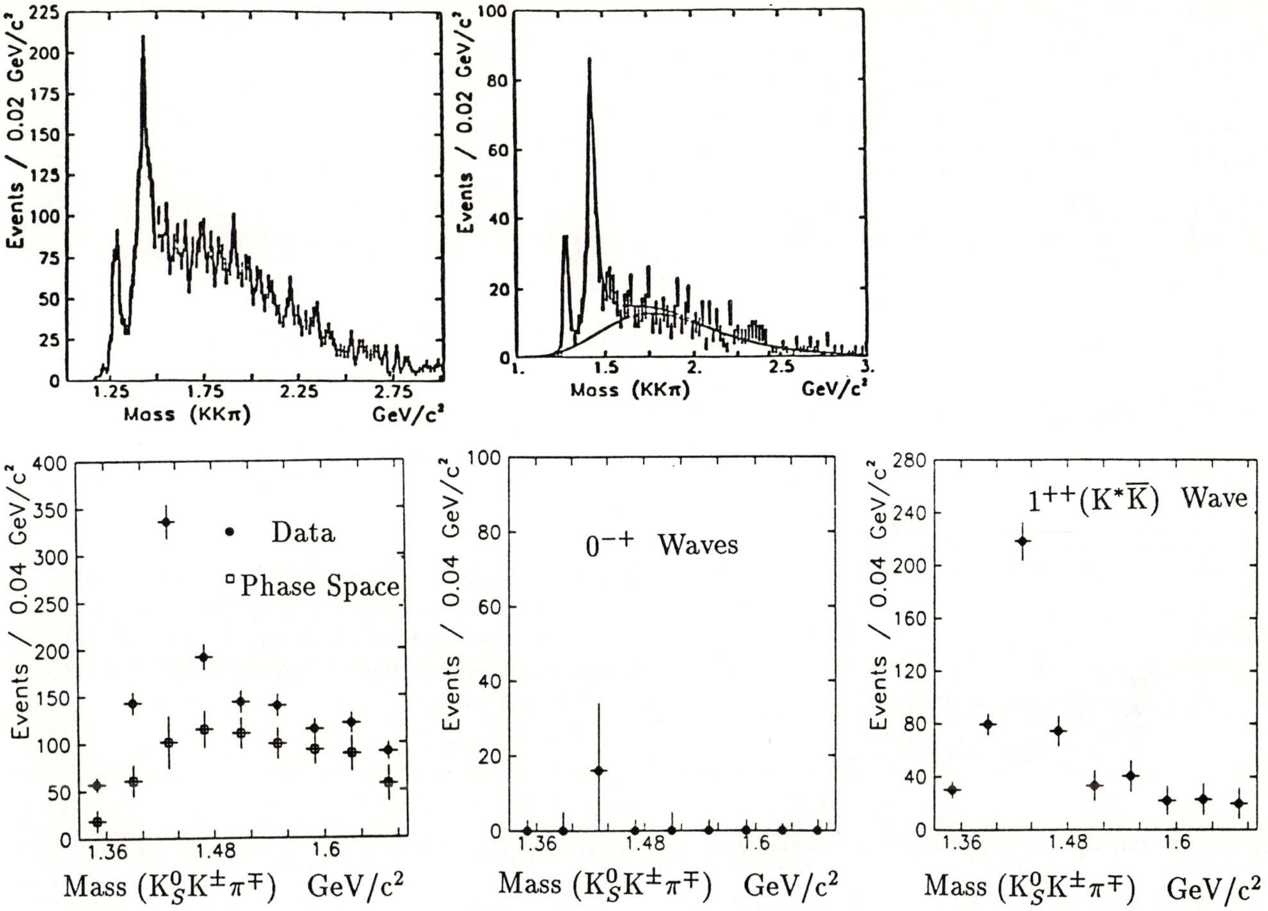

Figure 4: $K\overline{K}\pi$ Spectra, 0^{-+} and 1^{++} waves, and phase space, from Central Production (WA76) [16]. The upper right-hand mass spectrum results from selecting events in which K and/or p have been identified by the Cerenkov system.

The name $\iota/\eta(1450)$ generally refers to the broad $K\overline{K}\pi$ enhancement seen in the J/ψ radiative decays [14,15] (see fig. 3). It is possible that a portion of the $\iota/\eta(1450)$ is in fact the $\lambda/\eta(1420)$ depending on whether the $\iota/\eta(1450)$ is to be identified as a pure or mixed gluonic state. Nevertheless, the entire enhancement from 1.30 to 1.58 GeV has been established as a 0^{-+} state using as analyzer the normal to the decay plane of the $K\overline{K}\pi$ system [19]. Thus this conclusion is independent of the nature of intermediate states, e.g. $\delta/a_0(980)$ or K^*, the branching ratios of which are poorly known at the moment for the $\iota/\eta(1450)$. Recently, Burnett [20] presented new results from his Dalitz-plot analysis of the Mark III data. The partial-wave $0^{-+}(K^*\overline{K})$ exhibits a bump at mass (1447 ± 9) MeV and width (90 ± 11) MeV, while the fit had difficulty separating $0^{-+}[\delta/a_0(980) + \pi]$ from $1^{++}[\delta/a_0(980) + \pi]$; the combined partial waves show a broad structure in the mass range between 1.39 to 1.49 GeV (see fig. 5). New results [21] from DM2 have also become available regarding the spin-parity of the $\eta(1430)$ in the $K\overline{K}\pi$ system. The analysis takes into ac-

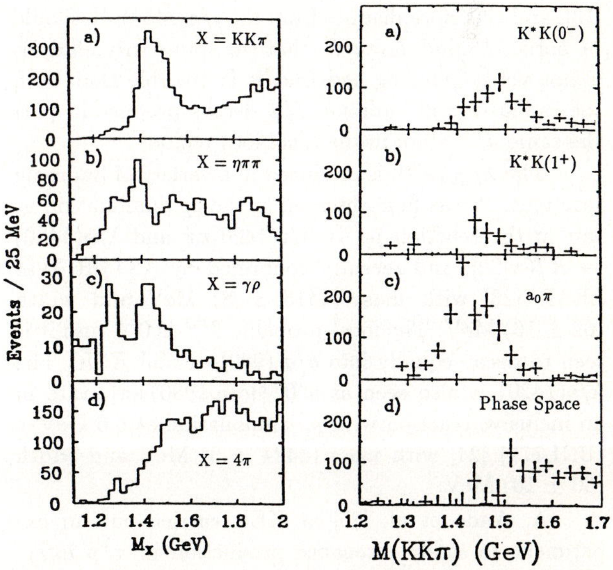

Figure 5: Results of the $K\overline{K}\pi$ Dalitz-plot fits (Mark III) [20]. Shown are the waves $0^{-+}(K^*\overline{K})$, $1^{++}(K^*\overline{K})$, $\delta/a_0(980) + \pi$ with 0^{-+} and 1^{++} combined, and phase space.

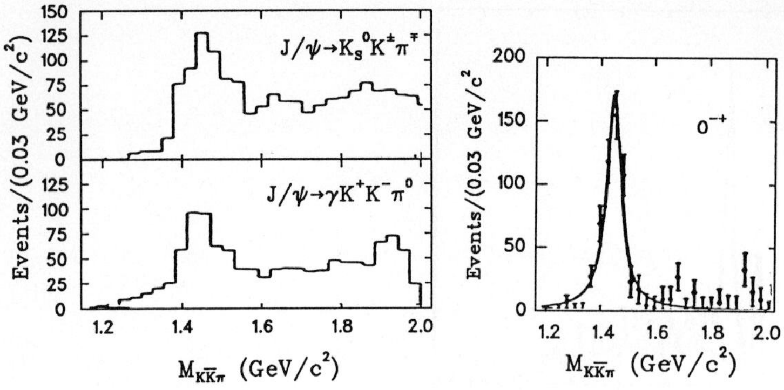

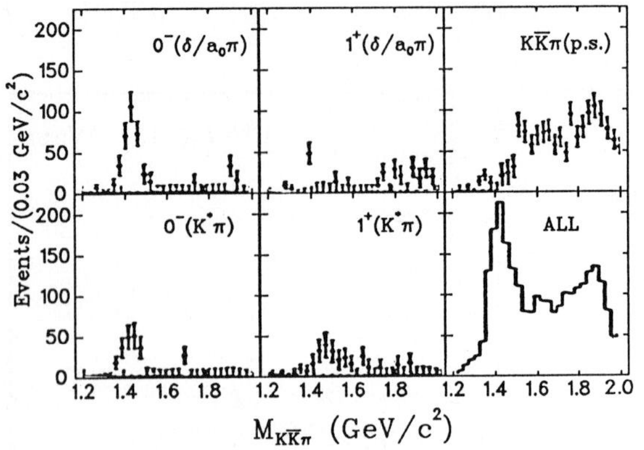

Figure 6: Results of the partial-wave analysis of the $K\overline{K}\pi$ system from the DM2 data [21]. The mass spectra and the fitted partial-waves are shown. Shown on the upper right-hand corner is the 0^{-+} wave with $\delta/a_0(980)+\pi$ and $K^*\overline{K}$ combined.

count not only the Dalitz-plot variables but also the production variables of a resonance in the J/ψ radiative process. The results, shown in fig. 6, clearly demonstrate the presence of a bump in the $0^{-+}(K^*\overline{K})+0^{-+}[\delta/a_0(980)+\pi]$ with mass (1449 $\pm$ 4) MeV and width (66 $\pm$ 7) MeV.

In the present review, the state seen recoiling off ω in the J/ψ hadronic decay is assumed to be a 1^{++} state [13], and therefore distinct from the $\iota/\eta(1450)$. It should be borne in mind, however, that the spin-parity analysis is not yet convincing and that it is possible that both the radiative and hadronic J/ψ decays produce in part the same J^{PC} state in the $\iota/\eta(1450)$ region.

The $\lambda/\eta(1420)$ is produced in a variety of hadronic reactions. It was first observed in the $\overline{p}p$ annihilation at rest in the reaction $\overline{p}p \rightarrow \lambda/\eta(1420)\pi\pi$ and $\lambda/\eta(1420) \rightarrow K\overline{K}\pi$[22], and recently confirmed by ASTERIX at LEAR [23] with mass (1413 $\pm$ 8) MeV and width (62 $\pm$ 16) MeV. The most probable J^{PC} is 0^{-+} and it is seen to decay equally into $\delta/a_0(980)+\pi$ and $K^*\overline{K}$. The $\lambda/\eta(1420)$ is also seen as a $0^{-+}[\delta/a_0(980)+\pi]$ state in an inclusive reaction with $\overline{p}p$ annihilation at 6.6 GeV/c (BNL771) [24] with mass (1424 $\pm$ 3) MeV and width (60 $\pm$ 10) MeV.

A. Ando et al. [25] at KEK carried out an experiment to study resonance production in π^-p interactions at 8.06 GeV/c. They have performed a full partial-wave analysis of the $\eta\pi^+\pi^-$ system in the reaction $\pi^-p \rightarrow \eta\pi^+\pi^-n$. They find from this analysis a 0^{-+} state $\zeta/\eta(1280)$ and in addition observe the $\lambda/\eta(1420)$ as a $0^{-+}[\delta/a_0(980)+\pi]$ state. The masses and widths are (1279 $\pm$ 5) and (32 $\pm$ 10) MeV for the $\zeta/\eta(1280)$, and (1420 $\pm$ 5) and (31 $\pm$ 7) MeV for the $\lambda/\eta(1420)$, respectively. Lack of a significant bump in the $\eta\pi\pi$ mass spectrum at 1.42 GeV is attributed to a destructive interference between the $0^{-+}[\delta/a_0(980)+\pi]$ wave and a $0^{-+}[\epsilon/f_0(1400)+\eta]$ wave. Such an interference effect was previously hypothesized to explain absence of the $\iota(1460)$ in the $\eta\pi\pi$ channel in J/ψ radiative decays [26]. All three $\delta/a_0(980)+\pi$ states, $\zeta/\eta(1280)$ $\lambda/\eta(1420)$ and $D/f_1(1285)$, are seen to exhibit rapid phase motions characteristic of resonant states. The $\zeta/\eta(1280)$, first observed by Stanton et al. [27], is thus confirmed by Ando et al [25].

A high-statistics study of the E/ι phenomenon in the hadronic sector has been carried out by the BNL771 experiment with the MPS. The study involves the channel $K_SK^+\pi^-$ produced in π^-p, K^-p and $\overline{p}p$ interactions at 6.6 and 8.0 GeV/c. A Dalitz-plot analysis of the $K\overline{K}\pi$ system from the reaction $\pi^-p \rightarrow K_SK^+\pi^-n$ at 8 GeV/c has been published earlier, based on a portion of the data [28]. Since then, a full partial-wave analysis has been performed on the same channel with a computer program written expressly for this experiment at BNL, and the results have been presented at a number of international conferences [29] and also published [30] in 1988. The BNL partial-wave program takes into account in the

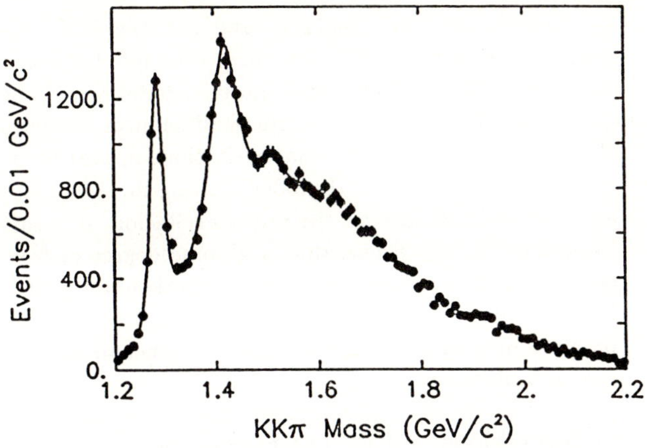

Figure 7: $K\overline{K}\pi$ Spectrum from the BNL771 data [32]. The solid curve is a fit with three resonances and a polynomial background.

most general way any degree of coherence between any partial waves through a parameterization of the general spin-density matrix [31].

The most recent data from the BNL771 experiment [32], with a sample of some 52,000 events for the reaction $\pi^- p \rightarrow K_S K^+ \pi^- n$ at 8 GeV/c, show two prominent peaks in the D and E/ι regions and, for the first time in a $\pi^- p$ interaction, a significant peak at around 1.52 GeV(see fig. 7). The $K\overline{K}\pi$ spectrum has been fitted with S-wave Breit-Wigner forms over a smooth polynomial background. The areas, fitted masses and widths for the three peaks are: 4750 $\pm$ 100 events for the first peak with (1285 $\pm$ 1) and (22 $\pm$ 2) MeV, 8800 $\pm$ 200 events for the second peak with (1419 $\pm$ 1) and (66 $\pm$ 2) MeV and 600 $\pm$ 200 events for the third peak with (1512 $\pm$ 4) and (35 $\pm$ 15) MeV.

The results of a partial-wave analysis for three $-t$ bins show a clear 0^{-+} signal near 1.4 GeV in both the $\delta/a_0(980)+\pi$ and $K^*\overline{K}$ decay modes (see fig. 8). The $0^{-+}[\delta/a_0(980)+\pi]$ in addition exhibits near 1.42 GeV a proper phase motion characteristic of a resonant behavior with respect to the $1^{++}(K^*\overline{K})0^+$ wave, where the notation 0^+ for the reference wave stands for M = 1 and reflectivity (naturality) = +1 (see fig. 9). It is noted that a significant 0^{-+} state is seen at around 1280 MeV. This constitutes an independent confirmation of the $\zeta/\eta(1280)$ in an independent channel, $K\overline{K}\pi$. A hint of the 1^{++} $D'/f_1'(1530)$ is also detected for $-t < 0.2$ $(\mathrm{GeV/c})^2$. It is becoming increasingly clear that the presence of a 1^{+-} $h_1'(1380)$ in this data cannot be neglected. The $D'/f_1'(1530)$ and the $h_1'(1380)$ are much less significant, however, in the present BNL data than in the LASS data; one may conclude therefore that these are mainly $s\overline{s}$ states but contain nevertheless some $n\overline{n}$ components.

The BNL769 experiment [33] has accumulated data on the reaction $\pi^- p \rightarrow K_S K_S \pi^0 n$ at 21.4 GeV/c. Since the $K_S K_S \pi^0$ system is a pure $C = +1$ eigenstate, possible interference effects due to $C = -1$ states in other charged modes of the $K\overline{K}\pi$ system are eliminated. Indeed, the $K_S K_S \pi^0$ spectrum from the BNL769 data shows the $\eta(1430)$ structure different from other

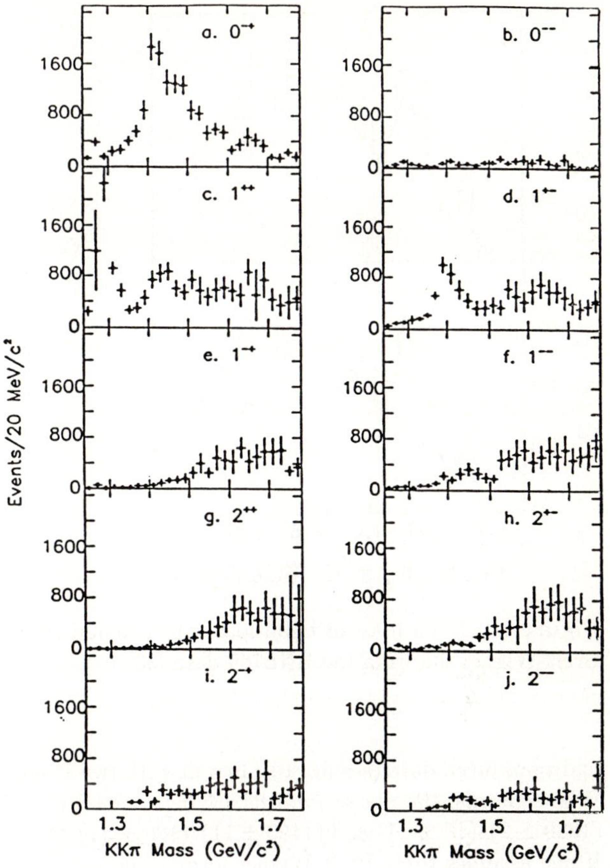

Figure 8: $K\overline{K}\pi$ Partial waves from the BNL771 data [32] with $0 \leq -t \leq 1.40(GeV/c)^2$. All significant partial-waves up to spin 2 are shown. Isobars K^* and $\delta/a_0(980)$ (where applicable) have been allowed in the fit.

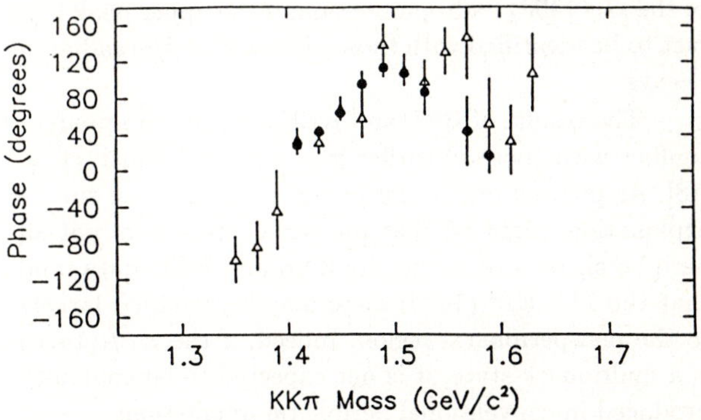

Figure 9: Phase motion of $0^{-+}[\delta/a_0(980) + \pi]$ against $1^{++}(K^*\overline{K})0^+$ for $0.14 \leq -t \leq 0.40(GeV/c)^2$ (black dots) and $0.40 \leq -t \leq 1.40(GeV/c)^2$ (triangles) from the BNL771 data [32].

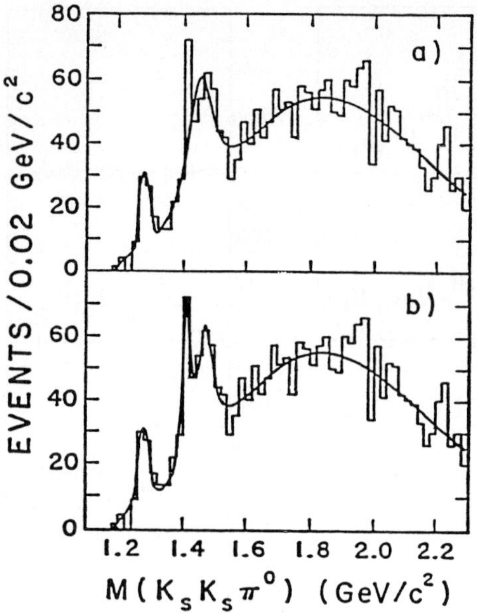

Figure 10: $K\overline{K}\pi$ mass distributions with two and three Breit-Wigner fits from the BNL769 data [33].

hadroproduced data (see fig. 10). In a fit with two Breit-Wigner forms, the masses/widths are determined to be $(1280 \pm 2)/(18 \pm 11)$ and $(1450 \pm 11)/(100 \pm 11)$ MeV. If one assumes three Breit-Wigner forms, one obtains a better fit with the χ^2 probability going from 3% to 29%. The masses/widths obtained are: $(1280 \pm 2)/(18 \pm 11)$, $(1412 \pm 4)/(25 \pm 5)$ and $(1488 \pm 5)/(82 \pm 8)$ MeV. The Dalitz-plot analysis shows that the 1280-MeV peak is a $1^{++}[\delta/a_0(980)+\pi]$ state, while the $\eta(1430)$ region is dominated by a 0^{-+} wave. This would be the first instance of the $\iota/\eta(1450)$ hadroproduction, if the upper peak is in fact to be identified with those observed in J/ψ radiative decays.

The results of BNL and KEK data are in apparent conflict with those of earlier π^-p data of Dionisi, et al. [18]. As pointed out in the previous section, one simple explanation might be that the earlier data were not all peripheral, in contrast to the BNL and KEK data, and that the 1^{++} $E/f_1(1420)$ state may be confined largely to the non-peripheral region. Indeed, if the $E/f_1(1420)$ is a multiquark state, it is not expected to be copiously produced in conventional peripheral interactions.

While the $\lambda/\eta(1420)$ is a significant component of the E/ι region in π^-p interactions, it is not seen in the K^-p LASS data. Therefore, the $\lambda/\eta(1420)$ should be mostly an $n\overline{n}$ state. Then, where is the $s\overline{s}$ state predicted at 1.55 GeV in the GI model ? The results of the LASS analysis show no 0^{-+} state up to 2.0 GeV. The situation is likely to remain confusing until a 0^{-+} state with significant $s\overline{s}$ component is discovered experimentally, and, one may surmise, its mass is likely to be much higher than 1.5 GeV or its width very large.

Since both the $\zeta/\eta(1280)$ and the $\lambda/\eta(1420)$ appear to be mainly $n\overline{n}$, one may attempt to assign the $\lambda/\eta(1420)$ as the second radial excitation of the $\zeta/\eta(1280)$; however, typical mass differences between the first and the second radial excitations appear to be around 550 MeV in the GI model, making the $\lambda/\eta(1420)$ an unlikely candidate for the radial excitation. It is safe to assume, in any event, that a gluonic degree of freedom needs to be introduced into the quarkonium model. It may even be that the presence of a hybrid may be required instead of, or in addition to, a glueball and the radial excitations.

5. Discussions and Future Prospects

A survey of established meson states, with the work of Godfrey and Isgur as reference, shows the basic validity of the 'relativized' quarkonium model. The survey highlights a need to go beyond the $q\overline{q}$ model, to include the gluonic degrees of freedom and multiquark states in the light-quark sector. Of the states discussed in previous sections, only three states $\zeta/\eta(1280)$, $h_1'(1380)$ and $D'/f_1'(1530)$ are seen to have reliable spectroscopic placements. In this sense, a new frontier of meson spectroscopy is being opened up in the mass region between 1.5 and 2.5 GeV, where a vast majority of the complications seem to occur, pointing beyond the quarkonium model.

Among the 1^{++} states, the $E/f_1(1420)$ is increasingly becoming a redundant state, difficult to place in the scheme of the SU(3) nonet families. It is noted that the recently confirmed $D'/f_1'(1530)$ is the likely candidate for being the $s\overline{s}$ member of the ground-state 1^{++} nonet with the $D/f_1(1285)$ as its $n\overline{n}$ isoscalar partner. Copious production of the $E/f_1(1420)$ in photon-photon collisions suggests that this state may be a multiquark state, perhaps a 1^{++} bound state of $K^*\overline{K}$ at threshold [34] or a P-wave π orbiting around an S-wave $K\overline{K}$ at the center [35]. These ideas are natural extensions of the concept that the $S/f_0(975)$ and the $\delta/a_0(980)$ are 0^{++} $K\overline{K}$ molecules [36].

There exist two 0^{-+} states which are not understood: the $\lambda/\eta(1420)$ seen in hadronic productions and the $\iota/\eta(1450)$ seen in J/ψ radiative decays. It is safe to assume that there should be a large overlap between the two states. It is even possible, however remote, that the two states are in fact two different manifestations of the same state. One may also ask: Could it be that the $\iota/\eta(1450)$ is a superposition of the $\lambda/\eta(1420)$ and a glueball with its mass around 1.5 GeV ? The $\lambda/\eta(1420)$ is itself a state difficult to place in the $q\overline{q}$ model. If the $\zeta/\eta(1280)$ is assumed to be the first radially excited $n\overline{n}$ state, then its $s\overline{s}$ partner is expected at 1.55 GeV in the GI model. The $\lambda/\eta(1420)$ is some 130 MeV below this mass and furthermore it has little $s\overline{s}$ component. And yet it decays into the $K\overline{K}\pi$ channel.

In this review a survey is given of the current status of meson spectroscopy in the light-quark sector. What are the future prospects of this field ? In the very near future, results of a partial-wave analysis of the $K\overline{K}\pi$ system from the K^-p BNL data will become available, which will provide additional information of the $s\overline{s}$ nature for the E/ι region. Additional data and/or analysis will be forthcoming from the GAMS-4000 at CERN and the GAMS-2000 at IHEP, Serpukhov. There exist also an approved experiment [37] at UNK, Serpukhov, the GAMS-10000, designed to study gluonic states up to 15 GeV with a beam in the range between 500 GeV/c to 3 TeV/c. Crystal-Barrel, OBELIX and JETSET experiments at LEAR [38] should also provide new data on meson spectra in the near future. The KEK experiment E179 [39] took data in 1989, to look for the exotic $1^{-+}\pi\eta$ resonance at 1.4 GeV. Another exotic meson search [40], E818 at BNL, will take place in 1990, concentrating mainly on the decay mode $D/f_1(1285) + \pi$. A similar exotic meson search in the same decay mode is being proposed at Fermilab, capitalizing on the Primakoff production of meson states in the mass range 1.0–2.0 GeV from a ~ 0.5 TeV/c beam [41].

A stage-one approved experiment E852 at BNL is designed to carry out a systematic search of all the decay channels involving photons as well as charged particles at the MPS, with a probable running time in 1993 [42]. Looking further into the future, one can expect to see data coming from the e^+e^- collider BEPC in China. In addition, with installation of the Booster at the AGS [43], BNL may commission an intense K beam, to be followed closely by the KAON factory at TRIUMF; these facilities will be able to produce K^-p data with statistics comparable to those of the present-day π^-p data. It is hoped that the tau-charm factory proposed at SLAC will become a reality in the near future, with statistics on the J/ψ data several hundred times greater than those available today.

In addition to the quest for the t quark and its attendant spectroscopy, another frontier for the meson spectroscopy lies with the task of deciphering the roles played by the gluonic and multiquark constituents as building blocks of mesons. Evidently, there is still a need for more high-statistics experiments designed to look for light-quark states, in particular with novel production or decay channels.

Acknowledgement

The author is grateful to the organizers of the International Meeting on the KAON Factory proposed at TRIUMF, Vancouver, Canada, for their kind invitation. The meeting was held at a splendid setting in Bad Honnef, W. Germany, with a large turnout of German physicists. It is hoped that the KAON factory at TRIUMF can be built in the near future, with an extensive international participation so characteristic of our endeavor.

He is indebted to W. Kern, S. Protopopescu, H. J. Willutzki and D. Zieminska for their helpful comments after reading a preliminary draft of this review.

References

[1] Proceedings of the BNL Workshop on Glueballs, Hybrids and Exotic Hadrons, BNL, August 29 - September 1, 1988 (AIP Conf Proceedings No. 185—Editor: S. U. Chung)

[2] S. U. Chung, Current Status of Light-quark Spectroscopy, BNL-preprint 40599, 1987

[3] S. Godfrey and N. Isgur, Phys. Rev. D32, (1985) 189

[4] Review of Particle Properties, Particle Data Group, Phys. Lett. 204B, (1988) 1.

[5] D. Aston et al., Phys. Lett. 201B, (1988) 573
B. Ratcliff, Ref. 1, p. 160
D. Aston, Ref. 1, p. 350
S. Suzuki, Proc. Second Hadron Spectroscopy Conf., KEK, (1987) 64.

[6] K. L. Au, D. Morgan and M. R. Pennington, preprint RAL -86 -076 (1986), Phys. Rev. D35, (1987) 1633.

[7] R. L. Jaffe, Phys. Rev. D15, (1977) 267; D15, (1977) 281; D17, (1978) 1444.

[8] J. Weinstein and N. Isgur, Phys. Rev. Lett. 48, (1982) 659; Phys. Rev. D27, (1983) 588.

[9] M. Boutemeur, Ref. 1 (p. 389)

[10] P. Gavillet et al., Z. Phys. C16, (1982) 119.

[11] G. Gidal, Ref. 1 (p. 171)

[12] M. S. Chanowitz, Phys. Lett. 187B, (1987) 409.

[13] J. J. Becker et al., Phys. Rev. Lett. 59, (1987) 186.

[14] U. Mallik, SLAC Summer Institute on Particle Physics, August 1986, SLAC -PUB -4238.

[15] S. Cooper, SLAC -PUB -3819 (1985); SLAC -PUB - 4139 (1986).

[16] T. A. Armstrong et al., Phys. Lett. 221B, (1989) 216;
Z. Phys. C34, (1987) 23; Phys. Lett. 146B, (1984) 273
A. Kirk, Ref. 1 (p. 340).

[17] O. V. Baille, Proc. Europhysics Conf. on H.E.P., Bari, Italy, (1985) 31.

[18] C. Dionisi et al., Nucl. Phys. B169, (1980) 1.

[19] J. D. Richman, Proc. 20th Rencontre de Moriond, Les Arcs, France, (1985) 471; Ph.D. Thesis (Caltech), CALT -68 -1231 (1985).

[20] T. Burnett, Ref. 1 (p. 102)

[21] L. Stanco, Ref. 1 (p. 318)

[22] P. Baillon et al., Nuovo Cimento 50A, (1967) 393;
Proc. Experimental Meson Spectroscopy, BNL, (1983) 78.

[23] R. Landau, Ref. 1 (p. 246)
[24] D. F. Reeves et al., Phys. Rev. D34, (1986) 1960.
[25] A. Ando et al., Phys. Rev. Lett. 57, (1986) 1296.
[26] W. F. Palmer and S. S. Pinsky, Phys. Rev. D27, (1983) 2219.
[27] N. R. Stanton et al., Phys. Rev. Lett. 42, (1979) 346.
[28] S. U. Chung et al., Phys. Rev. Lett. 55, (1985) 779.
[29] D. Zieminska, Proc. First Hadron Spectroscopy Conf., Maryland, (1985) 27,
J. Dowd, Proc. Int'l Europhysics Conf. on H.E.P., Bari, Italy, (1985) 318,
S. U. Chung, Proc. 20th Rencontre de Moriond, Les Arcs, France, (1985) 489,
S. Protopopescu, Proc. Annual Meeting of APS Div. of Particles and Field, Eugene, Oregon, (1985) 671,
S. U. Chung, Proc. Second Aspen Winter Particle Physics Conf., (1986) 77,
S. U. Chung, Proc. 23rd Int'l Conf. on H.E.P., Berkeley, Vol. I, (1986) 725,
S. Protopopescu, Proc. Second Hadron Spectroscopy Conf., KEK, (1987) 56,
D. Zieminska, Ref. 1 (p. 112)
S. Blessing, Ref. 1 (p. 363)
[30] A. Birman et al., Phys. Rev. Lett. 61, (1988) 1557.
[31] S. U. Chung and T. L. Trueman, Phys. Rev. D11, (1975) 633.
[32] D. Zieminska, Ref. 1 (p. 112)
[33] N. Cason, Ref. 1 (p. 334)
M. G. Rath et al., Phys. Rev. Lett. 61, (1988) 802.
[34] D. Caldwell, Ref. 1 (p. 465)
[35] R. Longacre, 'The $E(1420)$ meson as a $K\overline{K}\pi$ Molecule,' BNL preprint, to be submitted to Phys. Rev.
[36] J. Weinstein, Ref. 1 (p. 400)
[37] Yu. D. Prokoshkin, Proc. Workshop on the Experimental Program at UNK (1987), (p. 214)
[38] U. Gastaldi, General Detector Review, Ref. 1 (p. 50)
R. Eisenstein, JETSET, Ref. 1 (p. 636)
C. Amsler, Crystal Barrel, Ref. 1 (p. 653)
[39] T. Iwata, Ref. 1 (p. 678)
[40] J. Dowd, Ref. 1 (p. 631)
[41] M. Zielinski, Ref. 1 (p. 395)
V. Chaloupka and T. Ferbel, 'Physics at Fermilab in the 1990s', Breckenridge, CO (August 1989)
[42] A. Dzierba, Ref. 1 (p. 621)
[43] D. Lowenstein, Ref. 1 (p. 185)

Z. Phys. C – Particles and Fields 46, S121 (1990)

Zeitschrift für Physik C Particles and Fields

High statistics experiments: the LASS experience *

D. Aston [1], N. Awaji [2], T. Bienz [1], F. Bird [1], J. D'Amore [3], W. Dunwoodie [1], R. Endorf [3], K. Fujii [2], H. Hayashii [2], S. Iwata [2], W. Johnson [1], R. Kajikawa [2], P. Kunz [1], Y. Kwon [1], D. Leith [1], L. Levinson [1], T. Matsui [2], J. Martinez [3], B. Meadows [3], A. Miyamoto [2], M. Nussbaum [3], H. Ozaki [2], C. Pak [2], B. Ratcliff [1], P. Rensing [1], D. Schultz [1], S. Shapiro [1], T. Shimorura [2], P. Sinervo [1], A. Sugiyama [2], S. Suzuki [2], G. Tarnopolsky [1] T. Tauchi [2], N. Toge [1], K. Ukai [4], A. Waite [1], S. Williams [1]

[1] Stanford Linear Accelerator Center, Stanford University
[2] Department of Physics, Nagoya University
[2] University of Cincinnati
[4] Institute for Nuclear Study, University of Tokyo

Abstract

This paper summarizes the main data processing and analysis procedures employed by the LASS collaboration in the course of a Kp experiment involving 138 million triggers at an incident momentum of 11 GeV/c. The role of multi-vertex kinematic fitting at the coordinate level is described, and the importance of the $\sim 4\pi$ acceptance for both large and small statistics exclusive channels is discussed. The evolution in understanding of the amplitude structure of mesonic states which has resulted from the acquisition and analysis of large data samples is illustrated using the $K\pi$ $S-$wave and partial wave decomposition of the low-mass $K\pi\pi$ system. Implications for future studies of $s\bar{s}$ spectroscopy at the KAON factory, and for amplitude analyses of meson systems produced in J/ψ decay at a Tau-Charm factory are considered.

†Work supported by the Department of Energy under contract No.DE-AC0376SF00515; The National Science Foundation under grant Nos PHY82-09144, PHY85-13808, and the Japan U.S. Cooperative Research Project on High Energy Physics.

Based on invited talks presented by W. Dunwoodie at the Tau-Charm Factory Workshop, SLAC, May 23-27, 1989 and at the International Meeting on Physics at KAON, Bad Honnef, W. Germany, June 709, 1989

Z. Phys. C – Particles and Fields 46, S123–132 (1990)

Zeitschrift für Physik C Particles and Fields

Physics at SATURNE

J. Arvieux

Laboratoire National Saturne, F-91191 Gif-sur-Yvette Cedex, France

1. Introduction

Saturne is a synchrotron of maximum momentum P/Z=4 GeV/c which accelerates protons up to 3 GeV, deuterons up to 2.3 GeV, ^{3}He ions up to 5.2 GeV and heavier particles (from ^{4}He to ^{84}Kr) up to 1.15 GeV per nucleon. A strong emphasis is put on polarized beams of protons and deuterons which benefit from a high degree of polarisation (up to 90 %), large intensities (up to 2 x 10^{11} polarized ions per cycle, one cycle lasting 1 to 3 seconds depending on the final energy) and broad versatility (deuterons with vector and tensor polarization, easy change from protons to deuterons).

An important improvement of the machine qualities in intensity, reliability and range of available ions has been made possible by the dedication in oct. 87 of a new injector and storage ring "MIMAS" which now replaces the former linac injector for all particles except when the largest intensities (5 x 10^{11} to 10^{12} ions per cycle) of light unpolarized particles are required, a mode of operation which does not represent more than 10 % of the total machine schedule.

Due to the wide array of probes and extensive range of energy (from 100 MeV up) the physics program is very broad, ranging from "conventional" nuclear physics in which nuclear degrees of freedom only are taken into account (e. g. nuclear structure studies, dynamics of heavy-ions reactions below pion-production threshold, ...) to tentative discovery of quark phenomena in nuclear reactions (e. g. dibaryons). The main core of Saturne physics is meson-nuclear physics in which explicit mesonic degrees of freedom have to be taken into account. These degrees of freedom may manifest themselves in the production of real mesons (π , η , ρ , ω , k) or of excited nucleons (Δ 1232, N* 1440, N* 1535). They might also underlie reaction mechanisms as virtual excitations in which mesons or excited nucleons do not appear explicitely (in nucleon-nucleon elastic scattering for example).

I will present 3 areas of physics in which Saturne has recently made some important contributions :

- Nucleon-nucleon interactions,
- Creation and propagation of Δ-isobars in nuclei and studies of medium effects,
- Intense production of η-mesons in nuclear reactions.

2. The Nucleon-Nucleon program

In itself, the Nucleon-Nucleon (NN) problem is very important since it involves the most elementary interactions whether they are calculated in a meson-exchange model or in a quark model. Furthermore the NN interactions are the basic building blocks for the interaction of heavier nuclei. The NN forces are highly spin-dependent and their determination requires an extensive use of polarization. When one takes parity conservation and time reversal invariance into account, the interaction of 2 identical spin 1/2-particles is described by 5 complex amplitudes (pp interaction) or by 6 complex amplitudes if charge-dependence is assumed to be broken (np interaction). Since amplitudes are related via an arbitrary overall phase, the number of spin-dependent experimental quantities ("observables") to be measured is 9 for pp or nn scattering (I=1 amplitudes) and 11 for np scattering (I=0 amplitudes in addition to pp scattering which determines the I = 1 amplitudes). In fact observables are bilinear combinations of amplitudes and to avoid ambiguities one must measure more observables than strictly needed.

A dedicated beam linac has been set up at Saturne for the study of the NN interaction. In a first round of experiments the pp interaction has been extensively studied from 800 MeV to 2.7 GeV (data below 800 MeV are available from meson factories). A combination of dipoles and spin processing solenoids allows to align the beam spin into 3 directions along **k** (incident momentum), **s** (sideways) and **n** (normal or vertical). A frozen-spin polarized target is also used with spin aligned in the **n** and **k** directions. Target polarization is typically 85 % . Observables are labelled as X_{srbt} where the subscripts represent the scattered, recoil, beam and target spins respectively. For **pp** scattering a minimum of 11 observables (and up to 15 at selected energies) have been measured at 0.834, 0.874, 0.935, 0.995, 1.095, 1.295, 1.596, 1.796, 2.096, 2.396 and 2.696 GeV.

The scattering matrix is governed by 5 amplitudes called by convention **a, b, c, d, e (1)**. The real and imaginary part of each amplitude can be determined from the data in a model-independent way. Since one phase is arbitrary one sets $\phi_e = 0$. As examples the results for 0.874 GeV and 2.096 GeV are shown in fig 1 et 2 respectively **(2)**. The 0.874 GeV results are in good agreement with the predictions of a phase-shift analysis (solid line) for which the major

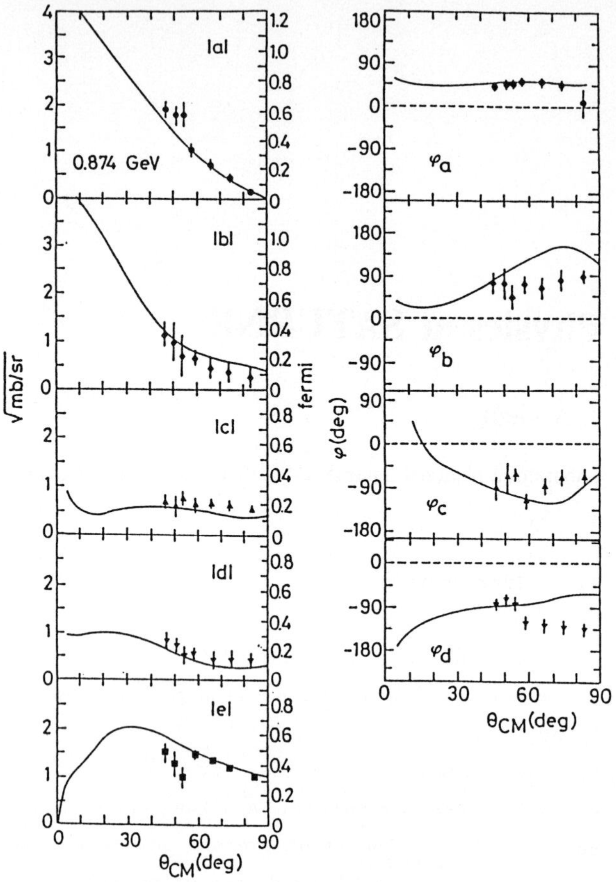

Fig. 1 - Modules and phases of pp elastic scattering amplitudes at 0.874 GeV.

part of the data used in the amplitude analysis were not available. The same smooth dependence is observed at all energies except at 2.1 GeV where 3 phases ϕ_a , ϕ_b and ϕ_c change sign abruptly (fig. 2). The energy dependence at 51° CM of the 4 phases is shown in fig. 3. A similar behaviour is observed at different angles. The effect is statistically very significant.

According to these results ϕ_a must go through zero at 2 energies, one between 1.7 and 2.1 GeV and one between 2.1 and 2.4 GeV. Since ϕ_a is very simply related to the analyzing power through

$$\sigma A_{oono} = \sigma A_{ooon} = |a|c \cos \phi_a ,$$

the analyzing power must show two extrema between 1.7 and 2.4 GeV. Existing data, with low momentum resolution (Δ **p/p** = 3.5 %) seemed to indicate a structure in A_{oono} around 2.2 GeV (3).

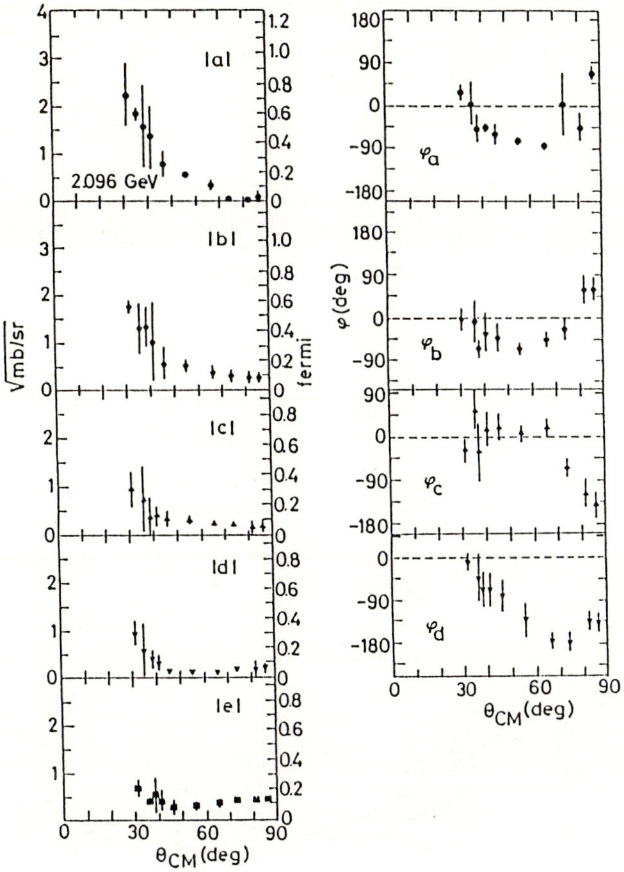

Fig. 2 - Modules and phases of pp elastic scattering amplitudes at 2.096 GeV.

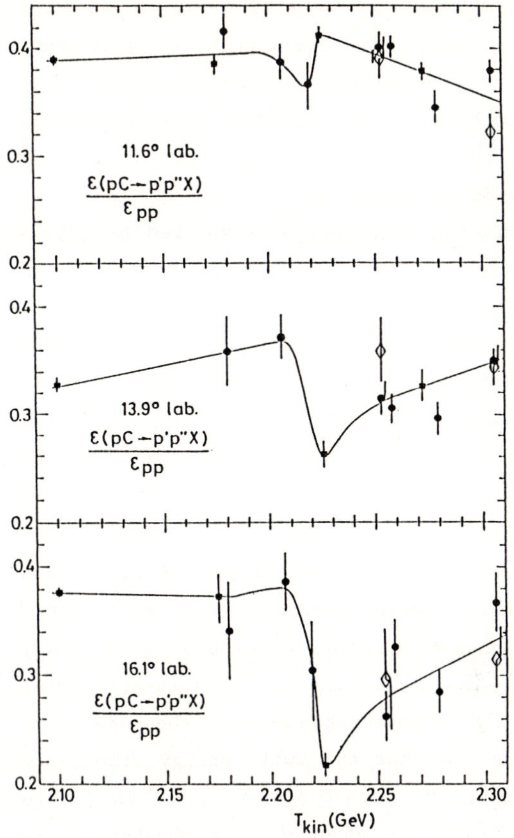

Fig. 4 - Energy dependence of the ratio R of p-c top-p analyzing power. The particle p' and p" are unidentified charge particles.

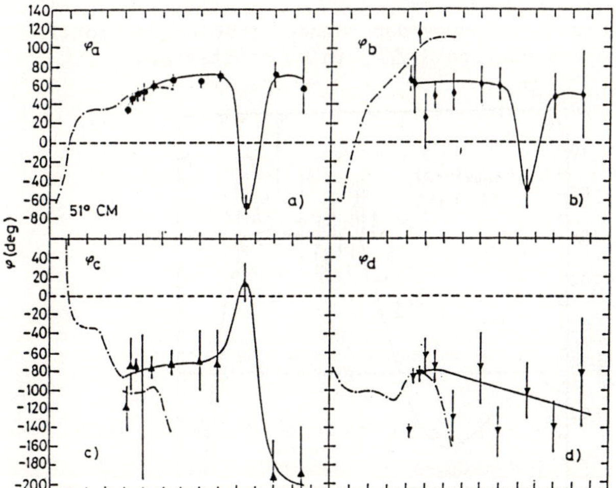

Fig. 3 - The energy dependence of phases at 51° The full lines are hand-drawn curve and the dashed-dotted lines are results of a phase-shift analysis (42).

We have investigated this observable in small energy steps at lab angles 11.6, 13.6 and 16.1 degrees (approximately 34, 40 and 46 degrees C.M.). Since there exists a so-called "imperfection" resonance for γ G = 6 at 2.20 GeV, the beam energy was changed by using Cu-absorbers. Preliminary results seem to indicate a very narrow structure between 2.20 and 2.25 GeV.

In another experiment the ratio R of the analyzing power on carbon and on hydrogen was measured, this time by varying the incident energy. Although the beam polarization may change through the depolarizing resonance region, the carbon to hydrogen analyzing power ration is unaffected . Since the width of the structure observed in pp-scattering is of the order of 10-20 MeV and since the Fermi energy spread of a proton in carbon is larger than 50 MeV one expects to see a structure in the R-ratio independent of the incident beam polarization. The results (fig. 4) show indeed a narrow structure around 2.22 GeV ($s^{1/2}$= 2.73 GeV). To confirm these results a third experiment has been done, this time by using an unpolarized incident beam scattered by a polarized target whose polarization is well known and independant of energy. Data are being analyzed.

Existing data showed unexpected behaviour in this energy region : spin-dependent total cross sections $\Delta\phi_L$ from Argonne (4) and pp=>dπ analyzing powers from Saturne (5). Even before the data became available there were NN calculations based on a P-matrix boundary condition theory in which the short-range behaviour is described by six-quark states calculated within the Cloudy-Bag Model and the long-range behaviour is due to meson-exchanges (6-7). These calculations predict narrow ($\Gamma \sim$ 50 MeV) dibaryon resonances at 2.63 ($^3S_1$-$^3D_1$), 2.70 ("S_0) and 2.88 ($^1D_2$) GeV. More recent calculations (8) predict also P-wave resonances in the same energy region. Whether the swift energy dependence of NN amplitudes around $s^{1/2}$= 2.7 GeV is due to "true" dibaryon resonance(s) or whether it is produced by an unknow mechanism (e. g. opening of highly excited N* resonance channels) is not yet sure and the answer will probably have to await a complete study of the NN interaction in this energy region.

The NN beam line has been presently converted to a polarized neutron beam of maximum energy 1.15 GeV, polarization 60 % and intensity up to 3 10^7 neutrons per cycle obtained by polarized deuteron break up. The beam polarization can be aligned in all directions in space.

The results for ϕ_{1t}= − $\Delta\phi_T$/2 and − $\Delta\phi_L$ are shown in fig. 5 and 6 for np scattering. The results agree qualitatively with the Argonne data (4) but the absolute value seems to be larger. The Argonne data were obtained by subtracting pp from pd total cross-sections and by correcting for 3 body final state interactions. This procedure might induce errors which are difficult to quantify.

An example of 2-spin data in n-p scattering at 0.63 GeV is shown in fig. 7 together with phase-shift analysis predictions. The new, very precise data, will have a significant im-

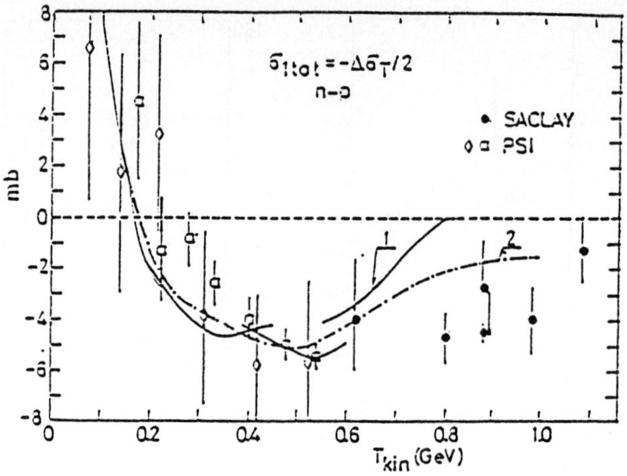

Fig. 5 - Energy dependence of polarized total cross-section σ_{1T} in n-p scattering.

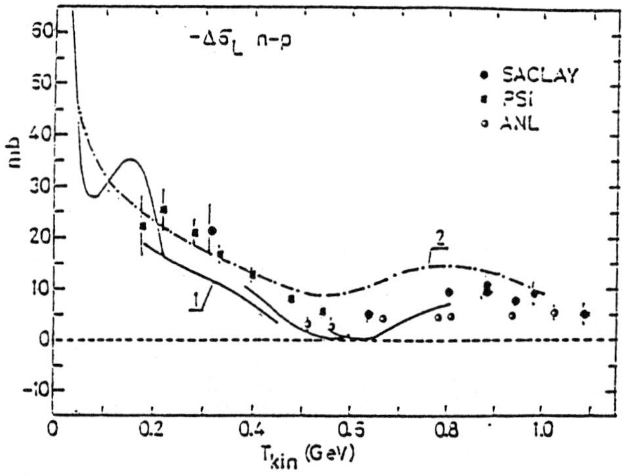

Fig. 6 - Energy dependence of polarized total cross section − $\Delta\sigma_L$ in np scattering.

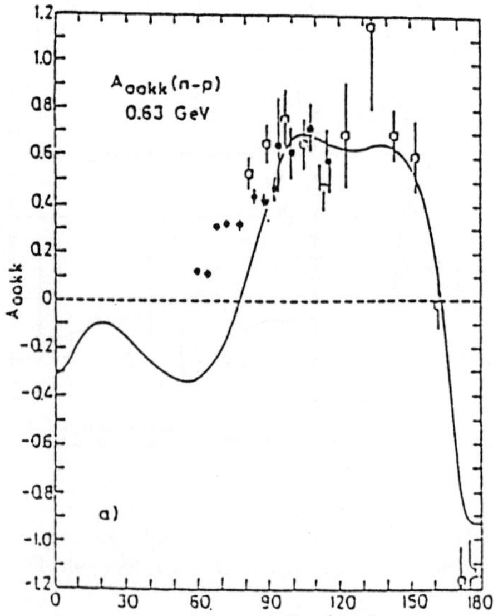

Fig. 7 - Two spin correlation parameter A_{ookk} in np scattering at 0.63 GeV.

pact on subsequent phase-shift analysis. Complete np data will be obtained at selected energies between 0.63 and 1.07 GeV (9).

3. Production and propagation of Δ-isobars in nuclei.

Single charge exchange reactions excite selectively the spin-isospin ($\Delta S = 1$, $\Delta I = 1$) response of a nucleus. At low energy one excites preferentially the L = 0 Gamov-Teller resonance, the L = 1 dipole resonance and the L = 2 (spin-isospin) quadrupole resonance. When the energy becomes larger than the Δ-excitation threshold (ω = 300 MeV in the center-of-mass system) the most prominent feature in the energy spectra is the excitation of a Δ-isobar in the nuclear medium through elementary mechanisms like

$$p(p, n)\Delta^{++} \text{ or } p(n, p)\Delta^{\circ}$$

The reactions involving neutron projectiles or targets are not easy experimentally and the use of composite particles allow to alleviate the problem. Δ-excitation in nuclei has been observed in (p, n) reactions at LAMPF (10, 11) with 800 MeV protons. At Saturne an array of complementary probes has been used to create Δ-isobars in nuclei and to study their propagation : (d, ^{2}He) reactions induced by tensor polarized deuterons, (^{3}He, t) and heavy-ion charge exchange reactions with projectiles ranging from carbon (A = 12) to Argon (A = 40) (12 - 16). The main characteristics of these spectra is a strong-excitation of Δ-isobars in all reactions, the shape of the Δ-peak being the same for C and heavier targets but its position being systematically shifted downwards by 70 MeV compared to free Δ-excitation on a nucleon as seen from the (^{3}He, t) data in fig. 8. Trivial kinematical effects (Fermi motion, finite size effects) may explain half of the observed shift (30-40 MeV out of 70 MeV) leaving about 30-40 MeV to be accounted for by dynamical effects.

The most interesting explanation relates to nuclear medium effects. Since hadronic probes couple mainly to be longitudinal response function one expects that the attractive Δ-hole

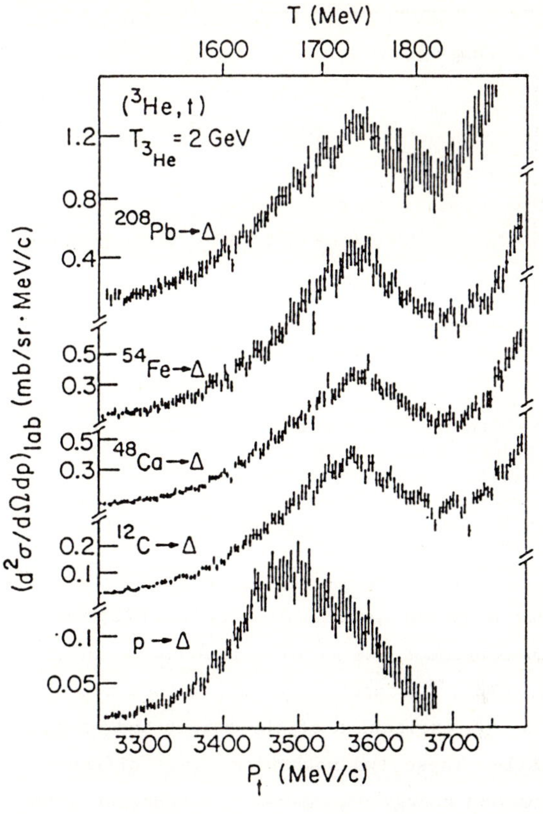

Fig. 8 - Triton spectra in the (He, t) reaction at 0° for a number of targets (13).

interaction produced by pion-exchange would induce strong correlations giving the Δ an apparent mass smaller than the free mass. This effect is density-dependent : large shifts (up to 150 MeV) are predicted at normal nuclear density (17) reducing to the observed 30-40 MeV in (^{3}He, t) reaction where the effective density is of the order of 0.3 ρ_0 as calculated in a Glauber-type calculation (18). A further confirmation of the surface-nature of these reactions comes from the fact that the mass-dependence of the reaction goes like $A^{1/3}$ instead of $A^{2/3}$ which would prevail in a volume-reaction. Several theoretical calculations explain the data by Δ-hole collective excitations (17, 19, 20). Distortion effects (21) can also reproduce the shift in light nuclei but since they predict that it has to be mass (or size) dependent, they are less successful for heavier targets (experimentally the shift is strictly constant from ^{12}C to ^{208}Pb).

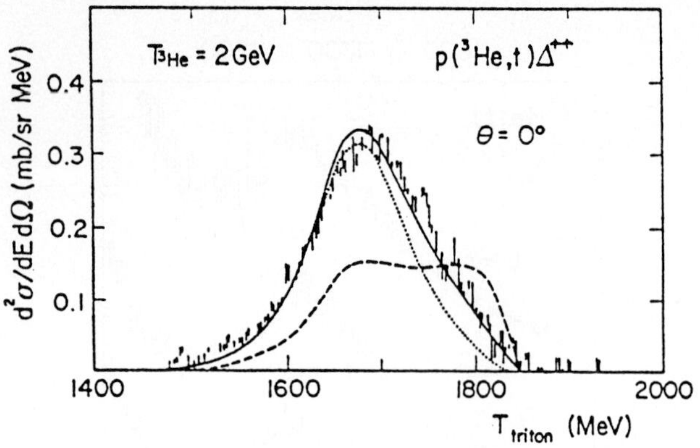

Fig. 9 - Cross-section for the elementary $p(^3He, t)\Delta^{++}$ reaction at 2 GeV. Dotted line : delta excitation of the target, solid line : with inclusion of delta-excitation of the projectile, dashed line : prediction for the $n(^3He, t)\Delta^+$ reaction.

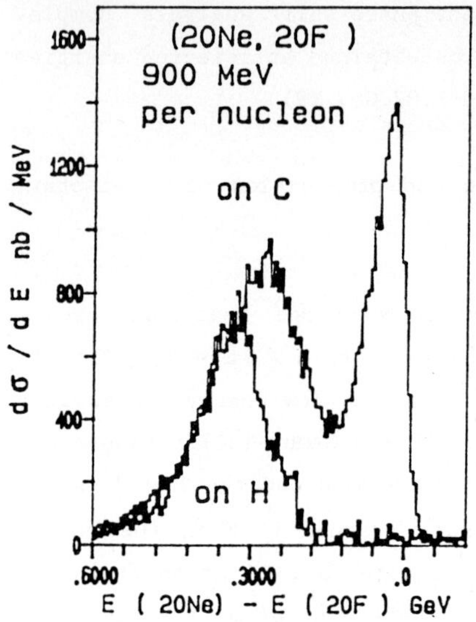

Fig. 10 - Zero degree spectra of the $H(^{20}Ne, ^{20}F)\Delta^{++}$ and $^{12}C(^{20}Ne, ^{20}F)^{12}N$ reactions at 900 MeV per nucleon.

In a recent paper, Oset et al (22) were able to reproduce the free Δ -production in the p $(^3He, t)$ Δ^{++} reaction by taking into account both the Δ-excitation of the target and of the projectile. These two mechanisms have different momentum and energy dependence. In reaction with complex nuclei they appear to be very different depending if the Δ is produced on a target-proton via the $p(^3He, t)\Delta^{++}$reaction or a target-neutron via the $n(^3He, t)\Delta^+$ reaction (fig. 9). However there are yet no quantitative predictions for heavy targets and it is still to be seen if this theory will explain the constancy of the shift and broadening from C to Pb.

In order to help understand medium effects we have enlarged the scope of our investigation by using different probes which test selective aspects of the phenomenon. By using heavy-ions at 900 and 1100 MeV x A from ^{12}C to ^{40}Ar, one tests the density dependence and peripherality of the interaction (23) ; ^{20}Ne induced spectra are shown in fig. 10 as an example. They appear to be dominated even more so by Δ -excitation, making heavy-ion charge-exchange reactions a very attractive tool to investigate Δ degrees of freedom in nuclei. The same shift between free and nuclear Δ -excitation is observed leading to the conclusion that in all cases the reaction proceeds essentially through a one-step mechanism exciting fundamental modes in the nucleus.

Due to the strong absorption of the projectile, the impact parameter is very sharply determined (within a couple of fermis) which in turn corresponds to a very narrow window for the momentum transfer. Then, by varying the target size and incident energy one can map the momentum dependence of the NN => NΔ interaction in medium (24).

Another interesting aspect of heavy ion charge exchange reaction is that they allow to study (p, n) - type or (n, p) - type reactions, e. g. in $(^{12}C, ^{12}B)$ and $(^{12}C, ^{12}N)$ reactions with the same projectile and the same detection set-up just by selecting the proper charge and momentum of the outgoing particle(23).

Polarized deuterons are probes of great interest to constrain the spin-isospin dependent part of the NN-interaction (26) through the p(d, 2p) n reactions where the 2 outgoing protons are left in the $^1S_0$ state. At higher energy the same reaction can be used to excite a Δ-isobar through the p(d, 2p) $\Delta°$ reaction and learn about basic properties of the NN => NΔ interaction. The polarization response denoted **P** is a combination of the T_{20} and T_{22} tensor analyzing power and tensor beam polarization ρ_{20}. Near 0° it reduces essentially to 1/2 ρ_{20} T_{20}. The results are shown for 2 angles (16) in fig. 11. They are compared with calculations based on me-

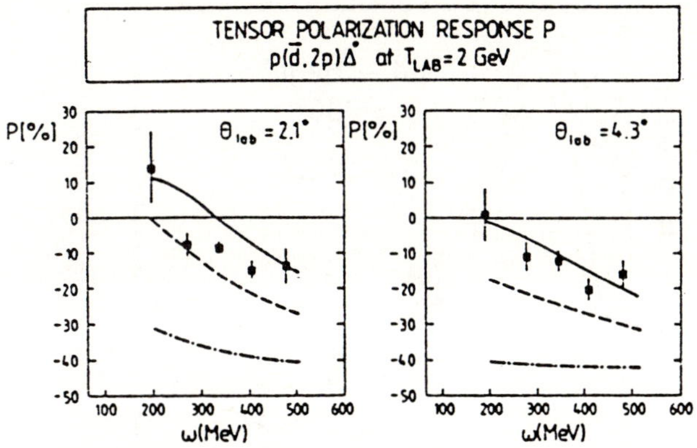

Fig. 11 - Dashed-dotted curve : pure OPE, dashed curve : OPE + minimal short-range correlations, solid curve : with ($\pi+\rho$) exchange.

son exchange (27). One-pion exchange (dashed-dotted line) which is pure spin- longitudinal excitation fails completely to fit the data. By introducing minimal short-range correlations (through removal of the δ-function arising from the point like OPE transition potential (28) and taking into account π NN and $\pi N\Delta$ vertex form factors, which add some spin-transverse component, the agreement is much improved (dotted-line). Finally by introducing 2 π-exchange through a ρ-exchange term the agreement becomes very good (solid line).

The next step is to produce the Δ into nucleus via the (d, 2p) reaction. The use of polarized deuterons allows to study the momentum dependence of longitudinal and transverse nuclear response function in a single experiment. These studies complement (e, e'p), (^{3}He, t) and heavy-ion charge exchange reactions since the mechanism of absorption will be different for each probe. The first study shows that the spin-transverse component is larger in ^{12}C than in deuterium. Since collective medium effects appear essentially in the spin-longitudinal response this result would indicate less collectively in ^{12}C than in deuteron which is difficult to understand. Experiments on heavier nuclei and more theoretical work will be needed to fully exploit this new tool.

4. **Meson-production.**

Together with the excitation of nucleon resonances in nuclear medium, the production of real mesons is an important part of the Saturne programs. With the energy available at Saturne one can create mesons up to mass 1.5 GeV (iota (1440) for example) in inclusive reactions or even higher in exclusive reactions like the pd => ^{3}He X reaction where the coherence of the meson-production mechanism increases the energy available in the center of mass system by 50 % (relativistic effects put aside). Another special feature already exploited in the (d, 2p) reaction (see section 3) is the availability of deuterons up to 2,3 GeV, whether unpolarized or polarized. Deuteron polarization can be of vector type (it_{11} component) or tensor type (t_{20} or t_{22} component). The t_{20} tensor polarization (or "alignment") plays a special role since it is directly related to the static quadrupole moment of the deuteron. Moreover the T_{20} tensor analyzing power is a combination of squares of amplitudes as is the cross-section and it is then less sensitive to minute effects which make the vector analyzing power difficult to calculate reliably.

A measurement of the cross-section and T_{20} analyzing power has been carried out in the pion production reaction dp => ^{3}He $\pi°$, where the recoil ^{3}He ion is detected in the SPES 4 spectrometer at 0°. By changing the spectrometer field one can investigate the center of mass angles 0° and 180°. The T_{20} data are shown on fig. 12. Calculations with a meson-dominated model (29) indicate that the 2 nucleon diagrams of fig. 13 completely fail to reproduce the energy dependence of the cross-section. Addition of N* to Δ-excitation improves the fit at 0° but makes no difference at 180°. The secondary maximum above 1500 MeV can only be reproduced by taking 3-nucleon diagrams, in which mesons are exchanged between all 3 nucleons (2Δ -excitation) are taken into account (fig. 14) (30).

The same pd => ^{3}He X reaction has been used to produce, at threshold, mesons heavier than the pion (31). The corresponding cross-

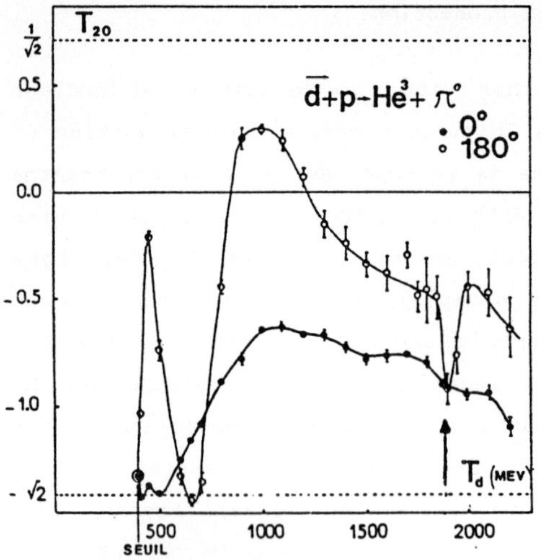

Fig. 12 - Tensor analyzing power T_{20} in the p(d, ^{3}He)π° reaction at center of mass angle 0° and 180°. The eta-production threshold is indicated by an arrow.

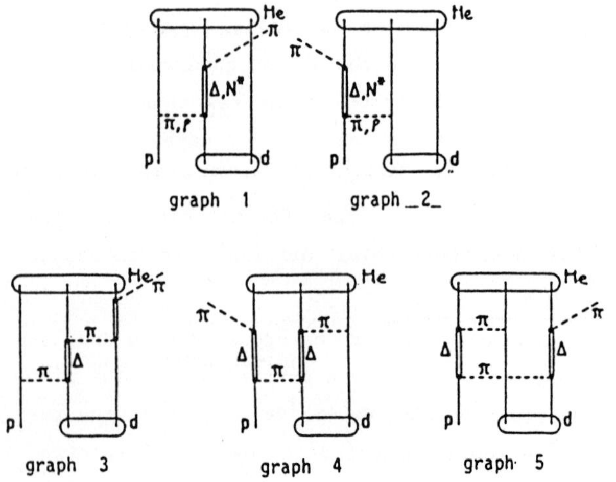

Fig. 13 - Mechanisms for the p(d, ^{3}He)π° reaction. 1-2 : two-nucleon diagrams, 3-5 : three-nucleon diagrams.

section as a function of incident proton energy up to 1.8 GeV, is shown in fig. 15. One sees clearly two prominent peaks (notice the log scale) corresponding to the π° (M=135.0 MeV, Γ=7.6 MeV) and the η (M=548.8 MeV, Γ=1.05 keV) mesons. One striking feature is the fact that η-production at threshold is more intense than π^0-production making this reaction most interesting to produce large quantities of clean eta-meson at rest (the background which comes mainly from 2π production is below 1 per cent).

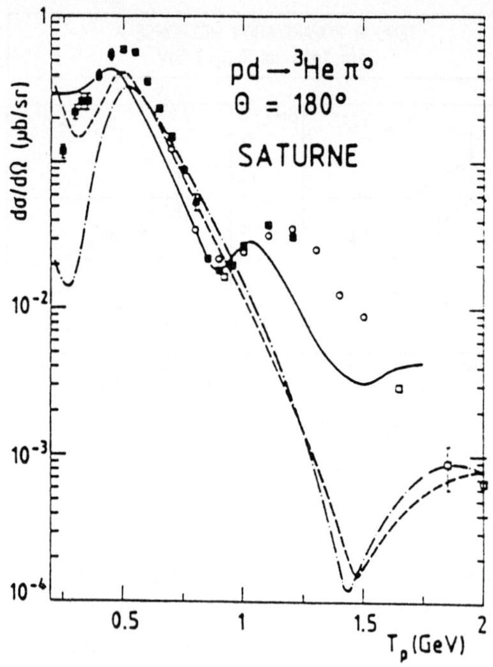

Fig. 14 - Excitation function of the pd→^{3}Heπ° reaction. The solid line includes the contribution of three-body mechanism dot.

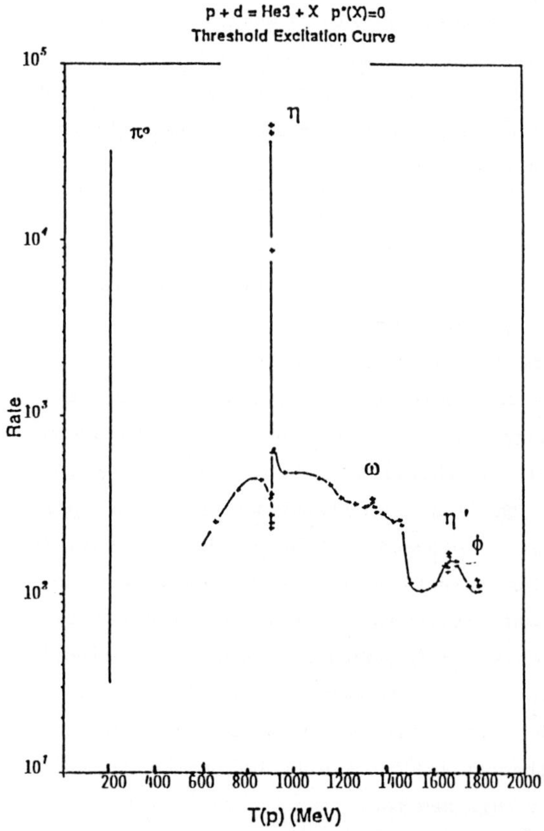

Fig. 15 : Threshold excitation of the pd→^{3}HeX. Event rates correspond approximately to particles per second with realistic running conditions (5 x 10^{11} protons per second, 5 cm LD_2 target).

Besides Π and η-production, the ω-meson (M=783 MeV, Γ=8,5 MeV) is clearly excited on top of the ρ (M=770 MeV, Γ=153 MeV) and a 3-Π background ; the η'-meson (M=958 MeV, Γ =0,2 MeV) is signalled by a cusp effect but no production peak is observed. Finally the Ø (M=1020 MeV, =4.4 MeV) is showing at the high energy end of the plot. The fact that the η' is barely seen compared to the η is reminiscent of what is observed in $\Pi p \rightarrow \eta$ n at higher energies (32) where the η'-production cross-section is only a few per-cent of the η'-production. This is attributed to the OZI quark line rule (33) which states that disconnected quark diagrams are forbidden. At Saturne energies the η' cross-section seems to be even more hindered compared to higher energies. This might be due to a kinematically enhanced meson-production of mass around 500-600 MeV. To test this hypothesis it would be interesting to study other nuclear reactions with different form-factors.

It seems interesting enough to remark that the same explanation based on quark-dynamics which suppresses the η' compared to the η at high energies, seems also to hold qualitatively at threshold.

Eta-production rates are of the order of 6 x 10^4 per second (or 5 x 10^9 per day) for realistic beam conditions (5 x 10^{11} protons per second and 5 cm LD_2 target). This makes the pd => ^{3}He η reaction the most powerful, known to date, to produce beams of η-mesons and opens the way for a study of η rare decays. The η-meson decays essentially into 2 (38,9 %) or 3 pions ($3\Pi^0$=31.9% ; Π^+,Π^-,Π^0 =23.7%). Amongst its rare decays only the $\mu^+\mu^-\gamma$ (BR = 3.1 +/- 0.4 x 10^{-4}) and $\mu^+ \mu^-$ channels (6.5 +/- 2.1 x 10^{-6}) are quoted with definite values (34). The latter value is slightly below the unitary limit of 4 x 10^{-6} (as quoted by Landsberg (35)) or 1.1 x 10^{-6} (as quoted by Amettler et al (36)). It does not agree with the standard-model prediction (BR =1.3 x 10^{-5}) by 3 standard deviations but without taking any uncertainty on the theory into account.

The first experiment which might be feasible with the Saturne η-beam could be to remeasure $\eta \rightarrow \mu^+ \mu^-$. Taking into account the rate of the Lepton-G experiment (34) of one tagged η per second and comparing with the 6 x 10^4 per second available at Saturne the same statistics could be obtained in 5 minutes running time but in a less pleasant environment produced by hadronic reactions. The next step could be a measurement of $\eta \Rightarrow e^+ e^-$ whose known BR limit is very high : BR < 3 x 10^{-4} against an unitary limit of 4.5 x 10^{-9} (38). The physics motivation would be a check of the Standard Model and a search for leptoquark (37). This experiment might be difficult owing to the $e^+ e^-$ background coming from ρ-decay (the ρ-meson has a large width overlapping the η region) (39).

Finally a very interesting experiment is $\eta \rightarrow \mu^+ \mu^-$ where the polarization of one muon is measured. This process is CP-violating and polarization of a few per cent are predicted (40) depending on the neutron dipole moment.

5. Conclusion

A series of experiments done at Saturne in the last 2-3 years has brought a wealth of data which should keep theorists at work for years to come. Some quite unexpected results like the intense production of η-mesons may open up now fields putting Saturne at the frontier of nuclear and particle physics. Another program not discussed in this talk, aims at developing a large detector for strangeness physics induced by high energy protons (41).

References

1 J. Bistricky, F. Lehar and P. Winternitz : J. Physique 39 (1978) 1

2 F. Perrot et al. : 3rd Int. Symposium on Pion-Nucleon and Nucleon-Nucleon Physics, Gatchina (USSR), April 17-22, 1989

3 J. A. Parry et al. : Phys. Rev. D8 (1973) 45

4 I. Auer et al. : Phys. Rev. D34 (1986) 2586

5 R. Bertini et al. : Phys. Lett. 162B (1985) 77 and 203B (1988) 18

6 P. Lafrance and E. Lomon : Phys. Rev. **D34** (1986) 1341

7 P. Gonzales, P. La France and E. Lomon : Phys. Rev. **D35** (1987) 2142

8 E. Lomon, private communication, May 1989

9 F. Lehar et al. : to be published

10 G. Glass et al. : Phys. Rev. **D15** (1977) 36

11 B. E. Bonner et al. : Phys. Rev. **C18** (1978) 1418

12 C. Ellegaard et al. : Phys. Rev. Lett. **50** (1983) 1745 ; Phys. Lett. **154B** (1985) 110

13 D. Contardo et al. : Phys. Lett. **168B** (1986) 331

14 I. Berquist et al. : Nucl. Phys. **A469** (1987) 648

15 C. Ellegaard et al. : Phys. Rev. Lett. **59** (1987) 974

16 D. Bachelier et al. : Phys. Lett. **B172** (1986) 23

17 G. Chanfray and M. Ericson : Phys. Lett. **141B** (1984) 163

18 C. Gaarde : Nucl. Phys. **A478** (1988) 475c

19 H. Ebensen and T. S. H. Lee : Phys. Rev. **C12** (1985) 1966

20 V. F. Dimitriev and T. Suzuki : Nucl. Phys. **A438** (1985) 697

21 T. Udagawa and F. Osterfeld : Nucl. Phys. **A482** (1988) 391c

22 E. Oset, E. Shiino and H. Toki : Phys. Lett. **B224** (1989) 249

23 M. Roy-Stephan : Nucl. Phys. **A482** (1988)373c

24 C. Guet, M. Soyeur, J. Bowlin and G. E. Brown : Nucl. Phys. **A494** (1989) 558

25 D. V. Bugg and C. Wilkin : Nucl. Phys. **A467** (1987) 575

26 C. Wilkin and D. V. Bugg : Phys. Lett. **154B** (1985) 243

27 S. Mundigl and W. Weise : Phys. Rev. **C39** (1989) 710

28 G. Brown and W. Weise : Phys. Rep. **C22** (1975) 279

29 C. Kerboul et al. : Phys. Lett. **B181** (1986) 28

30 J.-M. Laget and J.-F. Lecolley : Phys. Rev. Lett **61**(1988), 2069

31 J. Berger et al. : Phys. Rev. Lett. **61** (1988) 919

32 W. D. Apel et al : Phys. Lett. **83B** (1979) 198 - N. R. Stanton et al : Phys. Lett. **92B** (1980) 353

33 S. Okubo : Phys. Lett. **5** (1963) 165 ;

34 R. I. Dzhelyadin et al. : Phys. Lett. **97B** (1980) 471

35 L. G. Landsberg : Phys. Rep. **128** (1985) 301

36 L. Amettler, A. Bramon and E. Masso : Phys. Rev. **D30** (1984) 251

37 B. Nefkens : contribution to 5e Journées d'Etudes de Saturne (JES5), Piriac, May 15-20, 1989

38 A. Soni : In Proc. of Physics with Light Mesons
W. Gibbs and B. M. K. Nefkens eds : LANL Conference Proceedings **LA-11184c** (1987) p 55.

39 B. Mayer : private communication

40 C. Q. Geng and J. N. Ng : Phys. Rev. Lett. **62** (1989) 2646

41 R. Bertini : spokesman for letter of intent n0 213, LNS (1989)
R. Frascaria and R. Siebert : spokesmen for letter of intent n° 219, LNS (1989).

42 F. Lehar, C. Lchanoine-Leluc, J. Bystricky : J. Physique **48** (1987) 1273.

Z. Phys. C – Particles and Fields 46, S133–147 (1990)

Zeitschrift für Physik C Particles and Fields

High-$P_\perp^2$ spin dependent measurements*

A. D. Krisch

Randall Laboratory of Physics, The University of Michigan, Ann Arbor, MI 48109, USA

Abstract

The reason for studying the violent collisions of protons is to obtain information about the inner structure of the proton. To study the inner structure of the proton and to see what its constituents look like, one must observe large values of $P_\perp$ which is the canonically conjugate variable to the impact parameter.

Now recall that, when you study proton-proton elastic scattering with an unpolarized beam, you can only measure one quantity, the spin average differential cross section. However, if either the beam or the target is polarized then you can measure one-spin quantities such as A, the analyzing power, which is sometimes incorrectly called the polarization. If both the beam and the target are polarized then you can measure two-spin quantities such as A_{nn}, the spin-spin correlation parameter. It is useful to now define the polarization:

$$P = \frac{N(\uparrow) - N(\downarrow)}{N(\uparrow) + N(\downarrow)}$$

This definition applies to both the beam polarization and the target polarization.

1. Proton Proton Elastic Scattering

I will now review unpolarized proton-proton elastic scattering. Figure 1 is a slightly dated graph [1] summarizing all then-existing pp elastic scattering data which was plotted against my favorite scaling variable $\rho_\perp^2 = \beta^2\, P_\perp^2\, \sigma_{TOT}(s)/38.3$ which removes most or the energy dependence of "shrinkage of the diffraction peak". This graph shows several things. First it shows that the

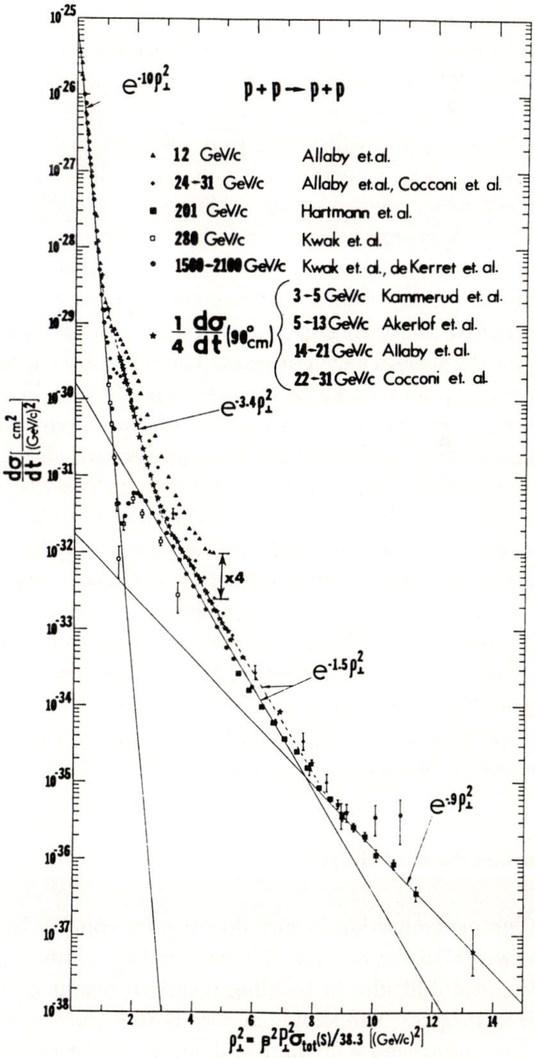

Figure 1: Unpolarized proton-proton elastic scattering plotted against scaled $P_\perp^2$ variable.

*Supported by a research grant from the U.S. Department of Energy

pp elastic cross section has been measured from about $10^{-25}\mathrm{cm}^2/(\mathrm{GeV/c})^2$ down to about $10^{-37}\ \mathrm{cm}^2/(\mathrm{GeV/c})^2$, a range of 12 orders of magnitude. I believe that this graph has more decades than any other process that has been studied in high energy physics.

You see from this graph that there are clearly several regions with different slopes and rather sharp "breaks" between the regions. First, you can easily see the small–$P_\perp^2$ "diffraction peak", which drops rapidly as $e^{-10P_\perp^2}$. Fourier transforming this slope of 10 gives the size that a proton sees when it scatters from another proton; this size is about 1 fermi. You can also see the large–$P_\perp^2$ hard scattering region extending from $P_\perp^2$ of about 3 to 8 $(\mathrm{GeV/c})^2$. This scattering between the constituents of the proton is characterized by a slope of about $e^{-1.5P_\perp^2}$; some people may remember that around 1966 when we first found this at the ZGS, I thought that this was the third region of proton-proton scattering in what was called the "onion model". However, as you can clearly see, when people did higher energy experiments at Fermilab and at the ISR, this region disappeared completely. This medium–$P_\perp^2$ region is thus a low energy effect probably due to the direct scattering of the protons as opposed to diffractive scattering.

The small–$P_\perp^2$ "diffraction peak" seemed to be totally independent of energy from about 3 GeV up to 2000 GeV when plotted against this scaling variable in figure 1; such energy independence is a characteristic of diffractive scattering. The large–$P_\perp^2$ region is also independent of energy, which suggests that this may also be diffractive scattering, but now the diffraction scattering of the proton's constituents. There is a hint that there may be another break at very large–$P_\perp^2$, but here the errors are quite large. Note that the last point on this figure is the largest $P_\perp^2$exclusive process which has ever been observed; it has a $P_\perp^2$ of about 15 $(\mathrm{GeV/c})^2$ and comes from my AGS thesis experiment of 1963. I stress this point because with the proposed kaon facility, one should have several orders of magnitude more intensity than was then available; thus it might be possible to extend these fundamental measurements to larger $P_\perp^2$. It would be most interesting to get high precision $p + p \to p + p$ data at $P_\perp^2 = 20\ (\mathrm{GeV/c})^2$. This concludes my overview of unpolarized proton-proton elastic scattering in the context of a geometrical model.

2. Importance of Spin

The next question is why do we want spin? Why should we go to the considerable intellectual, technical, and financial difficulty of building polarized beams and polarized targets? The simple answer is that pure spin cross sections are more fundamental, although, for many years most people thought that spin would not be important. But instead spin has turned out to be quite important. I think that people were driven to accept the importance of spin quite properly by experimental observation. Experimenters found large and quite unexpected spin effects beginning in the middle 1970's; much of this began with the start up of the ZGS polarized proton beam, but there were also many other important results. Some experiments that I consider especially important were:

1. $\Delta\sigma_T$ and $\Delta\sigma_L$: These spin dependent p-p total cross section measurements have resulted in an industry known as the Dibaryon Phenomenon [2]. As a fellow involved in the first measurement [3] of $\Delta\sigma_T$ around 1974, I must say that I still don't have a clear opinion of whether or not dibaryons exist. However, I certainly believe that there are huge spin effects in a region where people in 1974 thought there would be none.
2. Inclusive Λ polarization: This spin experiment is especially interesting because it requires neither a polarized beam or a polarized target. People find that when you produce Λ's inclusively at large–$P_\perp^2$ there is a large polarization [4]; but there is not much polarization at small–$P_\perp^2$. It is remarkable that this polarization seems totally independent of energy from KEK energies [5] of 12 GeV up to the ISR equivalent [6] of perhaps 2000 GeV. In my opinion, this total energy independence puts a lot of pressure on the belief that spin effects must disappear at higher energy.
3. High – $P_\perp^2$ $p_\uparrow + p_\uparrow \to p + p$: Our group using a polarized proton target with the ZGS polarized proton beam found large and unexpected spin-spin effects [7] in violent pp elastic collisions at large $P_\perp^2$. I will discuss these later.
4. Parity violation in σ_{TOT}: Groups at SIN and LAMPF found at energies of 45 and 800 MeV a parity violation [8] of about 1 in 10^7 which is consistent with the weak contribution to proton-proton scattering. However, a group containing many of the same LAMPF people did the same experiment[9] at the ZGS at 6 GeV/c; they found a violation of $(+2.65 \pm 0.60 \pm 0.35)\ 10^{-6}$. This 6 GeV/c result is difficult to reconcile with our understanding of the size of such weak interaction effects.
5. High – $P_\perp^2$ $p + p_\uparrow \to p + p$: Our group did a low priority measurement [10] of A in a region where perturbative Quantum Chromodynamics said that A should be 0. Instead A was found to be quite large. I will discuss this in some detail later.
6. $e_\uparrow + p_\uparrow \to e$ + anything: This well known experiment [11] at SLAC scattered a polarized electron beam roma polarized proton target. The experiment made an important contribution to the electroweak theory.

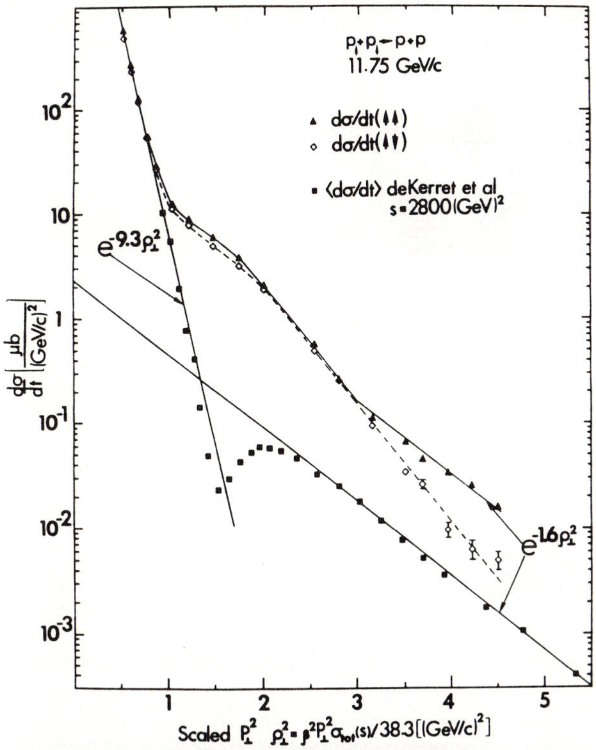

Figure 2: Proton-proton elastic scattering cross-sections plotted against the scaled $P_\perp^2$ variable.

7. p + p↑ → p + p at 45 GeV: This Serpukhov polarized target experiment [12] played an important role in showing that Regge theories could not adequately explain high energy scattering.
8. New EMC experiment: This experiment [13] has recently obtained a surprising spin result. We will certainly hear more about his later; it is clearly an important and controversial result.

Now I will turn to proton-proton elastic scattering with both protons polarized. Figure 2 is a graph of some 12 Gev/c data that we obtained at the ZGS [7]. The spin-parallel proton scatterings shown as triangles. The spin-anti-parallel data is shown as diamonds. For comparison we show as squares some ISR data at an equivalent momentum of about 2800 GeV/c. As you can see clearly, in the diffraction peak all 3 sets of data fall on top of each other; this shows that in small–$P_\perp^2$ diffraction scattering the proton doesn't much care what energy it is at or what spin state it is in. In the diffraction peak you just have 2 protons looking at each other and they each see an object with a size of about 1 fermi, much as Professor Hofstadter saw when he scattered electrons from protons and measured their charge distribution. Next notice the medium–$P_\perp^2$ region which still exists at 12 GeV/c but has clearly disappeared at high energy. Finally, consider the large–$P_\perp^2$ hard scattering region whose $e^{-1.6P_\perp^2}$ slope corresponds to the scattering of objects of about 1/3 of the fermi in size. The rather remarkable new result from the polarization experiment is that you see exactly the same $e^{-1.6P_\perp^2}$ slope, but only when the spins are parallel. When the spins are anti-parallel the cross section seems to just keep going down. It is as though the constituents of the protons have violent collisions only when they are spinning in the same direction; they just sort of pass through each other when they are spinning in opposite directions. This was a totally unexpected result and one that is now 11 years old, but it still has not been satisfactorily explained.

What can one learn about the constituents of the proton by scattering protons from each other? I believe that to really understand protons we need physicists studying both electron-proton scattering and proton--proton scattering. As many of you know, some strong evidence for the existence of the constituents in the proton came from the 1969 deep inelastic e-p scattering experiments at SLAC [14]. They found that by scattering an electron off a proton one could see that there were some point-like constituents in the proton. On the other hand, in 1966 we had studied 90°_{cm} proton-proton elastic scattering with an unpolarized beam at the Argonne ZGS [15]. We saw the sharp break at large–$P_\perp^2$ with a slope of about $e^{-1.6P_\perp^2}$ corresponding to the scattering from objects about 1/3 of a fermi in size. What is happening here? Are the constituents point-like or do they have some size of about 1/3 of a fermi?

A possible understanding of this paradox can be obtained from figure 3 by noticing that when an electron scatters off a proton, the constituents behave as if they are almost point-like. We certainly believe that the electron is point like at least at the level of perhaps 1/100 of a fermi. Let us next assume that each quark has a size of about 1/20 of a fermi as shown below, where two quarks are colliding with each other. Then you clearly would see a very small size for the constituents in e-p scattering. Next assume that when two quarks collide they have some strong interaction which has a range of about 1/4 of a fermi. Then when you scatter two quarks off each other by colliding 2 protons the characteristic range of the interaction is about 1/3 of a fermi. This is obviously an oversimplified picture; on the other hand it shows clearly that you must study both electron-proton scattering and proton-proton scattering if you want to understand the nature of the proton and its constituents.

3. AGS Polarized Beam

If one is going to have polarization capability at KAON, one must do a lot of work. It is important to understand that polarized proton acceleration is a non-trivial job. In fact, it took about 7 years to get polarization capability at the AGS starting with the Workshop in

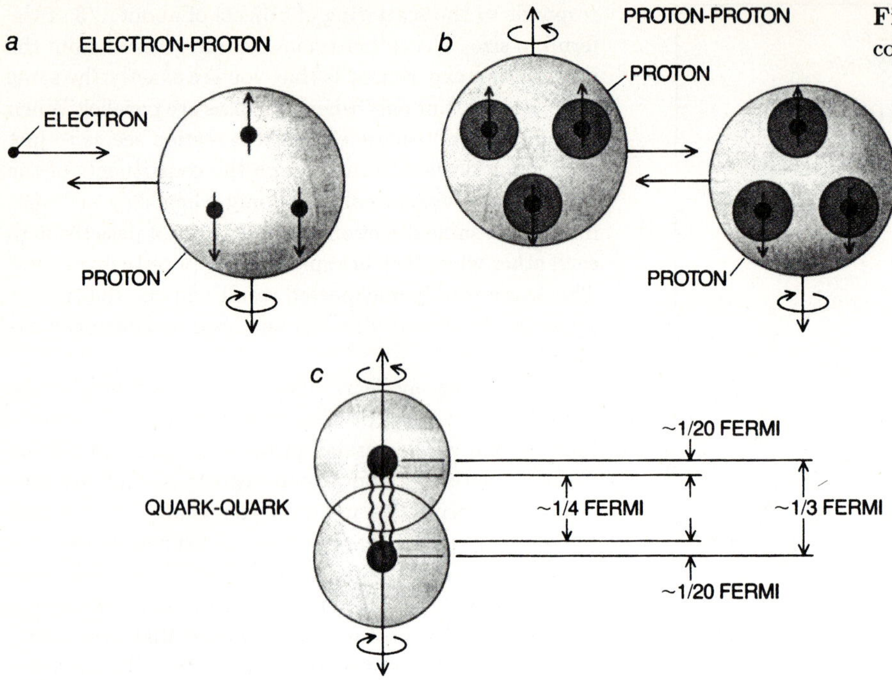

Figure 3: Quarks in e-p and in p-p collisions.

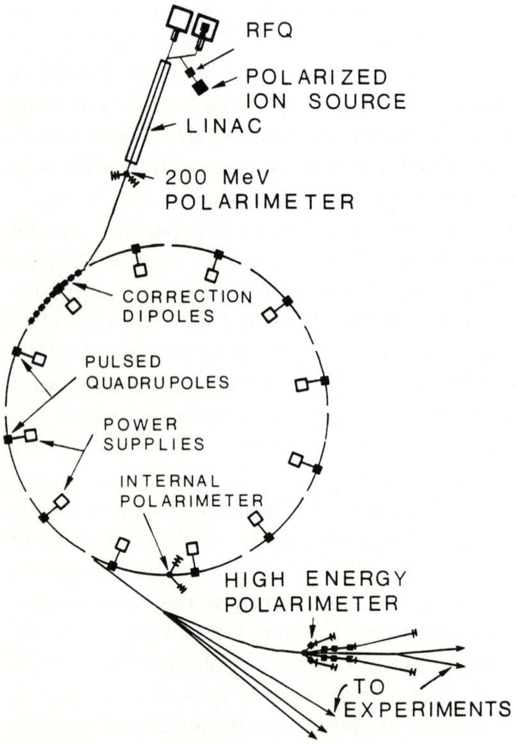

Figure 4: AGS layout for the operation of the polarized proton beam.

Figure 5: The AGS polarized H^- ion source.

Ann Arbor [16] in 1977, a Summer Study at Brookhaven [17] in 1978, and then funding approval in 1979; the polarized beam first operated at 16.5 GeV/c in 1984.

To allow the acceleration of polarized protons many modifications had to be made to the AGS [18], as shown in figure 4. These included the installation of: a polarized ion source, a new RFQ, a low energy beam transport line, 4 polarimeters, special power supplies for 95 tiny correction dipoles in the ring, 12 fast pulsed quadrupoles and their massive power supplies of 22 MegaWatts each, and much control computer hardware and software. To obtain a good polarized beam at KAON you should get a first class polarized ion source with as high an intensity as possible. Figure 5 shows the polarized H^- source at the AGS. The original atomic beam stage came from Argonne. It was transferred, along with some of us, from the ZGS to the AGS. The AGS staff added the cesium charge exchange unit; this enabled us to have polarized H^- ions which allowed fairly high intensity by using multi-turn injection at the AGS.

4. Polarimeters

A symbolic polarimeter is shown in figure 6. The beam comes in from the left and scatters from the target. There are identical detectors on the left and the right which allow you to obtain the quantity $A_{Measured}$ which is defined to be

$$A_M = \frac{L-R}{L+R}$$

The beam polarization P_B is obtained from A_M using the relation $P_B = A_M/A$, where A is the analyzing power which must be measured using a polarized target. One way to understand A is that it is what the measured left-right asymmetry would be if the beam were 100% polarized.

Figure 6: Generalized polarimeter.

The 200 MeV polarimeter at the end of the 200 MeV linac had a thin carbon target and two identical sets of scintillation counters on both sides; it is shown in figure 7. The polarimeter looks at p-carbon scattering at 200 MeV. It had an effective analyzing power of about 60%. The target was about the same size as a piece of pencil lead and it continuously intercepted 1% of the linac beam. The polarimeter had about 500 counts per AGS pulse so that you could get a few percent measurement of the beam polarization in about a minute. It only reduced the beam by 1% because that is all that the carbon target absorbed. This polarimeter was built by our colleagues at Rice, Jay Roberts and Gerry Phillips, and it was calibrated using the IUCF 200 MeV polarized beam. We at Michigan built a 760 keV polarimeter, an internal polarimeter placed inside the AGS ring, and a high energy polarimeter placed near our experiment; I will not describe them in detail [18].

Figure 7: The 200 MeV polarimeter.

5. Depolarizing Resonances

The main problem of accelerating a polarized proton beam in a high energy synchrotron is something called depolarizing resonances. A resonance occurs whenever the precessional frequency of spinning proton in a magnetic field somehow gets in phase with the frequency at which it sees horizontal magnetic fields as it goes around the accelerator. There are two types of these resonances:

<u>Intrinsic depolarizing resonances</u>: are caused by the quadrupole focusing fields which give the vertical beta oscillations.
<u>Imperfection depolarizing resonances</u>: are caused by the various imperfection fields in any manmade device.

Ernest Courant and his then student, Ron Ruth, calculated [19] the expected strengths of the depolarizing resonances at the AGS as a function of beam momentum as shown in the figure 8. Rather than explain the mysterious parameter, ϵ, let me just note that the lower dashed line corresponds to 1% depolarization, while the upper dashed line corresponds to 99% spin-flip. As you can see, there is a relatively small number of intrinsic resonances, but they appear to be quite strong. There are many imperfection resonances with one occurring about every 600 MeV, but they were calculated to be rather weak. Our general feeling was that if the depolarization was much below 1% you could ignore it. If it was much above 99% spin-flip there wasn't much you could do about it. If you were somewhere in between you might try to deal with it somehow. We thought that we could ignore most of these imperfection resonances. Unfortunately, the imperfections in the AGS were about an order of a magnitude worse than had been believed. The

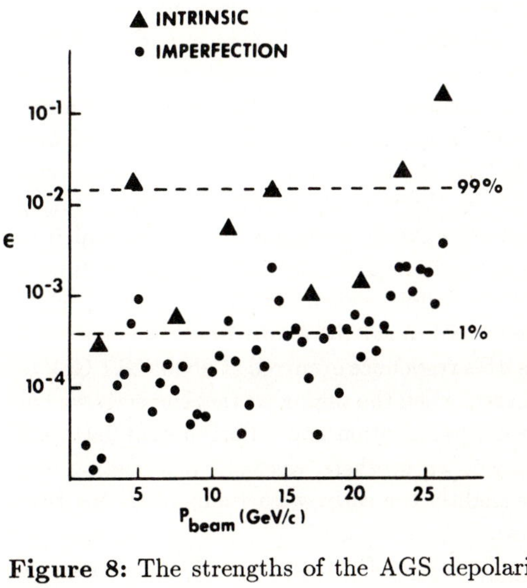

Figure 8: The strengths of the AGS depolarizing resonances.

Figure 9: A pulsed quadrupole.

imperfection resonances were all strong and we had to deal with every single one of them. Spin is a much more sensitive probe of imperfections and problems than any optical instruments.

I now have some advice for the KAON accelerator physicists; they should be cautious. Accelerating polarized protons at the AGS turned out to be a much bigger job than we had anticipated. Fortunately, we were cautious and built better hardware than we thought we would need. Some of you may know that KEK has tried to accelerate a polarized beam to 12 GeV and did not have enough financial support to allow caution; so far they have only been able to get the polarized protons to about 5 GeV with about 25% polarization.

Figure 9 shows one of the pulsed quadrupole magnets. We built a dozen of these at Michigan. They had to be made of a material called ferrite, so that they could be turned on in about 1.6 microseconds. They required hyperbolic shaped pole tips and each quadrupole needed a pulsed power supply of about 15000 volts at about 1500 amps; that is about 22 MegaWatts. There were a dozen of these, thus the total power is several hundred MegaWatts, which might drain Long Island. Fortunately, they were only on for a few microseconds so that the average power was fairly low. These quadrupoles were installed around the AGS ring and allowed one to jump through the intrinsic resonances. Figure 10 shows a timing curve for jumping an intrinsic depolarizing resonance. This is the $G\gamma = 48 - \nu_y$ intrinsic resonance, which occurs near 20 GeV/c. We were measuring the beam polarization at 22 GeV/c, while sweeping the time at which we turned on the pulsed quadrupoles. We measured time in some mysterious units called Gaussclock counts, which you divide by 200 to get the approximate momentum in GeV; thus, this resonance occurred at about 20.1 GeV/c. As you can see, when the timing was set correctly we had a rather good polarization and a fairly broad flattop. If the timing was wrong there was some polarization loss. If we were slightly too early, we actually made the resonance worse.

The other type of depolarizing resonance that we had to deal with was the so called imperfection reso-

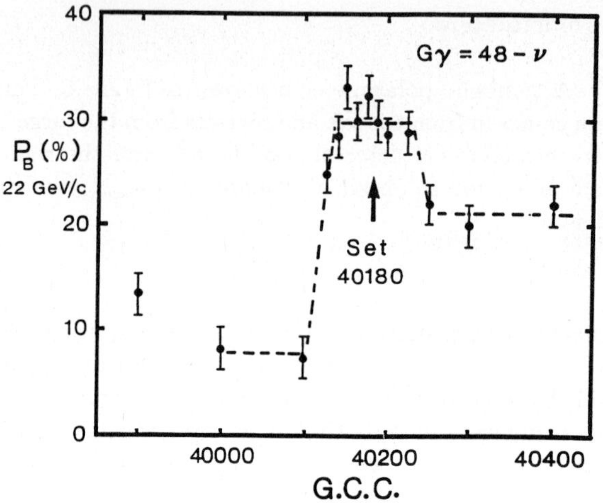

Figure 10: The $G\gamma = 48 - \nu$ intrinsic resonance timing curve.

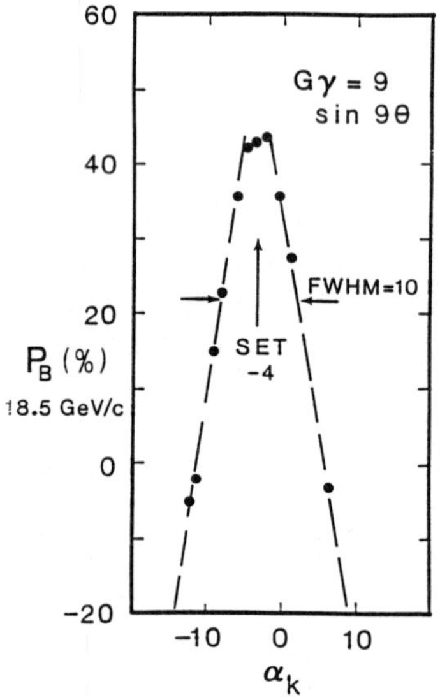

Figure 11: The $G\gamma = 9$ imperfection resonance correction curve.

nance. Figure 11 shows one of the strongest imperfection resonances, $G\gamma = 9$; as you can see, when the correction was turned off ($\alpha_k = 0$) we saw a polarization of about 26%. When the resonance was fully corrected we observed about 43% polarization. When we corrected in the wrong direction the polarization started going negative. In fact, in other curves we have demonstrated total spin-flip. If a resonance is very strong and is an imperfection resonance, then you can flip the spin rather than correct the resonance; there is then no depolarization at all. Unfortunately, for the intrinsic resonances the flipping doesn't seem to work very well. Therefore, you must jump through all the intrinsic resonances and can either correct or spin-flip the imperfection resonances. In the recent January 1988 run of the AGS polarized beam, several of the strongest intrinsic resonances were spin

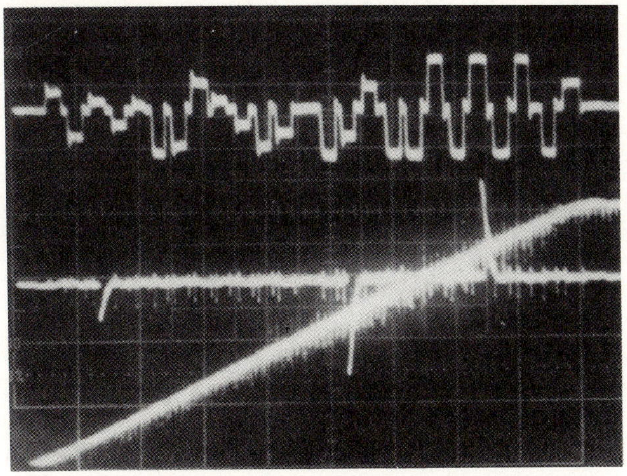

Figure 12: Scope trace of acceleration of polarized protons to 16.5 GeV/c at the AGS.

flipped rather than corrected. There are some beam orbit problems associated with spin-flipping, but you often get better polarization.

This concludes my overview of the work that had to be done to accelerate a polarized proton beam at the AGS. A scope trace of acceleration to 16.5 GeV is shown in figure 12. on the lower trace, you can see the 3 quadrupole pulses for the 3 intrinsic resonances which had to be jumped: $G\gamma = 0 + \nu, 12 + \nu, 36 - \nu$. On this scale the superfast rising edge of each quadrupole pulse looks like a vertical line. The middle sloping line is the AGS ring magnetic field going from zero up to full energy where it goes into flattop. On the upper trace, you can see the field in one of the 95 dipoles that collectively make pulses with the correct wave of horizontal field for correcting each imperfection resonance. The first pulse corresponds to $G\gamma = 6$; one had to program the 95 magnets into a harmonic wave, which made six oscillations in one turn around the ring; then we had to experimentally adjust the amplitude of that wave to correct the $G\gamma = 6$ resonance. This first pulse shows the amplitude of the 6th harmonic wave, which is the wave of the horizontal magnitude field with 6 oscillations. After that, we went on to $G\gamma = 7$ by triggering the program in the computer, which gave 7 oscillations. For each resonance we had to adjust the amplitude and the phase of the appropriate wave. Then we went on to $G\gamma = 8$ and so on. Using this painful but successful process, it took about 7 weeks to get the AGS polarized proton beam fully working [18].

6. Polarized Proton Target

To do spin-spin experiments, you also need a polarized proton target which is shown in figure 13. A polarized proton target using Dynamic Nuclear Polarization requires a high magnetic field of about 2.5 T and a very cold temperature of perhaps 1/2° K; the field and temperature together highly polarize the electrons in some special beads. Then you must transfer the electron polarization in these beads to the nearby protons using 70 GHz microwaves. You must also have 107 MHz NMR system, which measures the protons' polarization. Notice that the ratio of 70 GHz to 107 MHz is exactly 660, which is the ratio of the electron's magnetic moment to the proton's. These systems, which extremely complex, now work rather well. The basic idea of Dynamic Nuclear Polarization was developed in the late 1950's by Professor Abragam of the College de France and Profes-

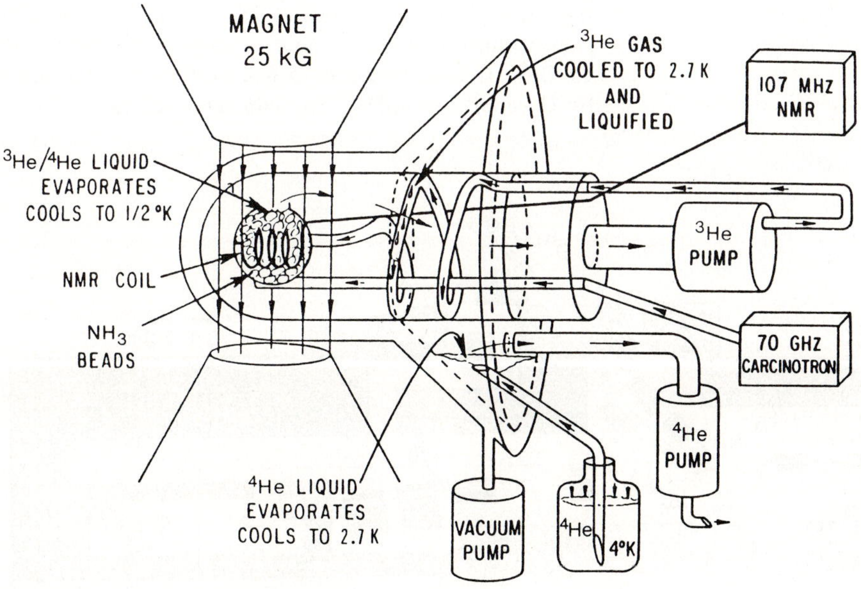

Figure 13: Michigan polarized proton target.

sor Jeffries of Berkeley, two distinguished atomic physicists. Many laboratories including TRIUMF now have successful polarized targets. Our group has specialized in high cooling power targets to allow the use of high intensity beams; so far we have been able to run with beams on our target of 10^{11} protons per 2.5 sec pulse. We are now working on a new 5 T/1° K polarized target which should allow considerably high beam intensities. This development [20] might be quite interesting to the people designing KAON.

7. Elastic Proton Proton Spin Experiments

The experimental layout that we have been using to study large–$P_\perp^2$ spin effects at the AGS is shown in figure 14. The beam comes from the left and is first scattered in the high energy polarimeter; this has a liquid hydrogen target and 2 double arm spectrometers which look at pp elastic scattering from the target. By determining the left-right asymmetry in pp elastic scattering one gets an absolute measurement of the beam polarization by using the pp elastic analyzing power, A, which has been measured by other groups; we have also measured A ourselves using our polarized target.

The polarized beam then strikes the polarized target and one can look at pp elastic scattering in all four pure initial spin states. We have a long string of spectrometer magnets to bend the forward particle which typically has a momentum of 20 or 30 GeV and a shorter arm for the few GeV recoil particle. We detect the protons using the F and B scintillation hodoscopes. We use scintillation hodoscopes because when our beam of intensity 10^{11} hits our rather thick target, then the inside of this tunnel glows with radiation. Wire chambers would not operate very well in this tunnel, which has a high intensity radiation environment.

8. Measurements of A_{nn}

The data from our most recent AGS run last January, which is shown in figure 15, was recently published [21]. At 18.5 GeV/c we measured both the analysing power, A, and the spin-spin correlation parameter A_{nn}, as a function of $P_\perp^2$. This is the first good precision data above ZGS energies on 2 spin effects.

Notice that in the medium–$P_\perp^2$ region there is a sharp dip in A near $P_\perp^2 = 3$. This is not a surprise because this dip had been seen twice before in unpolarized beam experiments, by a CERN group at 24 GeV [22] and by our group at 28 GeV [10]. But it is still a nice result, since our error bars are somewhat better than anyone else has achieved; this is because with both a polarized beam and a polarized target you can simultaneously measure the analyzing power in two independent ways. You then have an absolutely firm test of most systematic errors by using rotational invariance of space. As you can see we have rather tiny errors in A and we are pleased with that.

We also obtained some totally new data by measuring A_{nn}, the spin-spin correlation parameter at 18.5 GeV/c. This 2-spin parameter also has a dip, but this dip is near $P_\perp^2 = 2.3$ (GeV/c)^2 which seems rather surprising; in the lower energy data there was a dip near $P_\perp^2 = 1$ (GeV/c)^2 and possible near $P_\perp^2 = 3.5$ (GeV/c)^2 but there was only a broad shoulder near $P_\perp^2 = 2.3$ (GeV/c)^2. This new dip was certainly not expected. This was the first time that anyone was able to study these spin-spin effects above ZGS energies and spin appears to have brought another new surprise.

Another way to look at this new data is shown in figure 16, where we plot on the upper graph the ratio of the various spin cross-sections, such as $\sigma_{\uparrow\uparrow} \equiv (d\sigma/dt)_{\uparrow\uparrow}/(d\sigma/d\sigma)_O$. We also plot $\sigma_{\downarrow\downarrow}$ and the anti-parallel cross section ratio. There is clearly a great deal of structure in the spin-spin cross sections at 18.5 Gev/c. For comparison we have plotted the spin averaged cross section at 19 GeV; this precise data is from a CERN experiment by Cocconi's group[23]. You can see that the medium–

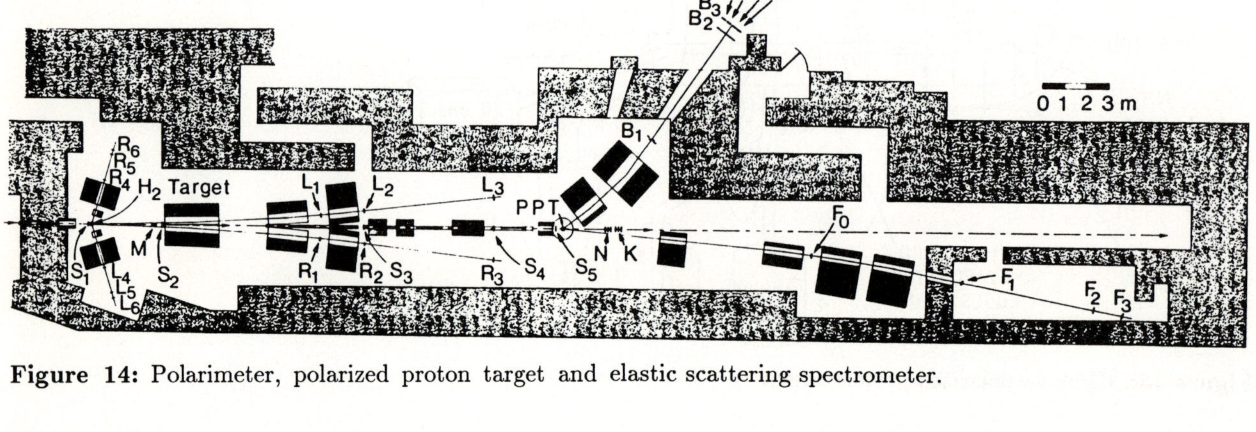

Figure 14: Polarimeter, polarized proton target and elastic scattering spectrometer.

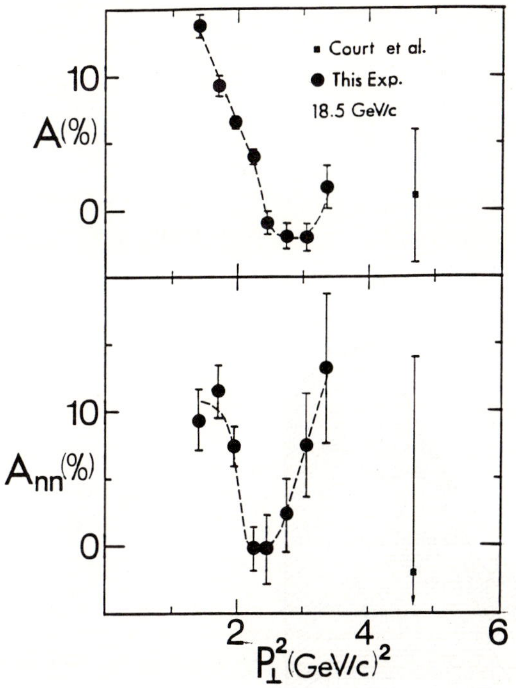

Figure 15: Plot of A and A_{nn} against $P_{\perp}^2$ for proton-proton elastic scattering at 18.5 GeV/c.

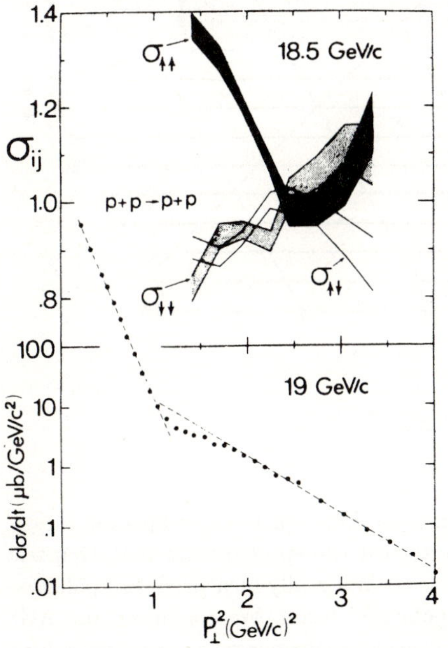

Figure 16: Pure spin p-p cross section ratios and the spin average p-p cross section near 18.5 GeV/c.

$P_{\perp}^2$ region of my old onion model [1] is already disappearing at 19 GeV. What do these two plots show? In the medium–$P_{\perp}^2$ region the spin average cross section is a smooth line with no real structure at all. However, in the same $P_{\perp}^2 = 2.3$ $(\text{GeV/c})^2$ region the pure spin state cross sections have an enormous amount of structure. At small–$P_{\perp}^2$ there is about 60% difference between $\sigma_{\uparrow\uparrow}$ and the other two cross sections. Next they all come together at medium–$P_{\perp}^2$ and then appear to be moving apart again at large–$P_{\perp}^2$. This clearly demonstrates that large and important strong interaction effects can be totally obscured if you have only unpolarized beams and unpolarized targets.

In figure 17 we have a compilation of A_{nn} plotted against both P_{lab} and against $P_{\perp}^2$ in a 3-dimensional plot. Notice the set of ZGS data where the angle was kept constant at 90° c.m. and the incident energy [24] was varied. When plotted against $P_{\perp}^2$ and compared with our earlier fixed energy 12 GeV data [2], which I showed in figure 2, the two sets of data fall on top of each other at large–$P_{\perp}^2$. We have not yet reached such large–$P_{\perp}^2$ at 18.5 GeV/c. We concentrated on medium–$P_{\perp}^2$ because we unfortunately had a run of only about 3 weeks; it would have taken longer to get good precision at large–$P_{\perp}^2$. The AGS data [21] is mostly at 18.5 GeV/c, but we have a few points at 13 and 16.5 GeV/c obtained in the 1984 [25] and 1986 runs [26].

As you can see in figure 17, there seems to be a broad shoulder at 12 GeV/c in the same $P_{\perp}^2$ region near 2.3 $(\text{GeV/c})^2$ where we see the dip at 18.5 GeV/c. At 12 GeV/c there is a dip at $P_{\perp}^2 = 1$ $(\text{GeV/c})^2$ and perhaps a dip near $P_{\perp}^2 = 3.5$ $(\text{GeV/c})^2$, but there appears to be no dip at medium–$P_{\perp}^2$. This new medium–$P_{\perp}^2$ dip at 18.5 GeV/c is quite surprising; we don't really understand what it means or why it is there.

I also have plotted on this graph the points that we plan to measure in our next polarized beam run, which we hope will be soon. We want to make many measurements at very small–$P_{\perp}^2$; this would let us see if there is still a dip near $P_{\perp}^2 = 1$ $(\text{GeV/c})^2$. It is possible that for some strange reason this small–$P_{\perp}^2$ dip has moved out to $P_{\perp}^2 = 2.3$ $(\text{GeV/c})^2$; we really don't know. We must move some of our magnets to make the small–$P_{\perp}^2$ measurements, but the rate is enormous at small–$P_{\perp}^2$, so we can easily measure all these points in 2 weeks. Then we want to measure the larger $P_{\perp}^2$ points shown in figure 17; these large–$P_{\perp}^2$ measurements will take more time because of the smaller cross section.

Further in the future, we are eagerly looking forward to the operation of the AGS booster which is wonderful for the polarized beam because it will give a factor of 20–25 increase in intensity. Then we will be able to start doing high precision 18.5 GeV/c measurements at very large $P_{\perp}^2$. Moreover we plan to extend our 90°_{cm} experiment from the ZGS up to these higher AGS energies. This may turn out to be an important experiment; somehow 90°_{cm} seems to be a special place for p-p scattering. We believe that we can reach about 18.5 GeV with fairly good precision; this assumes that the booster works as advertised and one gets a factor of about 2 increase in the polarized ion source intensity, but that seems more

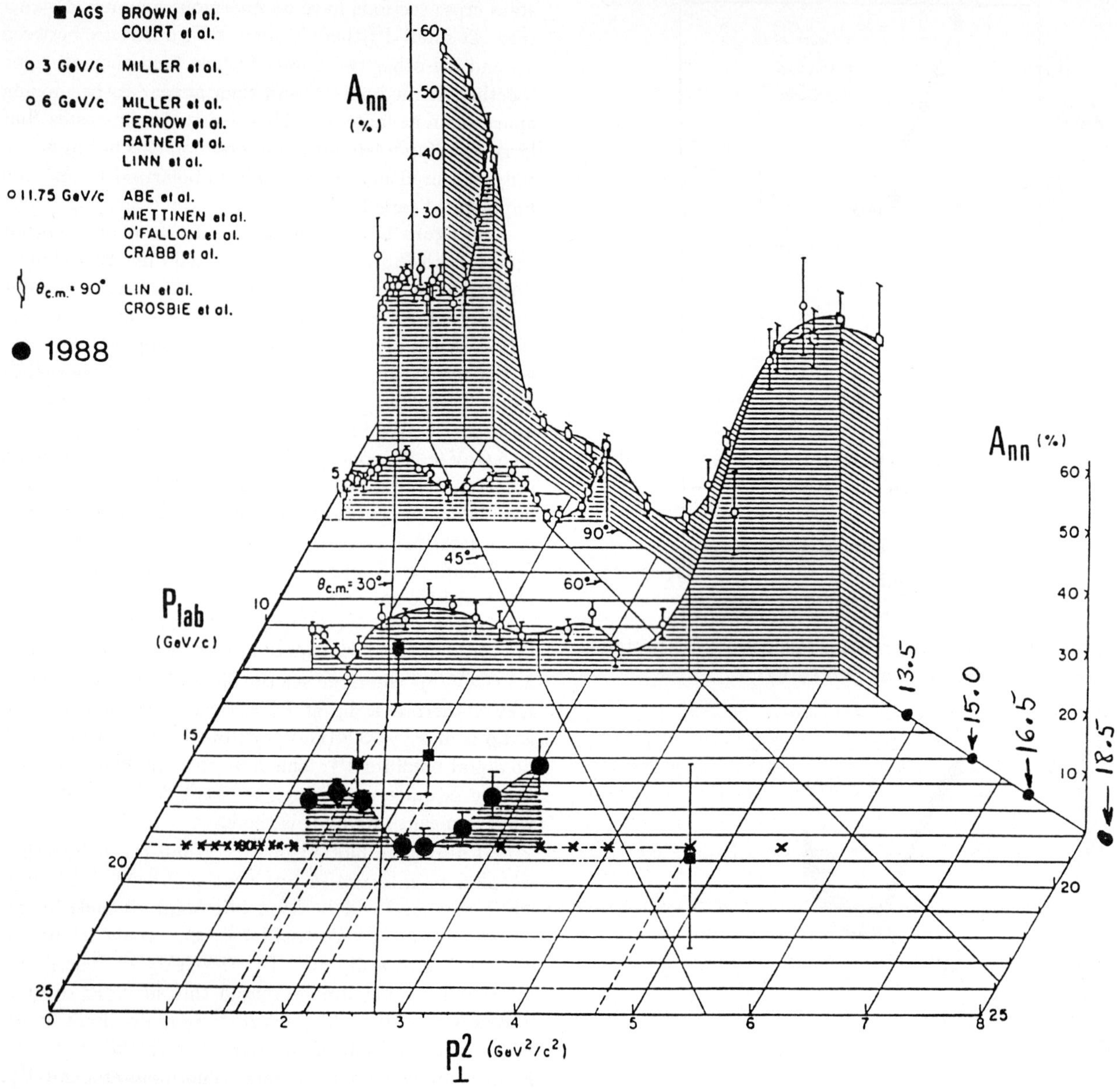

Figure 17: Compilation of A_{nn} data plotted in 3 dimensions against P_{lab} and $P_{\perp}^2$.

speculative. This 90°_{cm} experiment would work well with the booster plus this factor of 2. There are many different theoretical predictions about the 90°_{cm} behaviour of A_{nn} at higher energy and we are eager to see what A_{nn} really does at 90°_{cm}.

9. Measurements of A

We did a low priority measurement of the analyzing power, A, while we were getting ready to do these A_{nn} experiments. It took a number of years to get the polarized beam working and we decided that it would be sensible to test the apparatus by scattering the AGS unpolarized proton beam from our polarized proton target. This allowed us to test the spectrometer and even test the polarimeter to see if it really measured zero polarization for the unpolarized beam. We convinced the AGS program committee and management to approve us for a low priority experiment to measure the analyzing power in pp elastic scattering at 28 GeV, which is where the AGS normally runs. The experiment could easily run with low priority because we could only take at most 1/2% of the beam without burning our target away. The other reason for its low priority was that we were assured by most of our theoretical colleagues that, as we went to larger $P_{\perp}^2$, surely A would be zero because perturbative quantum chromodynamics had a rather firm prediction

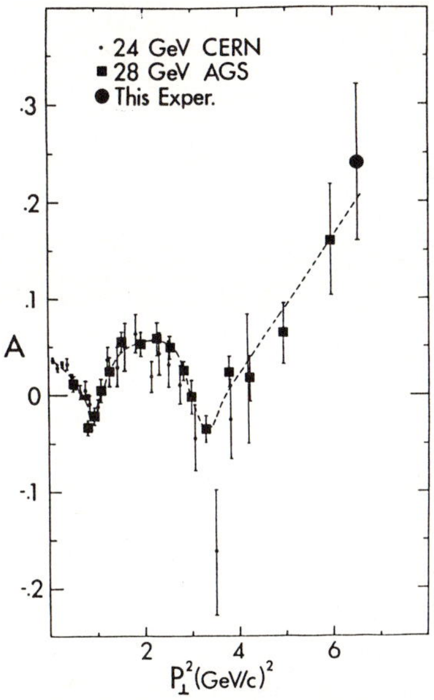

Figure 18: Plot of A against $P_\perp^2$.

that A should be zero. Moreover, this prediction should become more reliable at higher energy and larger $P_\perp^2$.

When we started doing this A experiment we obtained interesting but unexciting results [10] at small– and medium–$P_\perp^2$; we confirmed with better precision the 24 GeV CERN data shown in figure 18. Then we got out to $P_\perp^2$ of about 5 $(GeV/c)^2$ where there was a suggestion that A might be increasing, but it was not yet convincing. After a bit of excitement the then Director, Bob Palmer, overruled (for the first time in a decade) the AGS Program Committee which had voted to reject our extending these large–$P_\perp^2$ measurements. We then continued and found [10] that indeed A seems to be growing as shown in figure 18. Rumor has it that some theorists on the committee were so sure that A would be zero here that they didn't even want us to look. In any case, we did these experiment and found that A clearly seems to be growing. Many of our perturbative QCD friends now say, "But we meant that A would be zero at a much larger $P_\perp^2$ or a much higher energy".

Indeed one should try to pursue these measurements to larger $P_\perp^2$ and higher energy. This Fall or Winter, we hope to run at the AGS at a larger $P_\perp^2$ of about 7.2 $(GeV/c)^2$. This very large–$P_\perp^2$ region is where KAON could make a great contribution assuming that one can make a good enough polarized target. With the proposed KAON polarized beam one would have enough intensity to extend this A measurement to quite large–$P_\perp^2$. I think that this measurement should certainly be done; it would be most interesting if A either stays large or keeps growing.

10. Higher Energy Spin Experiments

Extending these spin experiments to higher energy also seems quite interesting and important. The prospects for higher energy at the AGS are clearly limited; there is no immediate plan to significantly upgrade the AGS energy. Therefore, we instead decided to look for other facilities. There is in fact a higher energy proton accelerator only a few hundred miles from Michigan, but for various reasons that didn't work out. Instead, we are going considerably further to a new accelerator called UNK being built in Protvino about 100 kilometers south of Moscow. This facility which is shown in figure 19 will come on in stages. Starting in late 1991, UNK will have a 400 to 600 GeV conventional ring. Starting around late 1993 or so, the 3 Tev superconducting ring will begin operation and eventually, around late 1995, there should be a second 3 TeV superconducting ring, which will make UNK a 3 TeV on 3 TeV collider. The injector for UNK is the 70 GeV accelerator at Serpukhov or Protvino which is the name of the town where the lab really is. Figure 19 shows the ring, which is 21 km in circumference, and will have an extracted beam area for the 3 TeV ring. UNK sits about 50 meters below ground.

We plan to do our experiment in the underground internal target station SS-3; where we would study A in $p + p \to p + p$ at large–$P_\perp^2$ first at 400 GeV and then 3 TeV. In March 1989 we signed the final agreement for our experiment [27], which is called NEPTUN-A. The architectural and engineering drawings for the 50 m long 5 m diameter special underground enclosure for NEPTUN-A were finished in February 1989. The SS-3 cave is now being excavated. The UNK beam will, of course, be unpolarized so we will need a polarized target to measure A. A solid polarized target would not work as well as our planned internal polarized jet target for a number of reasons.

We are therefore building a polarized gas jet target using a new principle called spin polarized ultra-cold atomic hydrogen, which was developed by Dan Kleppner, a distinguished atomic physicist at MIT. The prototype jet which is shown in figure 20 operates at about 1/3° K and uses a 5 to 8 T superconducting solenoid magnet which sucks in all the hydrogen atoms in one spin state. It is energetically favorable for these atoms to be in a high field; they are called high field seekers. The magnet rejects all those atoms that are in the wrong spin site; they are called low field seekers. Notice that the energy associate with the magnetic moment, $\mu \cdot B$, due to the 5 T field is equivalent to kT at about 4° K. We make the target very cold so that the atoms' energy spread is only kT at about 1/3° K; therefore, their energy spread is less than 10%. It is then easy to store these spin polarized hydrogen atoms in this "magnetic bottle".

We next try to extract the hydrogen atoms using a new idea called microwave extraction, which I think was

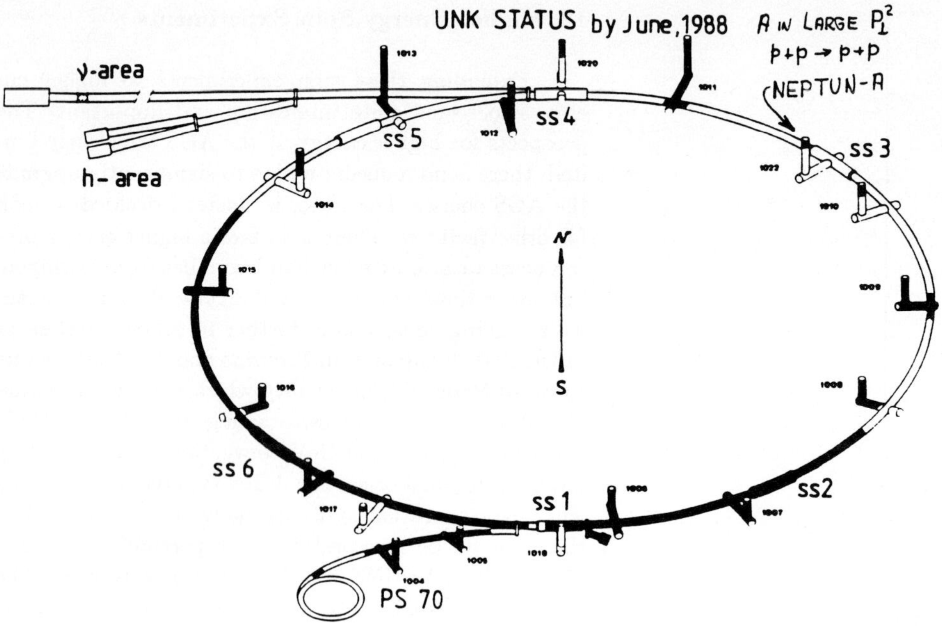

Figure 19: The UNK facility of Protvino.

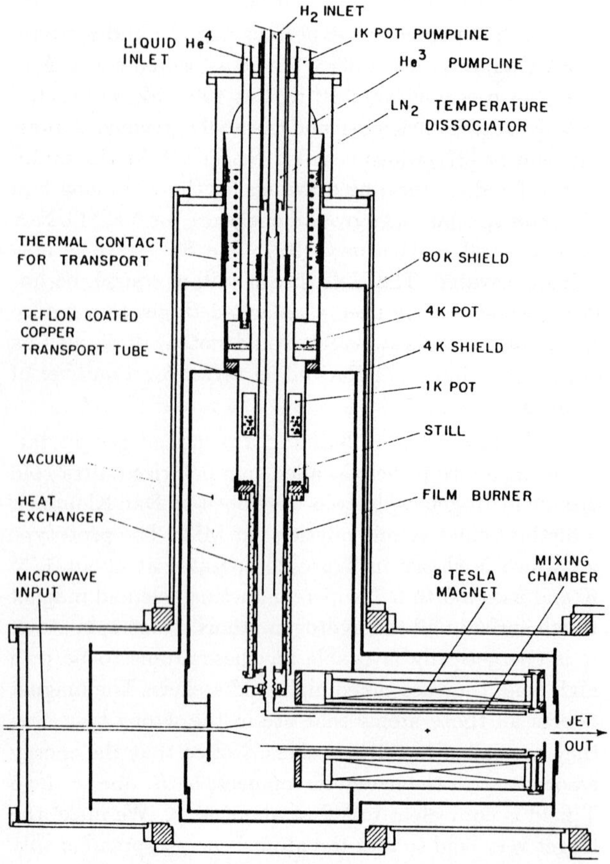

Figure 20: Michigan–MIT spin-polarized ultracold atomic hydrogen jet.

first proposed independently by Dan Kleppner and Tapio Niinikoski. They suggested that one shine microwaves into the magnetic bottle and if the microwaves have the right frequency they will induce spin-flip transitions and make the hydrogen atoms come out. When the spin of a high field seeker is flipped it becomes a low field seeker; it is then energetically unfavorable to stay in the 5 T field and it jumps out.

I am pleased to say that we now have this complex device working. We have obtained an extracted beam of spin polarized hydrogen atoms by shinning in 140 GHz microwaves which match the magnetic field at 5 T. We have extracted a beam of about 4 10^{16}/s spin polarized hydrogen atoms using our prototype jet in a pulsed mode; we have also had a dc beam of about 10^{15}/s. Our ultimate goal is a dc beam of about 10^{18}/s, but we could already do a respectable experiment with our present intensity.

A polarized jet is something that TRIUMF may want to think about. I don't know if there are any plans for internal target experiments at KAON. The largest $P_\perp^2$ point in figure 1, in fact, used an internal polyethylene target at the AGS. We did this experiment [28] around 1963 when the internal intensity was only about 3 10^{11} per pulse.

I next want to discuss the NEPTUN-A spectrometer [27] which is shown in figure 21 in our underground UNK station. The UNK proton beam, which will first be 400 GeV and then 3 TeV, comes in from the left. There will also be a magnet in the forward direction which is part of the large Russian NEPTUN collaboration [29]. Some of the people from NEPTUN, are participating in NEPTUN-A and all of the NEPTUN spectrometers will

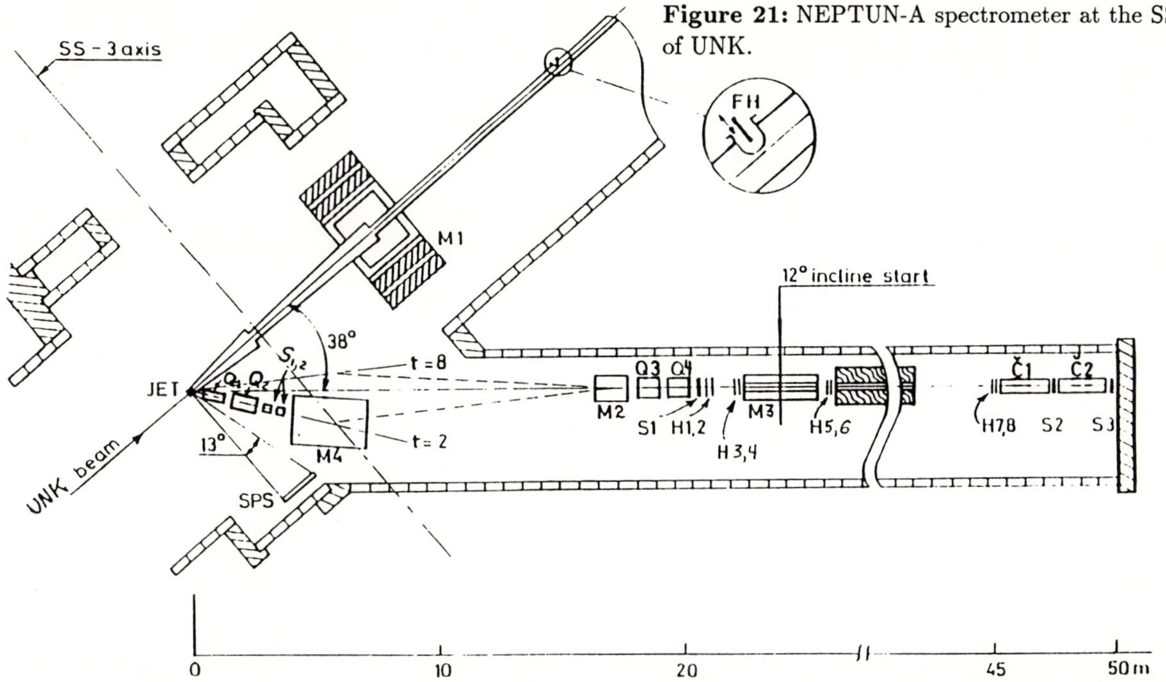

Figure 21: NEPTUN-A spectrometer at the SS–3 cave of UNK.

look at our polarized gas jet. Our focusing recoil spectrometer takes advantage of the fact that the angle and momentum of the recoil protons in pp elastic scattering are almost totally independent of the incident energy. For example, at $P_\perp^2 = 6$, the recoil proton comes out at about 37° with a momentum of about 4 GeV independent of whether you are at 12 GeV or 400 GeV or 3 TeV and that is a remarkable fact.

We are using this fact heavily in designing this spectrometer. We have done many studies of how one can momentum analyze elastically scattered forward protons at 400 GeV or at 3 TeV. We discovered a serious problem while considering a similar Fermilab experiment, where we were trying to momentum analyze the 800 GeV forward scattered protons. We tried to design a 300 m long spectrometer with about 60 m of strong magnetic field, but it was difficult and expensive. It appears even more difficult to momentum analyze 3 TeV particles; I don't think 20 TeV particles will be much easier. Fortunately, the recoil particle has a momentum of about 4 GeV, so it is easy to momentum analyze. We also plan to have some tiny forward hodoscopes in something like a Roman Pot. At 400 GeV the whole forward hodoscope will be about 4 cm × 1 cm; at 3 TeV it will be much smaller.

The Neptun-A recoil arm is an elaborate spectrometer containing dipoles, quadrupoles, skew-quadrupoles, and probably some sextupoles. This focusing spectrometer subtends a rather large solid angle, of about 200 miliradians vertically by 20 miliradians horizontally. It focuses this large $\Delta\Omega$ down to a small detector of about 20 cm × 20 cm, which is about 50 m away; moreover, it focuses mostly only elastic scattering events. We noticed that for pp elastic scattering there is a unique relation between the momentum and the scattering angle and we designed this spectrometer to focus only the elastic events. It doesn't much focus the inelastic events, thus it gives a large improvement in the signal to background ratio. If we had no focusing, a detector placed at 50 m would have a size of 10 m × 1 m since the spectrometer subtends 200 miliradians × 20 miliradians; instead our detector size will be about 20 cm × 20 cm. We hope to have momentum resolution of perhaps 1/10%. By using this good momentum resolution, we hope to be able to cleanly distinguish elastic scattering from inelastic scattering. We aren't sure that this will work perfectly at 3 TeV, but it seems our best hope. Our Russian colleagues and we are working hard on designing this spectrometer. We plan to do our momentum analysis vertically in this special underground cave, which has a 12° vertical blend at its mid-point. We plan to do the angle analysis in the horizontal plane.

In figure 22 we have plotted the ratio σ_{parallel}: $\sigma_{\text{antiparallel}}$ against $P_\perp^2$ for two of those sets of data [7,24] that I showed you earlier in the three dimensional plot (figure 17). The open squares are the 90°_{cm} data; the black circles are the 12 GeV/c data. Clearly at small–$P_\perp^2$ the two sets of data do not fall on top of each other; this is not surprising because this small–$P_\perp^2$ data at 90°_{cm} is a rather low energy. there are clearly some low energy effects, which are not related to large–$P_\perp^2$ scattering. However, for $P_\perp^2 \geq 1.5\ (\text{GeV/c})^2$ it is obvious that these two sets of data fall exactly on top of each other; this clearly shows that these large spin effects are indeed large–$P_\perp^2$ hard-scattering effects.

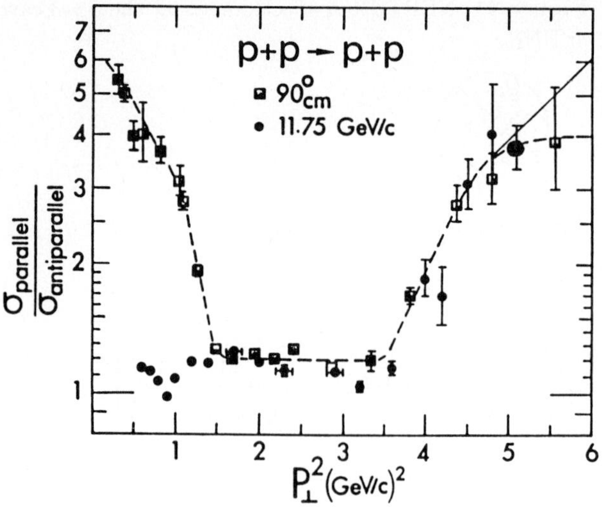

Figure 22: The ratio of the pure spin cross sections for fixed angle and fixed energy.

Some of you may know that Bethe and Weisskopf [30] proposed around 1977 that these effects might be related to particle identity because our 12 GeV large–$P_\perp^2$ data was close to 90°_{cm}. In recent years Lipkin [31], Tomozawa [32], and others have also pursued this particle identity question. However, I believe that it is unlikely that these large spin-spin effects are caused by particle identity alone. Although particle identity could contribute, it certainly isn't the only thing happening. The fact that these two totally different sets of spin data fall exactly on top of each other seems fairly remarkable.

Another view of large angle pp elastic scattering spin effects is shown in figure 23. This is a graph of the measured 90°_{cm} cross sections in different initial spin states plotted against $P_\perp^2$; the upper scale shows the laboratory momentum. At any given P_{lab} the largest $P_\perp^2$ occurs at 90°_{cm} so the $P_\perp^2$ value on the lower scale is the upper limit at each momentum. The figure shows clearly the low energy spin effects at small–$P_\perp^2$ and the absence of any spin effects in the medium–$P_\perp^2$ region. At large–$P_\perp^2$ the two curves are clearly moving apart up to our maximum value of $P_\perp^2 = 5.5\ (\text{GeV/c})^2$. Maybe they will come back together at larger $P_\perp^2$ as some people propose, but so far they are still moving apart. We are eager to see what happens at large–$P_\perp^2$; perhaps they will continue to move apart or perhaps the ratio will stay fixed at about four.

You can also see from the figure how rapidly the cross section decreases with $P_\perp^2$. This is important to stress, because if you want to measure these rare high–$P_\perp^2$ collisions you clearly need a large solid angle, a high intensity, and a polarized target that can take this high intensity. KAON seems an ideal place to extend these 90°_{cm} measurements because of the many orders of magnitude that occur in this curve. To do this experi-

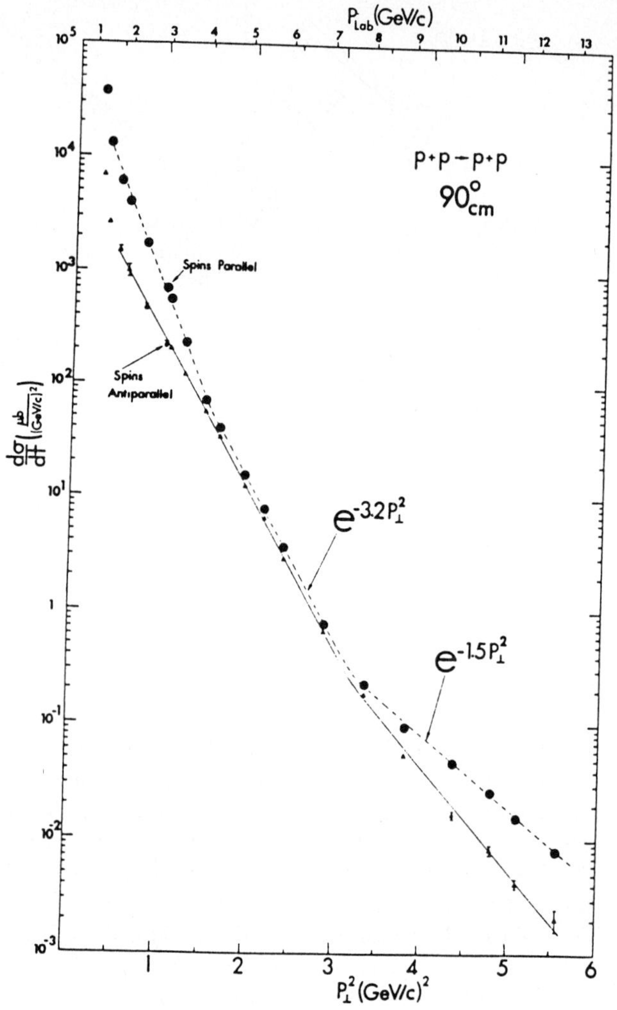

Figure 23: Pure spin cross sections at 90°_{cm}.

ment, KAON will need a high intensity polarized beam, good detectors, and high cooling power polarized target and/or a polarized internal jet target.

Figure 24 is a compilation of the spin-spin correlation parameter, A_{nn}, and 90°_{cm} plotted as a function of energy over the entire range of where it has been measured. I have never seen this plotted before and have used this talk as an excuse to prepare this figure. To make this graph more readable, I used log paper. On the right is the high energy data which I showed you before in the regime from about 1 GeV up to 13 GeV. On the left is the low energy regime which is normally plotted separately. I thought that such an overview might be especially appropriate for this talk at a medium energy lab which hopes to soon become also a high energy lab. One thing that I find quite striking is that at the lowest energy A_{nn} goes down very close to –1. The lowest energy point comes from a tabulation of low energy data from my friend, Willi Haeberli. While this last point doesn't quite go to –100%, it is about $-98 \pm 1\%$; thus, most reasonable peo-

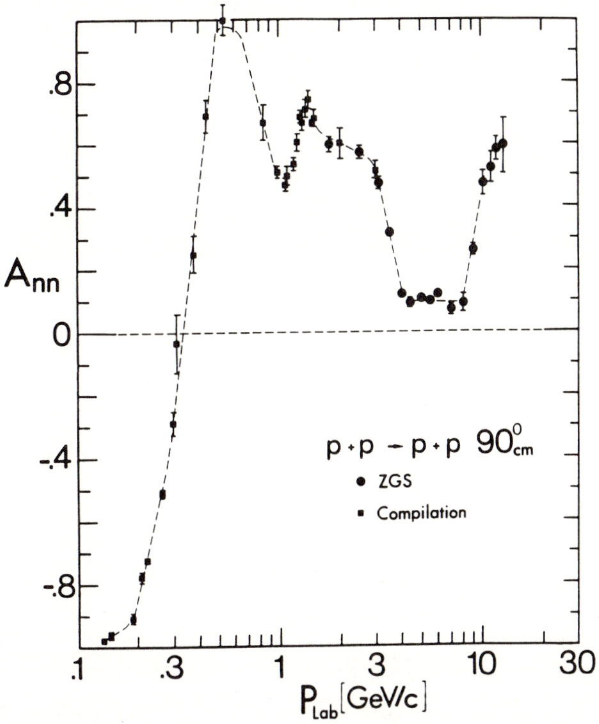

Figure 24: Compilation of the spin-spin correlation parameter at 90°_{cm} .

ple would agree that A_{nn} is headed to −100%. Another remarkable thing is that A_{nn} then increases rapidly and then goes to about +100%; the measured point is 100 ± 5%. A_{nn} then drops down to about 45% and then increases sharply again; I think that this structure between 1 and 2 GeV/c may be related to the dibaryon phenomena. Independent of the existence of dibaryons, there is certainly some interesting structure at low energy; I discussed the high energy structure earlier. In any case, I think that this is a rather remarkable plot. I also believe that one of the main goals of TRIUMF's KAON should be to extend this plot of A_{nn} at 90°_{cm} to higher energy.

References

[1] A.D. Krisch, Phys. Rev. Lett. 11, (1963) 217; Phys. Rev. 135, (1964) B1456; Phys. Rev. Lett. 19, (1967) 1149; Phys. Lett. 44B, (1973) 71; P.H. Hansen and A.d. Krisch, Phys. Rev. D15, (1977) 3287

[2] A. Yokosawa, Phys. Rev. 64, (1980) 47

[3] E.F. Parker *et al.*, Phys. Rev. Lett. 31, (1973) 783; W. de Boer *et al.*, Phys. Rev. Lett. 34, (1975) 558

[4] K. Heller, Proceedings 6th International Symposium on High Energy Spin Physics; Marseille Sept. 1984, ed. J. Soffer; Journal de Physique 46, (1985) C2121

[5] F. Abe *et al.*, Phys. Rev. D34, (1985) 1950

[6] P.E. Schlein, Jour. de Physique 46, (1985) C2-131

[7] D.G. Crabb *et al.*, Phys. Rev. Lett. 41, (1978) 1257; J.R. O'Fallon *et al.*ibid. 389, (1977) 733

[8] S. Kistryn *et al.*, Phys. Rev. Lett. 58, (1987) 1616; V. Yuan *et al.*, Phys. Rev. Lett. 57 (1986) 1680

[9] N. Lockyer *et al.*, Phys. Rev. D30, (1984) 860

[10] P.H. Hansen *et al.*, Phys. Rev. Lett. 50, (1983) 802; D.C. Peaslee *et al.*, Phys. Rev. Lett. 51, (1983) 2359; P.R. Cameron *et al.*, Phys. Rev. D32, (1985) 3070

[11] C.Y,. Prescott, Proc. 5th Intern. Symposium on High Energy Spin Physics; Brookhaven Sept. 1982, ed. G. Bunce; AIP Conf. Proc. 95 p28 (AIP New York 1983).

[12] A. Gaidot *et al.*, Phys. Lett. 57B, (1975) 389; 61B, (1976) 103

[13] J. Ashman *et al.*, Phys. Lett. B206, (1988) 364

[14] E. Bloom *et al.*, Phys. Rev. Lett. 23, (1969) 930

[15] C.W. Akerlof *et al.*, Phys. Rev. Lett. 17, (1966) 1105; Phys. Rev. 159, (1967) 1138

[16] A.D. Krisch and A.J. Salthouse, eds., Higher Energy Polarized Beams (AIP Conf. Proc. No. 42) (AIP New York 1978).

[17] B. Cork *et al.*, Proc. of the 1978 Summer Study on Polarized Protons in the Brookhaven AGS, ed. by A.D. Krisch (University of Michigan, Ann Arbor 1978), p1.

[18] F.Z. Khiari *et al.*, Phys. Rev. D39, (1989) 45

[19] E.D. Courant and R.D. Ruth, Report No. BNL 51270, 1980 (unpublished).

[20] D.G. Crabb *et al.*, in Proc. of the 8th Intern. Symp. on High Energy Spin Physics, (Minneapolis 1988).

[21] D.G. Crabb *et al.*, Phys. Rev. Lett. 60, (1988) 2351

[22] J. Antille *et al.*, Nucl. Phys. B185, (1981) 1

[23] J.V. Allaby *et al.*, Nucl. Phys. B52, (1973) 316

[24] E.A. Crosbie *et al.*, Phys. Rev. D23, (1981) 600

[25] K.A. Brown *et al.*, Phys. Rev. D31, (1985) 3017

[26] G.R. Court *et al.*, Phys. Rev. Lett. 57, (1986) 507

[27] A.D. Krisch, Experiment NEPTUN-A to be published in Proc. of the Workshop on UNK (March 1989).

[28] G. Cocconi *et al.*, Phys. Rev. Lett. 11, (1963) 499; Phys. Rev. 138, (1965) B165

[29] V.D. Apokin *et al.*, Proposal of Experiment NEPTUN (Serpukhov 1988).

[30] H.A. Bethe and V.F. Weisskopf, Independent comments at Copenhagen and CERN Seminars (June 1978).

[31] H.J. Lipkin, Phys. Lett. B181, (1987) 164

[32] Y. Tomozawa, Phys. Rev. D36, (1987) 2856

Z. Phys. C – Particles and Fields 46, S149–155 (1990)

The ΔS = 0 hadronic weak interaction

S. A. Page

University of Manitoba, Winnipeg, Manitoba, Canada R3T 2N2

Abstract

Our present understanding of the ΔS=0 hadronic weak interaction is based on a series of high precision experiments in the two-nucleon system and light nuclei which isolate the weak interaction via its parity violating signature. At low energy, the link between parity nonconserving (PNC) observables and the standard electroweak theory is via a set of weak meson-nucleon coupling constants which are predicted using a quark model of the physical hadrons. One striking feature that emerges is that the isovector pion coupling, which is almost entirely carried by neutral currents at the quark level, is strongly suppressed relative to theoretical predictions. At higher energy, a surprisingly large PNC asymmetry in p-p scattering has been found at 6 GeV/c. Direct QCD-based calculations which explain the data have been the subject of recent theoretical debate. New measurements at energies of 5 GeV and higher are needed to confirm the effect and guide theoretical efforts. With the availability of polarized proton beams at higher energy, as recently achieved at Brookhaven and planned for the KAON facility in Canada, the prospect for performing such experiments is very good.

1. Introduction

Despite the many successes of the Standard (GWS) Model, a quantitative understanding of weak hadronic interactions is still lacking due to the difficulty of treating QCD effects in the nonperturbative regime. Factors of 3 uncertainty are typically associated with calculations of weak hadronic matrix elements, ultimately limiting our ability to interpret measurements of parity and CP violation. An understanding of the ΔS=0 interaction is particularly important, because it provides a unique window on hadronic weak neutral current processes. Already, much has been learned from studies of parity violation in the nucleon-nucleon interaction and in nuclei, where all measurements to date are consistent with somewhat imprecise predictions based on the standard model. The present experimental situation is incomplete, however. Several complementary high precision measurements in systems for which the 'background' strong interaction physics is well understood are needed in order to constrain realistic calculations based on the standard weak interaction theory and QCD.

2. Theoretical Approach at Low Energy

At low energy, the weak nucleon-nucleon interaction has been successfully described in terms of a meson exchange model. The link between short range W and Z exchanges between quarks and the longer range π, ρ, and ω exchanges between nucleons is via a set of six weak meson-nucleon coupling constants, which have been predicted, e.g. by Desplanques, Donoghue and Holstein [1] (DDH) using a SU(6) quark model and treating strong interaction enhancement effects using renormalization group theory. The couplings are denoted (f^1_π, h^0_ρ, h^1_ρ, h^2_ρ, h^0_ω, h^1_ω), where superscripts refer to isospin changes. Their predictions are summarized in table 1. Large uncertainties are typical, and in some cases, even the sign of the coupling strengths cannot be predicted with confidence.

In a recent review [2], Adelberger and Haxton fitted the most significant nuclear parity violation data to a 2 parameter expression based on the quark model formalism of DDH. Unfortunately, the resulting values of the weak meson-nucleon couplings were only marginally better constrained than the 'reasonable range' estimates

Table 1: Weak Meson-Nucleon Couplings

Coupling	Theoretical Range ($\times10^{-7}$)	Theoretical Best Value ($\times10^{-7}$)	Experimental Best Fit ($\times10^{-7}$)
f_π^1	$0 \rightarrow 11.4$	4.6	2.1
h_ρ^0	$-31 \rightarrow 11.4$	-11.4	-6.7
h_ρ^1	$-0.38 \rightarrow 0$	-0.19	-0.21
h_ρ^2	$-11.0 \rightarrow -7.6$	-9.5	-7.3
h_ω^0	$-10.3 \rightarrow 5.7$	-1.9	-6.7
h_ω^1	$-1.9 \rightarrow -0.8$	-1.1	-2.3

from quark model predictions, due to the lack of independent, high-precision data. The results of the Adelberger-Haxton fitting procedure, which have been updated to include the recent 45 MeV measurement in $\vec{p}-p$ scattering, are shown in table 1.

3. Experimental Situation I : Nuclei

A thorough review of nuclear parity violation was given by Adelberger and Haxton [2] in 1985; only a brief summary will be presented here, with emphasis given to the status of present and future work. In order to attain a complete understanding of the $\Delta S=0$ weak interaction at low energy it is essential to determine the complete set of weak meson-nucleon couplings. A minimum of 6 independent experiments must be found. The parity-violating observables must be predicted with minimal model dependence in terms of the weak couplings of interest, which effectively restricts the selection to two-nucleon systems and light nuclei. In addition, the experiments must be technically feasible at the level of accuracy implied by the 'best fit' values of the weak coupling constants as determined from previous data (see table 1).

To date, many parity violation experiments have been performed in nuclear systems, but relatively few have achieved these goals. At present, there are only 4 independent constraints on the 6 weak meson-nucleon couplings; these are obtained from electromagnetic decays of parity mixed doublets in nuclei of the mass-20 region and from the helicity dependence of elastic proton scattering from protons and other light nuclei. The experimental situation is summarized in table 2. Experimenters face the challenging problem of extracting a tiny parity violating signal from a relatively enormous parity conserving background, the scale being set by the ratio of weak to strong amplitudes in the basic nucleon-nucleon interaction which is of order 10^{-7}. PNC observables are enhanced by nuclear structure in certain finite nuclei, but at the expense of introducing additional complexity into the nuclear wavefunctions required to deduce weak meson-nucleon coupling constants from the data.

Recently, some very large enhancements have been discovered in low energy neutron resonances in nuclei of the rare earth region, the largest effect being about a 10% admixture [3] between s- and p-wave resonances at 0.734 eV in ^{139}La. These large effects have raised the possibility of using the parity mixed neutron resonances to search for t-violation in either 3-fold (p-odd, t-odd) or 5-fold (p-even, t-odd) spin-momentum correlations – a thorough discussion of these ideas is found in the proceedings of the 1987 workshop on Tests of Time Reversal in Neutron Physics [4].

A compromise is reached in several nuclei of the mass-20 region, in which PNC observables are enhanced and the underlying nuclear structure is reasonably well understood (see figure 1). A common feature is the existence of a low-lying doublet of states of the same spin but opposite parity which are mixed by the weak interaction. Wavefunction admixtures lead to circular polarization of the γ-decays from the two states due to E1-M1 mixing. If the states have very different lifetimes, the circular polarization is enhanced for one member of the doublet and suppressed for the other; this additional enhancement is given by the square root of the lifetime ratio of the two states. As emphasized by Haxton, the nuclear matrix elements needed to extract weak meson-nucleon couplings from the γ-ray observables have been found to be extremely sensitive to radial shape of nuclear wavefunctions, short range correlations, and truncation of the model space. A major breakthrough occurred, when it was realized that an independent test of the shell model wavefunctions could be obtained from a measurement of first forbidden β-decay in two cases: the PNC matrix element in ^{18}F is related by an isospin rotation to the pion exchange contribution which dominates the first forbidden β^+ decay of ^{18}Ne to the upper member of the ^{18}F doublet [6], and a similar situation occurs for the isovector contribution to PNC in ^{19}F. Hence, a measurement of the forbidden beta decay rate, in addition to the ^{18}F γ-ray circular polarization, enables the isovector pion coupling f_π^1 to be deduced with minimal dependence on nuclear structure.

Parity violation experiments in light nuclei determine effective isovector (f_π^1) and isoscalar ($h_\rho^0 + 0.6h_\omega^0$) coupling strengths. The isovector pion coupling f_π^1 is unique in that 95% of its predicted value is attributed to neutral current diagrams in the quark model. Recent measurements [7] of γ-ray circular polarization in ^{18}F have shown f_π^1 to be 3.5 standard deviations smaller than the theoretical 'best value' from DDH. Holstein [8] has reexamined the quark model calculations and finds that

Table 2: Significant Experimental Constraints on Weak Meson-Nucleon Couplings

Case	Coupling	Expt. Performed	Additional Requirements	Proposed Expt.
1	f_π^1	^{18}F: P_γ (1081 keV)	None	
2	$h_\rho^0 + 0.6h_\omega^0$	^{19}F: A_γ(110 keV) $\vec{p}+\alpha$: A_z(46 MeV) ^{21}Ne: P_γ(2789 keV) ^{14}N: A_z(0^+; 8624 keV)	Case 1	
3	$h_\rho^0 + h_\rho^1 + \frac{h_\rho^2}{\sqrt{6}}$	None	None	$\vec{p}+p$: A_z, 222 MeV TRIUMF
4	$h_\omega^0 + h_\omega^1$	$\vec{p}+p$: A_z (15 MeV) $\vec{p}+p$: A_z (45 MeV)	Case 3	
5	$h_\rho^1 + h_\omega^1$	$\vec{p}+d$: A_z (15 MeV)	Cases 1,2,3,4	
6	$h_\omega^0 + 2.8h_\rho^2$	None	Cases 1,2	$\vec{n}+p$: ϕ_{PNC}/λ (Wash/ILL)

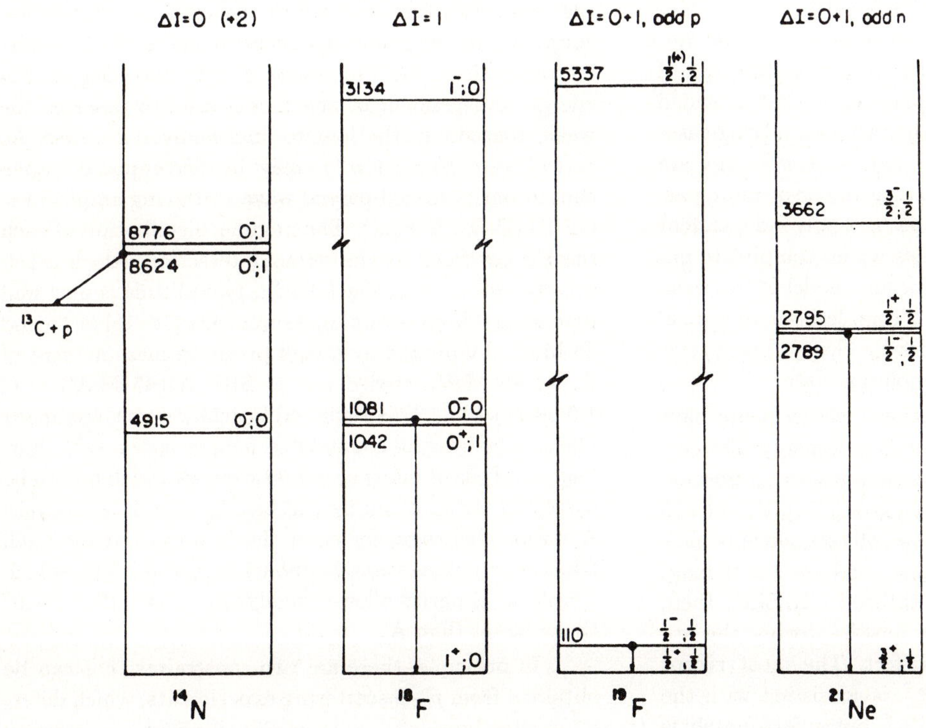

Figure 1: Parity mixed doublets in light nuclei [5].

the small measured value of f_π^1 cannot be understood unless the current algebra quark mass values are increased by about a factor of 2 over the original Weinberg values, which tends to produce a similar suppression of theoretical estimates in other areas, e.g. the $\Delta I=1/2$ rule.

In contrast to f_π^1, the weak isoscalar coupling ($h_\rho^0+0.6h_\omega^0$) is not well determined. A recent measurement of parity violation in $\vec{p}-{}^{13}C$ scattering by the University of Washington-Wisconsin collaboration was undertaken to settle a long-standing discrepancy between the interpretation of circular polarization measurements in ^{21}Ne and ^{19}F, which has been attributed to uncertainty in the shell model wavefunctions used to describe ^{21}Ne. The ^{14}N experiment is sensitive only to the isoscalar interaction. A positive result at the 1.5 σ level was obtained [9], which yields a constraint of the opposite sign to the DDH 'best value' prediction and agrees better with the ^{21}Ne experiment than with ^{19}F. Unfortunately, the shell model calculations for this system have been shown to have less predictive power than previously thought; a precise interpretation of the experiment therefore awaits improved shell model calculations of the ^{14}N system which are currently in progress [10].

4. Experimental Situation II : The 2-Nucleon System

Measurements of parity violation in the two nucleon system are extremely difficult, due to the lack of an enhancement mechanism analogous to that provided by nuclear structure effects in the two-level mixing cases discussed above. However, N-N data at low energy are straightforward to interpret in terms of one or more parity mixed partial wave amplitudes, virtually independent of the wavefunction uncertainties which complicate nuclear data. The ideal test case for any model of the weak nucleon-nucleon interaction is a complete set of partial wave amplitudes, since they afford a cleaner interpretation than most nuclear parity violation data.

A number of parity violation measurements have been performed in the n-p system at low energy. The earliest measurement of the circular polarization in thermal neutron capture by the Leningrad group [11], $P_\gamma=(-13 \pm 4.5)\times 10^{-7}$ excited considerable interest, since it indicated an enhancement of roughly three orders of magnitude over what was expected theoretically. Since then, a great deal of effort has gone into improving the apparatus and remeasuring the effect. The most recent result [12], $P_\gamma=(1.8 \pm 1.8)\times10^{-7}$, is consistent with the theoretical expectation of 0.6×10^{-7}, but unfortunately is not precise enough to significantly constrain predictions of the weak meson-nucleon couplings. A closely related observable is the circular polarization dependence of the cross-section for photodisintegration of deuterium, which was measured at Chalk River [13]. The result, $A_z(E_\gamma \leq 4.1$ MeV$) = (27 \pm 28)\times10^{-7}$, is consistent with theory, but unfortunately falls two orders of magnitude short of testing the meson exchange model prediction, which is less than 0.5×10^{-7}. Finally, the gamma ray asymmetry A_γ in polarized thermal neutron capture, which is sensitive to different partial wave amplitudes from P_γ, has been measured at Grenoble [14]. The result, $A_\gamma=(-0.15\pm0.48)\times10^{-7}$, is consistent with zero and would need to be improved significantly in order to challenge the theoretical prediction of -0.5×10^{-7}. The collaboration reports that their result was statistics limited and that the higher flux of neutrons presently available from the new cold neutron source at ILL should make possible a new generation of experiments at the $\pm3\times10^{-9}$ level. This possibility presents an intriguing challenge to experiment. A new experiment to measure a parity violating neutron spin rotation in liquid parahydrogen at ILL has recently been proposed [15]. The expected effect is of order 10^{-7} radians and is mainly sensitive to f_π^1; the collaboration expects to achieve a sensitivity of 4σ in a running time of 30 days. If the experiment is successfully carried out, it will be very interesting to compare the measurements of f_π^1 in the np system with that found in ^{18}F.

In contrast to the np system, where significant improvements in experimental precision are required to challenge theory, the pp system has provided one of the most significant constraints on weak meson-nucleon couplings to date through measurements of the helicity dependence (A_z) of elastic $\vec{p}-p$ scattering at low energy. Independent of any model used to describe the weak interaction, the longitudinal analyzing power, $A_z = (\sigma^+ - \sigma^-)/(\sigma^+ + \sigma^-)$, may be decomposed into a sum of parity mixed partial wave scattering amplitudes: ((S-P),(P-D),(D-F)...). The angular distribution of each term is governed by the strong interaction, which is relatively well known. The ($^1S_0$-$^3P_0$) amplitude is now well determined from several measurements [16–18] at 15 and 45 MeV, dominated by a high precision measurement of A_z at 45 MeV, carried out at SIN: A_z(45 MeV) = $(-1.5\pm0.2)\times10^{-7}$. This remarkable achievement was made possible by integral counting techniques and a very thorough analysis of many systematic errors which had to be corrected for on a run by run basis in the data analysis. A major challenge, for example, is presented by small transverse polarization components in the beam, which couple to a parity-allowed analyzing power that is 10^5 times larger than A_z.

In principle, there are two constraints that can be obtained from $\vec{p}-p$ scattering experiments, which determine effective ρ and ω couplings summed over isospin: $h_\rho^{pp} = h_\rho^0 + h_\rho^1 + h_\rho^2/\sqrt{6}$, and $h_\omega^{pp} = h_\omega^0 + h_\omega^1$. These can be evaluated from measurements of the two lowest partial wave contributions to A_z: the ($^1S_0$-$^3P_0$) amplitude discussed above, and the ($^3P_2$-$^1D_2$) amplitude which contributes significantly to the longitudinal an-

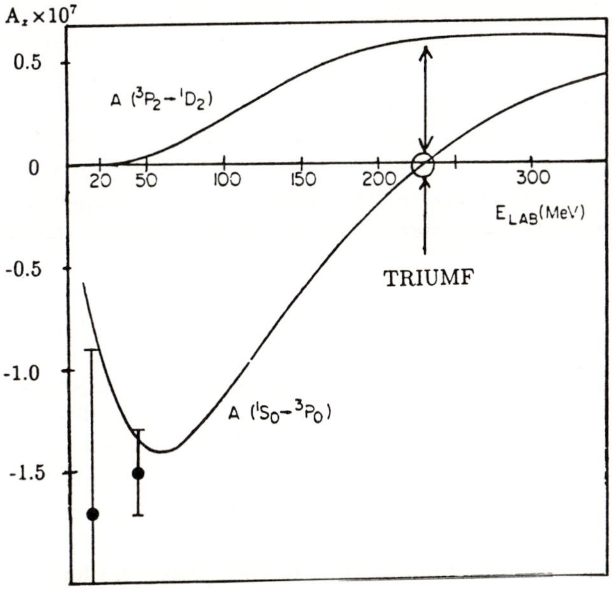

Figure 2: Partial wave contributions [19] to A_z.

alyzing power above 100 MeV. Simonius [19] has shown that the ($^3P_2$-$^1D_2$) amplitude depends only on weak ρ exchange in the meson exchange model, whereas ρ and ω contribute to the ($^1S_0$-$^3P_0$) amplitude with approximately equal weight.

An experiment is underway at TRIUMF [21] to determine the ($^3P_2$-$^1D_2$) amplitude, which may be measured independently at 230 MeV, where the $^1S_0$ and $^3P_0$ phase shifts exactly cancel (see figure 2). In practice, the angular distribution $A_z(\theta)$ must be weighted with the response function of the detectors, which has been simulated using Monte Carlo techniques; this shifts the desired beam energy to 222 MeV to cancel the ($^1S_0$-$^3P_0$) contribution in the TRIUMF experiment. The expected effect at 222 MeV [19,20] is $A_z \sim +0.6\times10^{-7}$. A_z will be determined to $\pm0.2\times10^{-7}$ or better in each of two simultaneous experiments, which will yield a value of h_ρ^{pp} to ±25% of the DDH 'best value' prediction.

A_z will be determined from simultaneous measurements of the helicity dependence of the transmission and scattering of longitudinally polarized protons from a 20 cm LH_2 target. The polarized proton beam will be provided by the new optically pumped polarized ion source at TRIUMF, which is currently being commissioned; this source is in principle ideal for parity violation measurements, since polarization reversal is achieved by changing the polarization and frequency of the laser light, with the minimum possible changes to beam current and emittance. The vertical polarization of the proton beam from the TRIUMF cyclotron will be transformed to longitudinal polarization by spin rotation in two dipole magnets, each preceded by a superconducting solenoid magnet, which are presently installed in the beamline. To minimize errors in rotation of the proton spin, which would lead to unwanted transverse components of polarization, a slow feedback system based on split-plate secondary emission monitors upstream of the first solenoid will control beam position and angle through these beamline elements.

The main components of the apparatus are a pair of transverse field parallel-plate ionization chambers similar in design to those used in the 800 MeV experiment at LAMPF [22], to measure the beam current before and after the target. In addition, a cylindrically symmetric axial field chamber 1 m downstream of the target will detect protons scattered from 6-43° (lab). The ionization chambers are operated in current mode to meet the high statistical accuracy requirements of the experiment.

The major systematic error which must be overcome is an apparent parity violating asymmetry generated by residual transverse polarization components in the beam that couple to a relatively large parity allowed transverse analyzing power $A_y \sim 0.3$. To minimize this and other systematic effects, beam properties must be identical in the two helicity states, and the detection apparatus must be highly symmetric. The additional apparatus to provide for the required systematic error control consists of: two dual-function multiwire beam profile/position centroid monitors, coupled to a fast feedback system which controls random beam excursions to less than 3 μm up to 1 kHz, and two scanning polarimeters which will monitor the profiles of residual transverse polarization components in the beam during the experiment. The sensitivity of the apparatus to transverse polarization components and beam displacement will be measured with a transversely polarized beam. Systematic error calculations predict a sensitivity of $\sim 2\times10^{-5}$ for a 100% transversely polarized beam offset by 1 mm from the symmetry axis of the apparatus; the effect scales linearly with the first moment of the transverse polarization distribution. A scanning polarimeter has been designed which is capable of measuring the first moment (xP_y) to $\pm5\times10^{-2}$ mm in 1 second, which will independently determine the false asymmetry from this source to a few times 10^{-9} during the course of the parity violation measurements. The TRIUMF experiment is currently in the second year of a 2-year instrumentation development programme; final data taking will begin following successful prototyping of the apparatus and demonstration that systematic errors can be controlled and understood at the 10^{-8} level.

Parity violation has also been studied in $\vec{p}-p$ scattering at higher energies (figure 3), where a simple meson exchange model loses much of its predictive power due to uncertainties in the treatment of inelastic processes. A measurement of A_z at 800 MeV was performed at Los Alamos [22]; the result, $A_z = (2.4 \pm 1.1)\times10^{-7}$, is in reasonable agreement with extrapolations of the meson exchange predictions, although the latter bracket a range of possible values roughly 5 times as large as the experimental uncertainty.

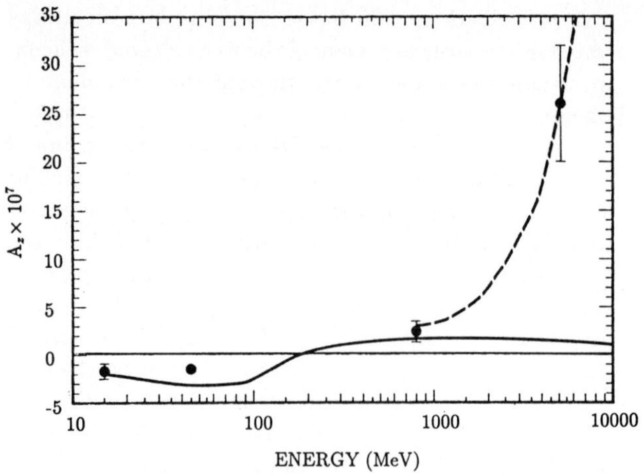

Figure 3: Energy dependence of A_z in $\vec{p}-p$ scattering [32]. The solid curve is a calculation based on the meson exchange model; the dashed curve is the diquark model prediction of ref. 27.

At still higher energy, there is only one additional parity violation measurement [23], which was performed as a transmission experiment with 5.1 GeV longitudinally polarized protons incident on a water target at the Argonne ZGS . The experimenters are confident in their analysis of possible systematic errors, which are of similar origin to those studied at lower energy. An additional background asymmetry due to polarized protons resulting from decays of polarized hyperons produced in the target was eliminated by installing a magnetic spectrometer to transmit only protons of the appropriate beam momentum. The result, $A_z = (2.6 \pm 0.7) \times 10^{-6}$, is an order of magnitude larger than expected based on a variety of models ranging from meson-exchange calculations to direct W and Z boson exchanges [24]. Furthermore, it has been shown [25] that the measured asymmetry for $\vec{p}$-H_2O scattering is expected to be only ~60% of the value for $\vec{p}$-nucleon scattering due to Glauber shadowing, implying an even larger effect for the $\vec{p}-p$ case.

Two independent models have succeeded at reproducing the large effect at 5.1 GeV; both have been the subject of debate in the literature. The calculations of Nardulli and Preparata [26], based on intrinsic parity violation in the nucleon wavefunction, predict an effect that is independent of energy above 800 MeV and consistent with the measured value at 5.1 GeV. In contrast, the diquark model calculations of Goldman and Preston [27] predict an asymmetry which rises rapidly with energy, by nearly two orders of magnitude from 800 MeV to 12 GeV. It has been shown [28] that the former model implies results at low energy which are too large and inconsistent with existing data. Simonius and Unger [29] have criticized the diquark model calculations, claiming that the absolute value of the predicted effects grow unphysically at high energy; the authors have responded [30] by arguing that their model provides a valid prediction of the energy dependence of A_z with an overall normalization to the 5.1 GeV data. Clearly, the enhanced parity violating asymmetry at 5.1 GeV represents an ongoing challenge, both for experimental confirmation and for renewed theoretical efforts. With further progress in this area, future studies of parity violation at high energy may some day provide an important test of models that describe the spin structure of the proton.

Detailed studies for the planned 30 GeV KAON Factory in Vancouver are being carried out to optimize the synchrotron lattice designs for acceleration of polarized beam with minimum losses due to depolarizing resonances. At present, it is expected that over 10 μA of protons can be injected from the TRIUMF cyclotron at 80% polarization, with 72% polarization at 3 GeV from the Booster, and 55% polarization or better surviving to 30 GeV [31]. If final approval for the project is obtained in 1990, this would make possible realistic plans for a 30 GeV parity violation measurement within the next 10 years.

5. Conclusions

After much careful experimental and theoretical study, the $\Delta S=0$ weak interaction has yielded some interesting physical results. At low energy, a weak meson-exchange model provides a consistent framework for the interpretation of experimental data in nuclear systems. Parity violating effects in light nuclei and the two nucleon system are consistent with standard model predictions, but a rigorous test of the latter awaits the outcome of several new experiments before a complete determination of the weak meson nucleon coupling constants is achieved. The strong suppression of the weak pion coupling f_π^1 has already provided important evidence of a dynamical neutral current suppression that was completely unexpected. While the enhanced parity violating effects in finite nuclei are, in most cases, too difficult to analyze unambiguously due to nuclear structure uncertainties, the extremely large effects seen in neutron resonances in the rare earth region hold promise for investigations of time reversal noninvariance which may rival sensitivities reached in the neutron electric dipole moment experiments. The greatest promise for precise tests of the standard model predictions lies in the two nucleon system, where recent advances in technology and systematic error control have made possible the measurement of effects at the 10^{-8} level with impressive accuracy, and the results are free of ambiguities in interpretation. The $\vec{p}-p$ experiments should be repeated at 6 GeV/c and higher, to confirm the tantalizingly large effect measured at the Argonne ZGS and to extend our knowledge of the energy dependence of this fundamental quantity.

References

[1] B. Desplanques, J.F. Donoghue, B.R. Holstein, Ann. Phys. (NY) 124, (1980) 449

[2] E.G. Adelberger, W.C. Haxton, Ann. Rev. Nucl. Part. Sci. 35, (1985) 501

[3] V.P. Alfimenkov et al., Nucl. Phys. A398,(1983) 93; C.D. Bowman et al., Phys. Rev. C39, (1989) 1721

[4] Workshop Proc.: *Tests of Time Reversal Invariance in Neutron Physics*, April 17-19 1987, Chapel Hill, N.C., eds. N.R. Roberson, C.R. Gould, C.D. Bowman, World Scientific (1987)

[5] W.C. Haxton, Proc. of the Symposium/Workshop on Parity Violation in Hadronic Systems, May 28-29 1987, Vancouver, B.C., Can. J. Phys. 66, (1988) 503; see also ref. 2.

[6] E.G. Adelberger et al., Phys. Rev. C27, (1983) 255

[7] H.C. Evans et al., Phys. Rev. Lett. 55, (1985) 791; M. Bini et al., Phys. Rev. Lett. 55, (1985) 795

[8] B.R. Holstein, Can. J. Phys. 66, (1988) 508

[9] V.J. Zeps, Ph.D. Thesis, University of Washington (1989)

[10] W.C. Haxton, V.J. Zeps, private communication

[11] V.M. Lobashov et al., Nucl. Phys. A197, (1972) 241

[12] V.A. Knyazkov et al., Nucl. Phys. A417, (1984) 209

[13] E.D. Earle et al., Can. J. Phys. 66, (1988) 534

[14] J. Alberi et al., Can. J. Phys. 66, (1988) 542

[15] E.G. Adelberger, B. Heckel, private communication

[16] J.M. Potter et al., Phys. Rev. Lett. 33, (1974) 1307; D.E. Nagle et al., AIP Conf. Proc. 51, (1979) 218

[17] R. Balzer et al., Phys. Rev. Lett. 44, (1980) 699; R. Balzer et al., Phys. Rev. C30, (1984) 1409

[18] S. Kistryn et al., Phys. Rev. Lett. 58, (1987) 1616

[19] M. Simonius, AIP Conf. Proc. 150, (1986) 185; M. Simonius, Can. J. Phys. 66, (1988) 548

[20] D.E. Driscoll, G.A. Miller, Phys. Rev. C39, (1989) 1951

[21] S.A. Page, J. Birchall, W.T.H. van Oers, TRIUMF Research Proposal E497, *Measurement of the Flavor Conserving Hadronic Weak Interaction* (1987)

[22] V. Yuan et al., Phys. Rev. Lett. 57, (1988) 1680

[23] N. Lockyer et al., Phys. Rev. D30, (1984) 860

[24] E.M. Henley, F.R. Krejs, Phys. Rev. D11, (1975) 605; T. Oka, Prog. Theor. Phys. 66, (1981) 977; A. Barroso, D. Tadic, Nucl. Phys. A364, (1981) 194; L.L. Frankfurt, M.I. Strikman, Phys. Lett. 107B, (1981) 99; P. Chiappetta et al., J. Phys. G 8, (1982) L93

[25] L.L. Frankfurt, M.I. Strikman, Phys. Rev. D33, (1986) 293

[26] G. Nardulli and G. Preparata, Phys. Lett. 117B, (1982) 445

[27] T. Goldman and D. Preston, Phys. Lett. 168B, (1986) 415

[28] J.F. Donoghue and B.R. Holstein, Phys. Lett. 125B, (1983) 509

[29] M. Simonius and L. Unger, Phys. Lett. 198B, (1987) 547

[30] T. Goldman and D. Preston, Proc. of the Symposium on Future Polarization at Fermilab, June 13-14 1988, Batavia, Illinois

[31] U. Wienands, Workshop on High Energy Spin Physics, TRIUMF, 15 Feb. 1989

[32] R.E. Mischke, Can. J. Phys. 66, (1988) 495

Z. Phys. C – Particles and Fields 46, S157–165 (1990)

Zeitschrift für Physik C Particles and Fields

Hypernuclear physics with the (π^+, K^+) reaction*

R. E. Chrien, D. J. Millener

Physics Department, BNL, Upton, NY 11973, USA

I. Introductory Background

The study of hypernuclei affords unique insights into the nature of hadronic forces beyond those obtainable from ordinary nuclear physics research. A hypernucleus consists of one or more hyperons bound to a nuclear core. In the SU(3) classification, the Λ, Σ, and Ξ hyperons occupy the same octet representation of spin–parity $\frac{1}{2}^+$ baryons as the neutron and proton, the familiar constituents of ordinary nuclei. The Λ and Σ possess strangeness $S = -1$ and isotopic spin $I = 0$, 1, respectively, while the Ξ has $S = -2$, $I = \frac{1}{2}$. In the underlying quark picture, the Λ and Σ have the flavor structure $s(ud)_{I=0,1}$, compared to the combination uud for the proton. The strange quark, s, carries the strangeness quantum number $S = -1$, which makes it distinguishable from the $S = 0$ u and d quarks. The lowest–lying hyperon is the Λ, with a mass of $1115.6\,\mathrm{MeV/c^2}$, some $177\,\mathrm{MeV/c^2}$ heavier than the proton. The study of the behavior of a hyperon embedded in the nuclear medium, through the theoretical analysis of level spectra, sheds light on the nature of the hyperon–nucleon (YN) effective interaction, i.e. on the role of the strange quark in strong interactions.

A single Λ behaves essentially as a distinguishable particle in the nucleus: there are no discontinuities in the binding energy B_Λ as a function of A, a signal of shell effects in ordinary nuclei. This property of the Λ implies additional dynamical symmetries for hypernuclear states that are not allowed for ordinary nuclei because of the Pauli principle. The Λ provides a superb example of single–particle structure in a many–body system. Deeply bound nucleon–hole states are very broad, a reflection of the large spreading width $\Gamma_\downarrow$ caused by admixtures with more complicated configurations. For the Λ, even for s–states in heavy system, $\Gamma_\downarrow$ is rather small, which indicates the rather weak ΛN residual interaction. The existence of a well–defined set of single particle states of different orbital angular momentum ℓ in a given hypernucleus enables one to extract information on the well depth, geometrical shape, and effective mass, all of which characterize the Λ–nucleus potential.

The principal production mechanisms for hypernuclei are strangeness exchange and associated production, corresponding to the elementary processes

$$
\begin{aligned}
&K^- + n \to \Lambda + \pi^- \\
&K^- + p \to \Lambda + \pi^0 \\
&K^- + n \to \Sigma^{0,-} + \pi^{-,0} \\
&K^- + p \to \Sigma^\pm + \pi^\mp,\ \Sigma^0 + \pi^0 \\
&\pi^+ + n \to \Lambda + K^+ \\
&\pi^+ + n \to \Sigma^{0,+} + K^{+,0} \\
&\pi^\pm + p \to \Sigma^\pm + K^+ \quad .
\end{aligned}
$$

Of these reactions, only those involving charged mesons in the exit channel have been studied extensively. In the case of nuclear species, the n, p above are replaced with the corresponding target nuclide.

The characteristic differences between the (K, π) and (π, K) production modes have been discussed by Dover, Ludeking & Walker [1] and by Bandō & Motoba [2,3], among others. The (K, π) reaction at small angles is a process with small momentum transfer q. It preferentially populates "substitutional" hypernuclear states, i.e. those states in which a nucleon in a shell model orbit with orbital angular momentum ℓ and total spin j is replaced by a Λ particle in the same orbit (ℓ, j). This transition is characterized by an orbital angular momentum transfer of $\Delta L = 0$. For the (K, π) reaction, there exists

*This manuscript has been authored under contract number DE-AC02-76CH00016 with the U.S. Department of Energy. Accordingly, the U.S. Department retains a non-exclusive, royalty-free license to publish or reproduce the published form of this contribution, or allow others to do so, for U.S. Government purposes.

a "magic momentum," i.e. a value of lab kaon momentum p_K for which q vanishes at $\theta = 0°$. This corresponds to $p_K = 530\,\mathrm{MeV/c}$ for Λ and $290\,\mathrm{MeV/c}$ for Σ^0 production. Note that q at $\theta = 0°$ remains less than $100\,\mathrm{MeV/c}$ for $p_K \leq 800\,\mathrm{MeV/c}$. At or near the magic momentum, the substitutional transitions with $\Delta L = 0$ are particularly enhanced with respect to those for which $\Delta L \neq 0$. The Λ hypernuclear states populated in $\Delta L = 0$ transitions tend to occur near, or above, zero Λ binding. The (K^-, π^-) process involves strong absorption of both the incoming and outgoing waves and consequently is localized in the nuclear periphery.

For the associated (π^+, K^+) reaction, there is no magic momentum, and $q \geq 350\,\mathrm{MeV/c}$ at the elementary cross section maximum near $p_K \approx 1.05\,\mathrm{GeV/c}$. The formation of high spin states is favored by the form factor of the (π^+, K^+) reaction. Furthermore, because of the long mean free path in nuclear matter for the K^+, the distortion for the outgoing wave is reduced; the reaction is less peripheral in nature. Fig. 1 nicely illustrates the comparison between (K^-, π^-) and (π^+, K^+) strength distributions. One sees the complementarity of these two reactions in exciting the low and high spin parts, respectively, of the hypernuclear spectrum.

The associated production of Λ hypernuclei by the (π^+, K^+) reaction was first studied theoretically by Dover et al. [1] and demonstrated in experiments carried on at the Moby Dick spectrometer at the AGS [4–7].

For (π^+, K^+) reaction studies, the restrictions imposed by the use of beam separators and decay backgrounds are not so severe. The principal problem in such studies is the relatively smaller cross sections for (π^+, K^+) as compared to the (K^-, π^-) production of substitutional states. To compensate for the lower cross section, incident beam intensities of up to 10^7 pions per second have been used in the Moby Dick configuration. The problem then reduces to the difficulties of handling particle intensities in that range. The rather significant contamination of pion beams with a positron component presents further difficulties with such beams, as it becomes difficult to separate positrons and pions at momenta of 1 GeV/c and above.

The experimental arrangement for (π^+, K^+) differs in several respects from the corresponding (K^-, π^-) reactions. The elementary (π, K) cross section has a maximum near $p_\pi = 1050\,\mathrm{MeV/c}$, which is at the upper limit for the LESB–1 line, and well above the $700-800\,\mathrm{MeV/c}$ usually chosen for the strangeness–exchanging (K^-, π^-) reaction. The roles of the kaon and pion spectrometers are interchanged, and the beam separator is tuned for pion transmission. Čerenkov counters are added after the target to veto pions in the rear spectrometer.

The cross sections obtained by Milner et al. [4] for the 1^- ground state and the 0^+, 2^+, 2^+ excited state multiplet in ${}^{12}_{\Lambda}\mathrm{C}$ near 11 MeV agree well with the DWBA calculated cross sections, after Fermi motion is folded in

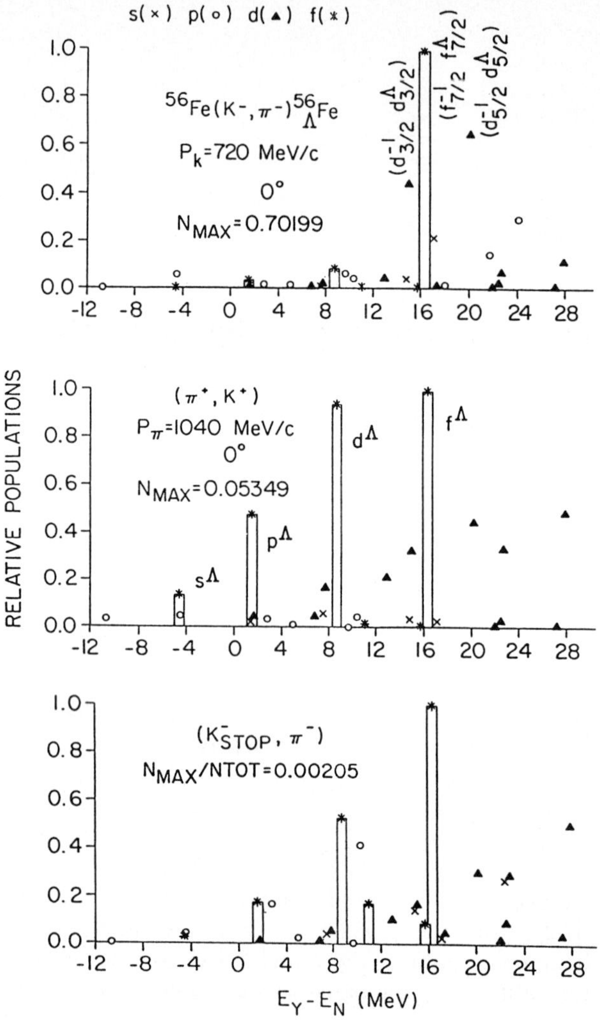

Figure 1: Relative strength for transitions to states in ${}^{56}_{\Lambda}\mathrm{Fe}$ induced by the (K^-, π^-), (π^+, K^+) reaction on a ${}^{56}\mathrm{Fe}$ target. The strength is in units of $N_{\mathbf{max}}$, the effective neutron number of the strongest transition. The selectivity of the (K^-, π^-) process at 0° in flight for low spin substitutional states and the tendency for the (π^+, K^+) reaction to populate a trajectory of high spin states $(f^{-1}_{7/2} \times j_\Lambda)$ is evident from the figure, which has been adapted from Bandō & Motoba [2]; the $f^{-1}_{7/2}$ series is highlighted.

and realistic optical model parameters for the π^+ and K^+ distorted waves are used. It is interesting to observe that the lower cross sections characteristic of the (π^+, K^+) reaction on light hypernuclei, compared to those for (K^-, π^-), are more than compensated by the increased particle flux available for pions as compared to kaons.

For hypernuclei beyond the p-shell, (K, π) reactions become less effective in populating bound states of hypernuclei because of the increasingly higher angular momentum of the valence shell neutrons. The coupling of the Λ to the high-spin neutron hole produces

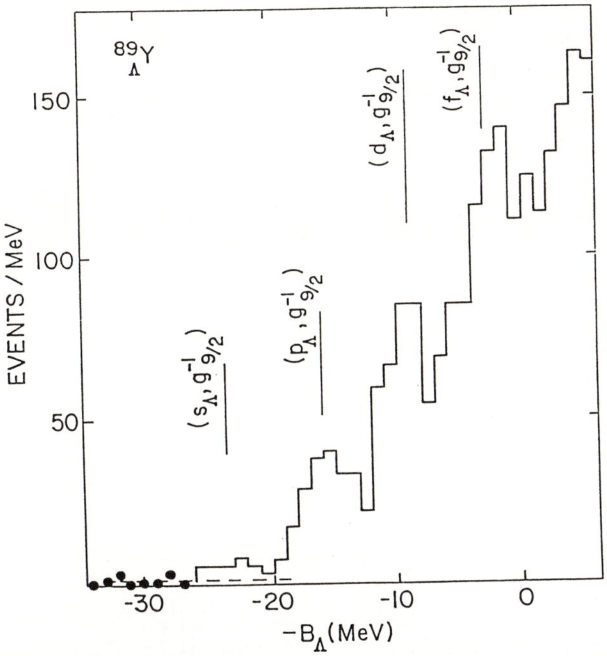

Figure 2: The excitation spectrum for the $^{89}Y(\pi^+, K^+)^{89}_{\Lambda}Y$ reaction at 1.05 GeV/c and $\theta_{K^+} = 10°$, from an AGS experiment [5–7]. The predicted Λ binding energies B_Λ for single particle configurations $(g_{9/2}^{-1} \times \ell_\Lambda)$ are indicated. The black dots below the ground state indicate the measured background.

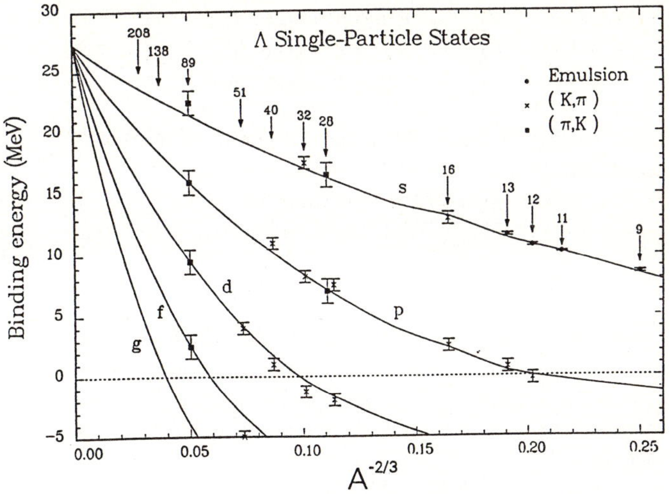

Figure 3: Data on binding energies of s, p, d, f single particle states of the Λ as a function of $A^{-2/3}$, from Millener et al. [8] The curves correspond to a nonlocal Λ–nucleus potential with ρ^2 density dependence, as explained in the text.

an advantageous momentum matching for the (π^+, K^+) reaction, as discussed by a number of authors. The high spin selectivity preferentially highlights a series of states in which the Λ–shell model orbitals $(s, p, d, f \cdots)$ couple to the valence neutron hole. The lack of spin–dependence in the Λ nuclear interaction produces a set of regularly spaced, narrow, single particle excitations that dominate over other more complicated particle–hole excitations. This strikingly simple sequence of levels was illustrated by Bandō & Motoba [2] for $^{56}_{\Lambda}Fe$ (see fig. 1). The cross sections can be calculated by a distorted–wave impulse approximation (DWIA) and are measurable, even for deeply–lying orbitals of heavy hypernuclei. In the second (π^+, K^+) experiment [5–7], eight targets were examined at $\theta = 10°$: ^{9}Be, ^{12}C, H_2O, Si, ^{13}C, ^{51}V, Ca, and ^{89}Y. For the latter four targets, beam particle intensities up to 10^7 pions per second were found usable with targets of reasonable size, namely 2 – 4 g/cm^2. The (π^+, K^+) spectra are observable over an irreducible background level of less than 100 n.b./sr/MeV, which allows a clear observation of the s_Λ ground state peak for $^{89}_{\Lambda}Y$, expected to be populated at a level of 0.5 μb/sr. Fig. 2 shows the data obtained for $^{89}_{\Lambda}Y$.

These two experiments have demonstrated that the (π, K) reaction is the method of choice for producing all but the lightest hypernuclei, preferable to (K, π) for directly accessing deeply lying hypernuclear states. They have made possible the interpretation of Λ single particle spectra in terms of the Λ–nucleus mean field.

Recently, through the (π^+, K^+) studies at BNL, it has become possible to track the evolution of Λ binding energies (ground and excited states) as a function of A, up to $A \approx 90$. The results for s_Λ, p_Λ, d_Λ and f_Λ single–particle binding energies, as obtained from (π^+, K^+), (K^-, π^-) and emulsion measurements, are plotted in fig. 3, taken from Millener et al. [8] The A dependence of the Λ level spacings enables us to constrain the geometry of the Λ–nucleus potential and also the well depth D_Λ.

A useful description of the data can be obtained in the Skyrme–Hartree–Fock approach, first used for hypernuclei by Rayet. Here, the nonlocality of the Λ–nucleus potential is parameterized in terms of an effective mass $m^*_\Lambda(r)$. The equivalent energy–dependent local potential $V_\Lambda(r, E)$ is of the form:

$$\begin{aligned} V_\Lambda(r, E) &= \frac{m^*_\Lambda(r)}{m_\Lambda} U(r) + \left(1 - \frac{m^*_\Lambda(r)}{m_\Lambda}\right) E \\ U(r) &= t_0\,\rho(r) + \frac{3}{8} t_3 \rho^2(r) + 1/4(t_1 + t_2) T(r) \\ T(r) &= \frac{3}{5}\left(\frac{3\pi^2}{2}\right)^{2/3} \rho^{5/3}(r) \\ \frac{h^2}{2m^*_\Lambda(r)} &= \frac{h^2}{2m_\Lambda} + \frac{1}{4}(t_1 + t_2)\rho(r)\,. \end{aligned} \tag{1}$$

The $t_3\rho^2(r)$ term can be used to adjust the potential radius, while the $(1 - m^*_\Lambda/m_\Lambda)E$ term serves to spread out the single particle levels; this enables one simultaneously to fit the spectra of light ($^{16}_{\Lambda}O$) and heavy ($^{89}_{\Lambda}Y$) systems. The choice

$$\begin{aligned} t_0 &= -402.6 \quad \text{MeV}\cdot\text{fm}^3 \\ t_1+t_2 &= 103.4 \quad \text{MeV}\cdot\text{fm}^5 \\ t_3 &= 3394.6 \quad \text{MeV}\cdot\text{fm}^6 \end{aligned} \tag{2}$$

leads to the fit displayed in fig. 3. These values correspond to

$$\begin{aligned} \frac{m^*_\Lambda(r=0)}{m_\Lambda} &\approx 0.8 \\ D_\Lambda &\approx 27.5\ \text{MeV} \quad . \end{aligned} \tag{3}$$

The calculations of Yamamoto et al. [9], which incorporate self-consistency and rearrangement energies (omitted by Millener et al. [8]), are consistent with the need for a strong repulsive $t_3\rho^2(r)$ term and a modest degree of nonlocality ($m^*_\Lambda(0)/m_\Lambda \approx 0.8$).

A strong $t_3\rho^2(r)$ term is also required to fit nucleon binding energies. However, within the Skyrme–Hartree–Fock model one cannot simultaneously describe nucleon levels near the Fermi surface (which requires $m^*_N(0)/m_N \approx 1$) and deeply bound levels (for which $m^*_N(0)/m_N \approx \frac{1}{2}$). For the Λ, on the other hand, a description of all levels is possible with a single $m^*_\Lambda(0)/m_\Lambda$.

The Λ behaves as a distinguishable particle in the nucleus. Unlike deeply bound nucleon–hole states, which are very broad, deeply bound Λ single particle states remain well defined. The possibility that strange quarks in the nucleus are partially deconfined is an intriguing one, but the signature of this effect in the Λ binding energies is likely to be subtle and easily masked by the complicated (but conventional) dynamics of density–dependent interactions.

An excellent example of the complementarity of the (K^-, π^-) and (π^+, K^+) reactions in investigation of hypernuclear structure is afforded by the case of ${}^9_\Lambda$Be. The ^{9}Be$(K^-, \pi^-){}^9_\Lambda$Be reaction at forward angles has been investigated at CERN. The spectrum at 720 MeV/c is shown in fig. 4, along with the excitation function of the ^{9}Be$(\pi^+, K^+){}^9_\Lambda$Be reaction at 1.05 GeV/c. The structure of ${}^9_\Lambda$Be was discussed in the shell model framework by Dalitz & Gal [10] and later by Auerbach et al. [11] The cluster model for ${}^9_\Lambda$Be has been extensively developed.

The essential features of the ${}^9_\Lambda$Be spectrum can be seen in a coupling scheme defined by $J = L + s_\Lambda$, $L = J_c + \ell_\Lambda$, where J_c is the spin of the nuclear core. For an interaction independent of the Λ spin s_Λ, L is a good quantum number, and states with $J = L \pm \frac{1}{2}$ (for $L \neq 0$) form a degenerate doublet. The structure of ${}^9_\Lambda$Be is similar to ${}^{13}_\Lambda$C in that the LS structure of the p^4 core of ^{8}Be with [4] and [31] symmetries resembles the [44] and [431] symmetries for ^{12}C. The neutron pickup strength goes mostly to the ^{8}Be ground state, the 2^+ state at 2.94 MeV, a group of states between 16 and 20 MeV and a 3^+ level above 19 MeV.

It is instructive to consider the ^{8}Be$(0^+, 2^+) \times p_\Lambda$ states in ${}^9_\Lambda$Be, in comparison with ${}^{13}_\Lambda$C. The 0^+ (ground state) and 2^+ (2.94 MeV) states in ^{8}Be have almost pure

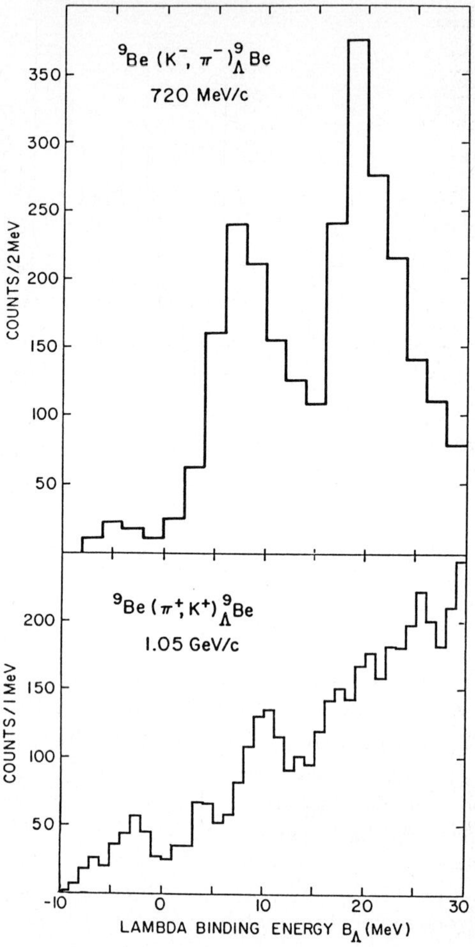

Figure 4: The measured (K^-, π^-) and (π^+, K^+) excitation spectra on a ^{9}Be target. The (K^-, π^-) forward angle data at 720 MeV/c are replotted from Bertini et al. [13] The (π^+, K^+) spectrum at 1.05 GeV/c, $\theta = 10°$, has been obtained from P. Pile (private communication).

[4] spatial symmetry with $S = 0$. Thus the states of ${}^9_\Lambda$Be can have [5] or [41] symmetry, respectively. Since the ^{8}Be core is prolate while ^{12}C is oblate, the matrix element of the quadrupole operator Q for ^{8}Be is opposite in sign to that for ^{12}C; this implies an inverted order in ${}^9_\Lambda$Be of the $L = 1, 2, 3$ states based on the 2^+ core state. There is then stronger mixing in ${}^9_\Lambda$Be than in ${}^{13}_\Lambda$C of $L = 1$ states based on the 0^+ and 2^+ core states. For $F^{(2)} = -3.2$ MeV, as derived earlier, this mixing is strong enough so that the eigenstate approaches the limit of good spatial symmetry $|[5]\ L = 1\rangle$. An important consequence is that the ^{9}Be$(K^-, \pi^-){}^9_\Lambda$Be cross section at 0°, leading to the lowest $3/2^-$ state ($L = 1$), becomes very small relative to the second $3/2^-$ state ($L = 1$), because the transition $[41] \to [5]$ is forbidden for $\Delta L = 0$. The remainder of the $\Delta L = 0$ strength is found in two states with $T = 0, 1$, about 12.5 MeV above the second $3/2^-$ level. This is in excellent agreement with the (K^-, π^-)

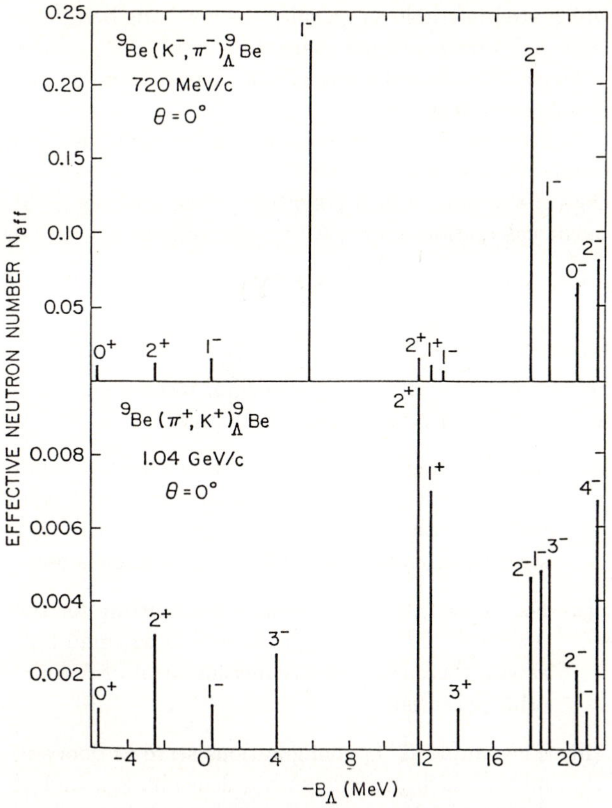

Figure 5: The effective neutron number N_{eff} for the (K^-, π^-) and (π^+, K^+) reactions on ^{9}Be, as calculated by Bandō [13]. The forward differential cross section for each hypernuclear state with binding energy B_Λ is given by $N_{\text{eff}} \cdot (d\sigma/d\Omega)_{\text{2-body}}$ in terms of the $K^- n \to \pi^- \Lambda$ or $\pi^+ n \to K^+ \Lambda$ two-body 0° cross section $(d\sigma/d\Omega)_{\text{2-body}}$. The states in $^9_\Lambda$Be are labeled by orbital angular momentum L and parity π, in a coupling scheme where $J = L + s_\Lambda$. Thus, each line with $L \neq 0$ corresponds to a doublet for which the spin splitting is neglected. Note the strong selectivity of both the (K^-, π^-) reaction, which has low q and favors $\Delta L = 0$ transitions, and the (π^+, K^+) process, which has $q \approx 350$ MeV/c and preferentially excites the higher spin $^9_\Lambda$Be states.

data shown in fig. 4. which display two strong peaks (at 7 and 19 MeV) separated by about 12 MeV.

The simplest version of the cluster model for $^9_\Lambda$Be, namely $\alpha + \alpha + \Lambda$, does not explain the third peak at 19 MeV in the (K^-, π^-) spectrum of fig. 4, since this involves a strong contribution of isospin one–core excited states of ^{8}Be. The cluster model was recently extended [12] to include $\alpha + \alpha^* + \Lambda$ configurations, where α^* is the intrinsic excited state of the α particle. The distribution of (K^-, π^-) strength for this model, shown in fig. 5, is rather similar to that obtained in the shell model.

The (π^+, K^+) reaction, in contrast to (K^-, π^-) at 0°, favors the excitation of the higher spin states in $^9_\Lambda$Be. The predicted strength distribution of ^{9}Be$(\pi^+, K^+)^9_\Lambda$Be reaction at 1.05 GeV/c is shown in fig. 5. In contrast to the (K^-, π^-) spectrum, one expects a measurable cross section to the "supersymmetric" 3^- state near $-B_\Lambda \approx 4$ MeV; these states were first discussed by Dalitz & Gal. Further, a peak near $-B_\Lambda \approx 12$ MeV, corresponding to ^{8}Be* $(2^+, 1^+, 3^+) \times s_\Lambda$, should be seen in (π^+, K^+), but not in (K^-, π^-). The experimental (π^+, K^+) spectrum of fig. 4 indicates three peaks below $-B_\Lambda = 15$ MeV, consistent with dominant excitation of $L^\pi = 2^+$, $(B_\Lambda = 2\,\text{MeV})$, $3^-(B_\Lambda = -4\,\text{MeV})$ and $2^+(B_\Lambda = -12\,\text{MeV})$ states. The hypernucleus $^9_\Lambda$Be affords an excellent example of the complementary use of both (K^-, π^-) and (π^+, K^+) to obtain a more complete picture of the hypernuclear spectrum.

II. Hypernuclear Physics for the Future

Some possibilities for a research program with high intensity (assumed 2×10^8 π^+/sec, which is about 20 times current AGS intensities) and good resolution (possibly 100 keV, to be compared with 2 – 3 MeV at BNL) are outlined.

1. Spectroscopy with the (π^+, K^+) reaction

(a) Λ *single-particle binding energies*

As noted in the introduction, the (π^+, K^+) reaction at $p_\pi \sim 1.05$ GeV/c is the optimal way to produce heavy Λ hypernuclei. This has been verified at Brookhaven [4–7] for targets up to $A = 90$. The presentation [5] of the results of a preliminary analysis of the data has produced a rash of theoretical papers [14–17].

On targets with last–filled neutron orbits of high spin, the (π^+, K^+) spectrum is dominated by the excitation of a series of high spin $j_N^{-1}\ell_\Lambda$ particle–hole states, as illustrated in fig. 2 for a ^{89}Y target. The energies of the $g_{9/2}^{-1}\ell_\Lambda$ states define the binding energies of Λ single particle states which fit into the regular pattern shown in fig. 3 and determine rather precisely the essential characteristics of the Λ–nucleus single particle potential [8].

Better data for the $f_{7/2}^{-1}\ell_\Lambda$ series states using at target with $A \sim 50$ would be welcome; the ^{51}V$(\pi^+, K^+)^{51}_\Lambda$V data from BNL suffer from poor statistics due to limited running time. Motoba et al. [14–16] present theoretical spectra for a ^{56}Fe target. The $h_{11/2}^{-1}\ell_\Lambda$ series should be easily observed using as a target any of the $N = 82$ nuclei, of which all exhibit $h_{11/2}$ pickup strength concentrated in a low–lying state [18] (e.g. ^{136}Xe, ^{138}Ba, ^{140}Ce, ^{142}Nd, ^{144}Sm).

For heavier targets the spacing of the Λ single–particle states decreases and good resolution is required. E.g., for ^{208}Pb the $i_{13/2}^{-1}\ell_\Lambda$ and $h_{9/2}^{-1}\ell_\Lambda$ series are interleaved, and the strongly excited states are separated by about 2 MeV [2,15].

Table 1: Ground state (π^+, K^+) cross sections (μb/sr)

Angle	^{28}Si	^{40}Ca	^{51}V	^{90}Zr	^{208}Pb$^{a)}$
0°	3.7	1.4	3.5	2.0	0.55
10°	1.8	0.5	1.4	0.8	0.15

a) $i_{13/2}^{-1}s_\Lambda$

Table 2: ^{89}Y(π^+, K^+)$^{89}_{\Lambda}$Y cross sections (μb/sr) for pure particle–hole configurations (summed over Λ spin–orbit doublets).

Configuration	ΔL	0°	5°	10°	15°	20°
$g_{9/2}^{-1}s_\Lambda$	4	2.05	1.67	0.82	0.16	0.013
$g_{9/2}^{-1}p_\Lambda$	5	8.70	7.52	4.54	1.52	0.17
$g_{9/2}^{-1}d_\Lambda$	6	18.8	17.0	12.0	5.6	1.2
	4	2.1	1.5	0.47	0.15	0.18
$g_{9/2}^{-1}f_\Lambda$	7	27.1	25.3	19.6	11.2	3.81
	5	5.4	4.3	1.9	0.40	0.19
$g_{9/2}^{-1}g_\Lambda$	8	7.5	7.4	6.7	5.1	2.7
	6	15.9	13.8	8.6	3.2	0.53
	4	0.75	0.49	0.21	0.32	0.23
$f_{7/2}^{-1}s_\Lambda$	3	0.60	0.42	0.12	0.01	0.01
$f_{7/2}^{-1}p_\Lambda$	4	3.4	2.7	1.2	0.19	0.03
$f_{7/2}^{-1}d_\Lambda$	5	9.3	7.9	4.4	1.3	0.13
$f_{7/2}^{-1}f_\Lambda$	6	15.8	13.9	9.0	3.6	0.66
	4	0.84	0.57	0.17	0.12	0.12
$p_{3/2}^{-1}f_\Lambda$	4	0.14	0.38	0.89	0.95	0.35
$f_{5/2}^{-1}g_\Lambda$	7	18.7	17.5	13.7	8.0	0.13
	5	3.6	2.9	1.3	0.28	0.13

(b) *(π^+, K^+) cross sections*

A representative sample of calculated cross sections for the production of hypernuclear states with the Λ in an s orbit are given in table 1. Angular distributions for various particle–hole configurations populated in the ^{89}Y(π^+, K^+)$^{89}_{\Lambda}$Y reaction are given in table 2.

These cross sections are calculated using optical potentials derived from the elastic scattering of 800 MeV/c π^+ and K^+ on ^{12}C (similar results are obtained for optical potentials derived from elastic scattering on ^{40}Ca, which are preferable for calculations involving heavy nuclei) and Fermi–average elementary $n(\pi^+, K^+)\Lambda$ cross sections. Such calculations give a good account of the Brookhaven data.

It would be useful to have measurements on pion elastic scattering at 1.05 GeV/c to improve our knowledge of the pion optical potentials which serve as input to the calculation of (π^+, K^+) cross sections.

(c) *Fine structure in (π^+, K^+) spectra*

The widths of the prominent peaks in the ^{89}Y(π^+, K^+)$^{89}_{\Lambda}$Y spectra are somewhat broader than can be accounted for by the experimental resolution. There are a number of possible reasons for this.

(i) The $g_{9/2}^{-1}$ strength in the ^{88}Y core nucleus is shared by 4^- and 5^- levels 233 keV apart, and the Λ couples to both states. For a ^{90}Zr target the $g_{9/2}^{-1}$ strength would be concentrated in a single state.

(ii) For $\ell_\Lambda \neq 0$, there is a spin–orbit splitting for the Λ orbits. This splitting is known to be small but, as yet, there is no precise measurement of Λ spin–orbit potential.

(iii) The simple $j_N^{-1}j_\Lambda$ configurations serve as doorway states which can mix with a dense background of Λ hypernuclear states (in analogy to an isobaric analog state which mixes with a dense background of $T_<$ states) leading to a spreading width.

(iv) The $A \approx 90$ nuclei have low–lying $f_{5/2}$ neutron hole states and a broad distribution of f hole strength at higher excitation energy. Hypernuclear configurations based on these states will give rise to some sharp states and a broad distribution of strength underlying the sharp states.

High resolution (π^+, K^+) studies could shed some light on the widths of the prominent peaks and on possible fine structure in the spectra.

(d) *Structure in light Λ hypernuclei*

For p–shell targets there are large (π^+, K^+) cross sections for $p_N \to s_\Lambda$, $\Delta L = 1$ and $p_N \to p_\Lambda$, $\Delta L = 2$ transitions to discrete states, as exemplified by the original ^{12}C(π^+, K^+)$^{12}_{\Lambda}$C experiment [4]. With good resolution, there is much detailed spectroscopy which could be done on a wide range of target nuclei. The energy separations of states and their relative (and absolute) (π^+, K^+) cross sections would help enormously in defining the effective ΛN interaction. A simple example is provided by two 2^+ states around 11 MeV excitation energy in $^{12}_{\Lambda}$C. These states, which are predominantly mixtures of Λ particles in $p_{3/2}$ and $p_{1/2}$ orbits coupled to the ^{11}C $3/2^-$ ground state are not resolved in the Brookhaven (π^+, K^+) experiments. However, emulsion data [19] shows two 2^+ states (and a 0^+ state) below the ^{11}C + Λ threshold, separated by 750 keV and with proton decay widths of about 500 keV. The energy separation of the 2^+ states is clearly sensitive to the spin–orbit splitting between the $p_{3/2}$ and $p_{1/2}$ Λ orbitals. Also, shell–model calculations show that the relative formation strength for the two states is a very sensitive function of the spin–orbit splitting and the detailed nature of the ΛN effective interaction. There are many other examples [11] where the energy separations and cross sections of dominantly $p_N p_\Lambda$ configurations could be measured in (π^+, K^+) experiments with good resolution.

In the case of a Λ in an s orbit coupled to a nuclear core state with non–zero spin, all the evidence suggests that whatever spin dependence exists in the ΛN interaction usually conspires to yield very small doubled splittings in the 100 keV range [20]. In any case, it is most often true that only one member of the doublet will be strongly populated in the (π^+, K^+) reaction. Nevertheless, precision measurements of the separation between states based on different nuclear core states would yield information on the ΛN effective interaction. To return to the case of ${}^{12}_{\Lambda}$C, three low–lying 1^- states based on the $3/2^-$ g.s., $1/2^-$ 2.00 MeV and $3/2^-$ 4.80 MeV states of the ^{11}C core can be formed in the (π^+, K^+) reaction. The upper two 1^- states are relatively weakly formed with strengths that are very sensitive to small admixtures of the weak–coupling basis configurations. The $p_N s_\Lambda$ interaction favored by Millener et al. [20] shifts formation strength into the lowest 1^- level (${}^{12}_{\Lambda}$C ground state); the sensitivity of both excitation energies and formation strengths to the ΛN effective interaction is clearly evident in table III of [20], as are the small doublet separations for the standard ΛN interaction.

2. Coincidence reactions

(a) $(\pi^+, K^+\gamma)$ *reactions*

Hypernuclear γ–ray transitions in ${}^{7}_{\Lambda}$Li and ${}^{9}_{\Lambda}$Be have been observed with NaI detectors using the $(K^-, \pi^-\gamma)$ reaction at Brookhaven [21]. Subsequently, attempts were made with Ge(Li) detectors to observed γ–ray transitions between the members of ground state doublets in ${}^{10}_{\Lambda}$B and ${}^{16}_{\Lambda}$O. These experiments proved to be very difficult and near the limit of experiments possible at the AGS. The non–observation of a transition from the 2^- member of the ground state doublet in ${}^{10}_{\Lambda}$B to the 1^- ground state suggests that the doublet splitting is considerably smaller than the predicted [20] 170 keV. For a sufficiently small separation, of the order of 100 keV, the 2^- state will preferentially weak decay as the M1 partial lifetime becomes longer than $\sim$ 200 ps.

The $(\pi^+, K^+\gamma)$ reaction could provide greater sensitivity to look for γ transitions between the members of ground state doublets. Also, it may be possible to utilize the resolution of Ge(Li) detectors to look for higher energy transitions (in the few MeV range) between the members of doublets based on different core states. In this way, it may be possible to measure doublet splittings indirectly and perhaps see several transitions in a γ–ray cascade. In some instances, the resolution of NaI detectors may suffice.

In many cases, hypernuclear states formed at high excitation energies will particle decay leading to the possibility of observing γ transitions in a daughter hypernucleus. For example, Majling et al. [22] have estimated the particle decay widths of hypernuclear states formed in (K^-, π^-) reactions and predicted the γ–ray lines that should be seen for specific cuts on the momentum of the outgoing pion (i.e. certain regions of excitation in the primary hypernucleus). Similar calculations could easily be made for primary hypernuclear production using the (π^+, K^+) reaction.

(b) (π^+, K^+x) *reactions*

The spectra of the decay products from the breakup of an excited Λ hypernucleus formed in the (π^+, K^+) reaction carry useful information on the structure of the parent hypernuclear state. Majling et al. [22] have urged the measurement of both particle and γ spectra as tools to investigate hypernuclear structure.

(c) *Weak–decays of* Λ *hypernuclei*

The ultimate fate of a Λ hypernucleus is to weak decay via one or other of the two processes

$$\begin{array}{lll} \Lambda & \rightarrow N + \pi & \text{(mesonic)} \\ \Lambda N & \rightarrow NN & \text{(non–mesonic)} \quad , \end{array}$$

with the non–mesonic mode dominating for all but the lightest systems. The decay can take place either from the ground state or from an excited state whose electromagnetic lifetime is long compared with that for weak decay (typically $\sim$ 200 ps). Weak lifetimes have recently been measured [23] for ${}^{12}_{\Lambda}$C and ${}^{11}_{\Lambda}$B (211$\pm$31 and 192$\pm$22 ps, respectively) and lifetime measurements for heavier systems are desirable. Of even more interest are measurements of the various partial decay rates, such as the division into mesonic and non–mesonic and especially the ratio for proton stimulated to neutron stimulated non–mesonic decay. The ratio of $\Gamma_{\Lambda n \rightarrow np}$ to $\Gamma_{\Lambda n \rightarrow nn}$ is sensitive to the relative contribution of different meson exchanges in the $\Lambda N \rightarrow NN$ weak decay process (see e.g. [24]). Measurements of this ratio have been made [25] for the non–mesonic decays of ${}^{12}_{\Lambda}$C and ${}^{11}_{\Lambda}$B, but poor statistics lead to large errors on this very interesting quantity.

Measurements of the neutrons and protons which result from non–mesonic decay, and share the large energy release of $(176 - B_\Lambda - B_N)$ MeV, for a range of light and a selection of medium and heavy nuclei would form a very interesting program. In many heavy nuclei, states involving the p_Λ orbit lie below the threshold for proton decay so advantage can be taken of the larger cross section for producing p_Λ states relative to s_Λ states (the produced hypernuclei then γ decay to the ground state).

Suggestions have also been made [26] to study the non–mesonic weak decay of p–shell Λ hypernuclei by looking at the delayed γ–rays from the $A-2$ daughter nuclei.

Recently, a very comprehensive study of the π–mesonic decay of p–shell Λ hypernuclei has been made [27]. The calculation includes π^0 and π^- total mesonic decay rates, the partial decay excitation functions and pion angular distributions from polarized hypernuclei. The actual decay rates are very sensitive to details of the pion optical potential [27–29]. However, the ratio $\Gamma_{\pi^0}/\Gamma_{\pi^-}$ for specific nuclear states are not. The π^0/π^- ratio can, in many cases, be used to distinguish the spin of the decaying member of a ground state doublet. The spectrum for decays to individual nuclear final states is generally an even more sensitive test of the spin of the initial hypernucleus and of hypernuclear structure.

(d) *Polarization of hypernuclei in the* (π^+, K^+) *reaction*

Bandō et al. [30] have calculated cross sections and polarizations for the (π^+, K^+) reaction on ^{12}C, ^{16}O, ^{28}Si, and ^{56}Fe targets using the elementary non–spin–flip f and spin–flip g amplitudes from a reanalysis of the $\pi^- p \to K^0\Lambda$ data. Large polarizations can be obtained for certain states which are populated with appreciable cross section; generally speaking the cross sections (for natural parity states) drop with increasing angle while the polarizations increase. E.g., the polarization for the $^{12}_{\Lambda}$C 1^- ground state reaches 50% at $\theta \sim 15°$ with a cross section of $\sim 3\,\mu$b/sr at $p_\pi = 1.04$ GeV/c.

The polarization is especially useful when combined with coincidence measurements of secondary decay particles, as in the weak decay examples discussed above.

III. Summary

Intense π^+ beams and high resolution spectrometers would open up a very broad field of hypernuclear structure studies with the (π^+, K^+) reaction. The coincidence detection of hypernuclear decay products including γ rays would provide severe tests of the ΛN effective interactions used in hypernuclear structure models and of the models themselves. Information on the weak–decay properties of hypernuclei could provide strong constraints on the meson exchange mechanism for the non–mesonic $\Lambda N \to NN$ decay mode. Observations of the mesonic decay mode would provide definitive information on hypernuclear structure and on pion distorted waves in the nuclear interior.

A large number of theoretical tools are at hand to interpret the experimental results. Detailed models of hypernuclear structure and the reaction mechanisms of hypernuclear production and decay have been developed and, in the absence of precise data, remain to be fully tested.

It will be interesting to see if a ΛNN three–body interaction [31] or a density–dependent two–body interaction, [8] which appear to be necessary to reproduce absolute binding energies of Λ hypernuclei, are necessary to reproduce the details of hypernuclear level schemes. G–matrices [32] obtained from the Nijmegen two–body potentials have some interesting characteristics which differ from the NN interaction. Bandō & Motoba [2] give as an example the repulsive pairing behavior for $Nd_{5/2}\Lambda d_{5/2}$ configurations in which the 4^+ state lies 1 MeV below the 0^+ state in $^{18}_{\Lambda}$O; the 0^+ state would be populated strongly by the (K^-, π^-) reaction and the 4^+ state by the (π^+, K^+) reaction, but it would require good resolution for both reactions to measure the $0^+ - 4^+$ energy difference.

References

[1] C. B. Dover, L. Ludeking and G. E. Walker, Phys. Rev. C22, (1980) 2073

[2] H. Bandō and T. Motoba, Prog. Theor. Phys. 76, (1986) 1321

[3] H. Bandō, Nucl. Phys. A478, (1988) 697c

[4] C. Milner et al., Phys. Rev. Lett. 54, (1985) 1237

[5] R. E. Chrien, Nucl. Phys. A478, (1988) 705c

[6] J. C. Peng, AIP Conference Proceedings, No. 176 (1988) 39

[7] P. H. Pile, in Proc. of Vth Int. Conf. on *Clustering Aspects in Nuclear and Subnuclear Physics*, Kyoto, Japan (1988), to be published

[8] D. J. Millener, C. B. Dover and A. Gal, Phys. Rev. C38, (1988) 2700

[9] Y. Yamamoto, H. Bandō and J. Zofka, Prog. Theor. Phys. 80, (1988) 757

[10] R. H. Dalitz and A. Gal, Ann. Phys. (NY) 131, (1981) 314

[11] E. H. Auerbach et al., Ann. Phys. (NY) 148, (1983) 381

[12] T. Yamada, K. Ikeda, H. Bandō and T. Motoba, Phys. Rev. C38, (1988) 854

[13] R. Bertini et al., Phys. Lett. 90B, (1980) 375

[14] T. Motoba, Nucl. Phys. A479, (1988) 227c

[15] T. Motoba, H. Bandō, R. Wünsch and J. Zofka, Phys. Rev. C38, (1988) 1322

[16] T. Motoba, in *International Symposium on Hypernuclear and low–energy Kaon Physics*, Padova (1988)

[17] R. Hausmann and W. Weise, Nucl. Phys. A491, (1989) 598

[18] R. K. Jolly and E. Kashy, Phys. Rev. C4, (1971) 1398

A. Chaumeaux et al., Nucl. Phys. A164, (1971) 176
S. Sen et al., Phys. Rev. C6, (1972) 2201

[19] R. H. Dalitz, D. H. Davis and D. N. Tovee, Nucl. Phys. A450, (1986) 311c

[20] D. J. Millener, A. Gal, C. B. Dover and R. H. Dalitz, Phys. Rev. C31, (1985) 499

[21] M. May et al., Phys. Rev. Lett. 51, (1983) 2085

[22] L. Majling et al., Nucl. Phys. A450, (1986) 189c; Phys. Lett. 130B, (1983) 235; Phys. Lett. 183B, (1987) 263

[23] R. Grace et al., Phys. Rev. Lett. 55, (1985) 1055

[24] J. F. Dubach, Nucl. Phys. A450, (1986) 71c

[25] J. J. Szymanski, A.I.P conf. proc. 150, (1986) 934

[26] L. Majling et al., Phys. Lett. 202B, (1988) 489

[27] T. Motoba, K. Itonaga and H. Bandō, Nucl. Phys. A489, (1988) 683

[28] K. Itonaga, T. Motoba and H. Bandō, Z. Phys. A330, (1988) 209

[29] R. Mach et al., Z. Phys. A331, (1988) 89

[30] H. Bandō, T. Motoba, M. Sotona and J. Zofka, Fukui University preprint, FUMP–1988–8

[31] A. R. Bodmer and Q. N. Usmani, Nucl. Phys. A477, (1988) 621

[32] Y. Yamamoto and H. Bandō, Prog. Theor. Phys. Suppl. No. 81 (1985) 9

Z. Phys. C – Particles and Fields 46, S167–181 (1990)

Zeitschrift für Physik C Particles and Fields

Theoretical investigations of reactions with strange particles

R. Büttgen, K. Holinde, D. Lohse, A. Müller-Groeling, J. Speth, P. Wyborny

Institut für Kernphysik (Theorie), Forschungszentrum Jülich, D-5170 Jülich, Fed. Rep. of Germany

Abstract:

We calculate within a generalized meson–exchange model observables for K^+N and K^-N scattering. We concentrate our investigations on the effect of the ω–meson exchange contribution. With a similar approach we investigate $\pi\pi$ and $K^+\pi^-$ scattering. Our main interest here is to find out to what extent the interaction between mesons can be understood in a meson–exchange frame work.

1. Introduction

It is generally accepted that quantum chromodynamics (QCD) is the fundamental theory of strong interaction. Therefore, in principle, the hadron–hadron interaction is determined by the quark–gluon dynamics. Unfortunately, little is known about QCD solutions in the low–energy (non–perturbative) regime, where most of the nuclear and medium energy physics experiments take place. In this region, however, there are indications [1] that most of the dynamics can be understood in terms of color–neutral objects, *i.e.* nucleons, mesons and isobars. Therefore in the low energy region (low compared to the perturbative QCD–regime) the strong interaction can probably be described to a very good approximation within the framework of meson–exchange interactions. The basic ingredients of such models are meson–baryon–baryon and meson–meson–meson vertices which represent an effective description of the very complicated, and mathematically yet intractible, multi–quark and gluon exchanges. These vertices contain coupling constants and form factors which paramtetrize the finite size of the hadrons. It is obvious that the meson exchange model is limited in its applicability, and that one expects corrections due to the underlying quark structure. The interesting question, however, is where do we see those "quark effects" in nuclear and medium energy physics? The range of its validity is clearly connected with the size of the confining region of the quarks. As the radii of hadrons are given not only by the size of the confining region of the quarks, but also by the extension of the surrounding meson cloud, the range of applicability of the meson–exchange models may be much larger than one expects from oversimplified estimates based on the hadron size.

In the past few years the Bonn NN–potential [2], which is based on suitably chosen meson–nucleon–nucleon and meson–nucleon–delta vertices, has been generalized to include the exchange of strange mesons [3], enabling the calculation of $N\Lambda$– and $N\Sigma$–potentials as well as the treatment of the K^+N–system which is the simplest meson–nucleon system [4] within the same framework.

In the systems which have been investigated previously, a repulsive core provided by the exchange of vector mesons, especially by ω–exchange prevents the two hadrons from coming too close to each other, so that one may argue that quark effects in those cases are not so obvious. Indeed it turned out, that one has to choose a coupling parameter for the ω–meson which is in all cases about 60% larger than the one given by SU(6) symmetry relations. This is necessary in order to obtain sufficiant short–range repulsion for a resonable discrition of the s–wave scattering phase shifts. This additional repulsion might be an indication that effects beyond the meson–exchange are important and that quark effects might be simply hidden behind the large ω–coupling parameter. In this contribution we report on calculations were we investigate especially this short range behaviour.

In the first part we compare K^+N and K^-N scattering. These two systems are especially appropriate for our purpose because in the K^+N–system the ω–meson exchange gives rise to a repulsive force, whereas it is attractive in the K^-N–system. Therefore a comparison of the two systems may give us information about the origin and the properties of the additional repulsion that is needed in order to reproduce the K^+N phase shifts. Indeed, we shall show that this additional repulsive contribution behaves differently in the K^-N–system compared to the ω–exchange.

In the second part we report on theoretical investigations of $\pi\pi$ and $K\pi$ scattering within the meson–exchange model. There exists a large amount of experimental data which can be compared with our theoretical results. The $\pi\pi$ scattering is of special interest in connection with the interaction between hadrons at short distances because here the repulsive ω–exchange is missing, and therefore quark effects might show up more clearly. Indeed, we will see that in order to understand theoretically the weak interaction at threshold in the spin–parity $(J^\pi) = 0^+$ and isospin (I)=0 channel of the $\pi\pi$ system, and in the $J^\pi=0^+$, I=1/2 channel of the $K\pi$ system, respectively, one has to introduce a repulsive piece of very short range into interaction that cannot be directly connected with experimentally known mesons. This repulsion is necessary in order to cancel at threshold the attraction provided by scalar meson exchange. However, we will also demonstrate that this effect depends on the way in which scalar mesons are coupled to the pseudoscalar mesons. If we use derivative coupling this repulsive contribution is no longer necessary.

In addition to this special question of the short range behaviour of the stong interaction a more general aim is to investigate to what extent, and how high in energy, the interaction between pseudoscalar mesons can be likewise understood in a meson–exchange framework. Thus we correspondingly formulate the strong effective interaction between those mesons and analyse the $\pi\pi$ and $K\pi$ data within the model. It turns out that for $\pi\pi$ scattering, the coupling to the $K\bar{K}$ channel is crucial, and is best dealt with in a coupled channel approach.

2. Comparison of K^+N and K^-N Scattering

2.1 The KN interaction model

Our strategy will be to treat the K^+N system in the meson exchange framework, in complete analogy to the Bonn NN model (see [2]). Therefore we start from a field–theoretical interaction Hamiltonian containing, apart from nucleon–nucleon–meson and nucleon–delta–isobar–meson couplings present in the Bonn model, additional K–K–meson and K–K*–meson vertices (K* being a strange vector meson with a mass of 895 MeV/c^2). Specifically, our total Hamiltonian is

$$H = H_0 + W \qquad (2.1)$$

with the free Hamiltonian

$$H_0 = \sum_\alpha E_\alpha a_\alpha^+ a_\alpha + \sum_\beta \omega_\beta b_\beta^+ b_\beta \qquad (2.2)$$

a_α^+, a_α and b_β^+, b_β are the creation, destruction operators for baryons (fermions) and mesons (bosons) obeying the usual anticommutator and commutator relations. Here α and β denote all quantum numbers which specify the respective state completely, *i.e.* momentum, spin, isospin and strangeness. The energies E_α and ω_β are the renormalized (physical) relativistic kinetic energies of baryons and mesons, respectively. Thus, apart from off–shell effects, self–energy contributions are automatically included and need not be evaluated explicitly. Furthermore, in complete consistency with the Bonn model, antinucleons are not included from the beginning although even with pseudovector coupling at the $NN\pi$ vertex, such contributions are not necessarily negligible. Rather, they depend sensitively on the NN vertex structure (form factor) which is not yet known with sufficient reliability to justify such an enormously involved calculation.

The interaction term W has the following structure

$$W = \sum_{\alpha\alpha' r} W_{\alpha'\alpha r} a^{+}_{\alpha'} a_{\alpha} b_r + \sum_{\beta\beta' r} W^{(1)}_{\beta'\beta r} b_{\beta'} b_{\beta} b_r + \sum_{\beta\beta' r} W^{(2)}_{\beta'\beta r} b^{+}_{\beta'} b_{\beta} b_r + \sum_{\beta\beta' r} W^{(3)}_{\beta'\beta r} b_{\beta'} b^{+}_{\beta} b_r + \sum_{\beta\beta' r} W^{(4)}_{\beta'\beta r} b^{+}_{\beta'} b^{+}_{\beta} b^{+}_r + \text{h.c.} \tag{2.3}$$

and thus contains baryon–baryon–meson as well as meson–meson–meson couplings. The vertex functions $W_{\alpha'\alpha r}$ and $W^{(i)}_{\beta'\beta r}$ can be derived in a straightforward way from suitable interaction Langrangians given explicitly in appendix A.

The next step is to define a transition matrix T for the kaon–nucleon system by relating it to the standard S–matrix,

$$\langle a_{n'} b_{m'} | S(z) | a_n b_m \rangle = \langle a_{n'} b_{m'} | a_n b_m \rangle - 2\pi i \; \delta(E_n + \omega_m - E_{n'} - \omega_{m'}) \langle a_{n'} b_{m'} | T(z) | a_n b_m \rangle \tag{2.4}$$

where $z \equiv E_n + \omega_m$ is the starting energy, $|a_n b_m\rangle \equiv a^{+}_n b^{+}_m |0\rangle$ (with) $|0\rangle$ the vacuum state.

Treating H in time–ordered perturbation theory, $\langle a_{n'} b_{m'} | T(z) | a_n b_m \rangle$ can be represented by a series expansion defined by all diagrams containing an incoming, $|a_n b_m\rangle$, and an outgoing, $|a_{n'} b_{m'}\rangle$, kaon–nucleon state,

$$\begin{aligned} \langle a_{n'} b_{m'} | T(z) | a_n b_m \rangle &= \langle a_{n'} b_{m'} | W \frac{1}{z - H_0 + i\epsilon} W | a_n b_m \rangle \\ &+ \langle a_{n'} b_{m'} | W \frac{1}{z - H_0 + i\epsilon} W \frac{1}{z - H_0 + i\epsilon} W \frac{1}{z + H_0 + i\epsilon} W | a_n b_m \rangle \\ &+ \ldots \; . \end{aligned} \tag{2.5}$$

This series can be partially summed by solving a (three–dimensional) integral equation of Lippmann–Schwinger type

$$\begin{aligned} \langle \alpha_{n'} b_{m'} | T(z) | a_n b_m \rangle &= \delta^{(3)}(\vec{p}_n + \vec{p}_m - \vec{p}_{n'} - \vec{p}_{m'}) \langle a_{n'} b_{m'} | V(z) | a_n b_m \\ &+ \langle \sum_{n''m''} \delta^{(3)}(\vec{p}_{n''} + \vec{p}_{m''} - \vec{p}_{n'} - \vec{p}_{m'}) \frac{\langle a_{n'} b_{m'} | V(z) | a_{n''} b_{m''} \rangle \; \langle a_{n''} b_{m''} | T(z) | a_n b_m \rangle}{z - E_{n''} - \omega_{m''} + i\epsilon} \end{aligned} \tag{2.6}$$

The kernel of this integral equation, *i.e.* the quasipotential $\langle a_{n'} b_{m'} | V(z) | a_n b_m \rangle$ consists of the (infinte) sum of all diagrams which are irreducible with respect to the kaon–nucleon channel. Diagrams involving at least one kaon–nucleon intermediate state are generated by scattering equation.

2.2 Determination of the quasipotential V(z)

For practical reasons, of course, one has to restrict oneself to those processes in V(z) which are of relatively low order in W. However, in order to be consistent with the Bonn NN model, our KN–interaction is based not only on single–particle exchanges but includes also fourth–order diagrams with N, Δ, K and K* intermediate states involving not only π– but also ρ–exchange, *i.e.* π– and ρ–exchange are grouped together. As in the Bonn model, σ–exchange effectively parametrizes correlated 2π–exchange processes in the s–channel. We mention that processes arising from the KK*ω coupling (having NK* intermediate states) are not considered; we feel justified to do so since their inclusion would inevitably require us also to include corresponding processes involving σ–exchange, because of the intimate interplay between attractive σ– and repulsive ω–exchange. We treat the K^+N and K^-N systems in an idential way as far as possible. The corresponding processes are shown in Fig. 2.1.

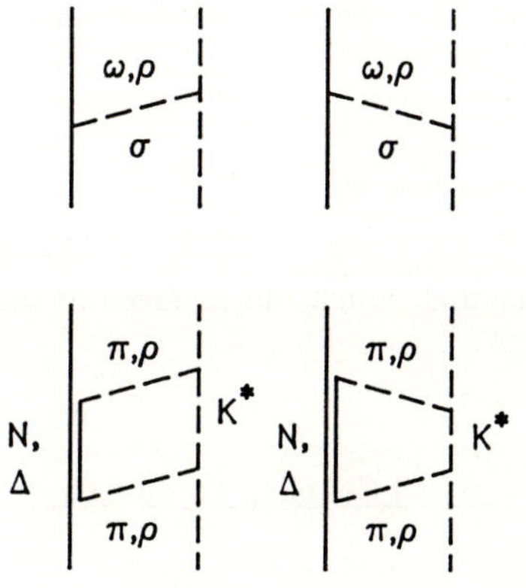

Fig.2.1: Contributions to K^+N– and K^-N–scattering.

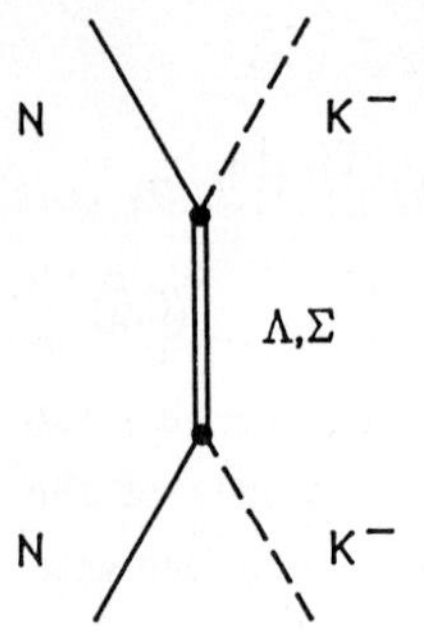

Fig. 2.2: "Pole"–contributions to K⁻N–scattering.

There is, however, one basic difference between the two systems which is connected with the quark–structure of the K^+ and K^- mesons. The K^+ consists out of an anti–strange quark $\bar{s}$ and an up quark u, whereas K^- includes a strange quark s and an unti up quark $\bar{u}$. In K^-N scattering the anti–up quark can annihilate with an up quark from the nucleon and the system can form a strange baryon or baryon resonance as an intermediate state. This gives rise to "pole graphs" as shown in Fig. 2.2.

In the calculation discussed here we consider only the Λ– and Σ–pole contribution. These poles are below the K^-N threshold and their influence is weak. On the other hand we analyse only K^-N data up to $P_{lab} = 300$ MeV (ϵ=1480 MeV), which is well below the first strange resonance at $E_{\Lambda*}$=1670 MeV (in the s–channel above the K^-N threshold). Therefore we expect that their influence on the present theoretical results are small. For reasons which we discuss later, the actual calculation of the K^-N system was performed with a coupled channel approach, where at energies below the K^-N–threshold the $\Lambda\pi$ and $\Sigma\pi$ channels are open. Here, indeed the $\Lambda*$ (1405) appears as a quasi–bound K^-N state in the $\Sigma\pi$ channel. The various coupled channels are shown in Fig. 2.3.

To calculate the contributions shown in Figs. 2.1 and 2.3 one needs the corresponding vertices, coupling parameters and form factors, which will be given in the next section.

2.3 Vertices, Form Factors and Coupling Parameters

First we list here the specific interaction Lagrangians (vertices) which dermine the interaction Hamiltonian

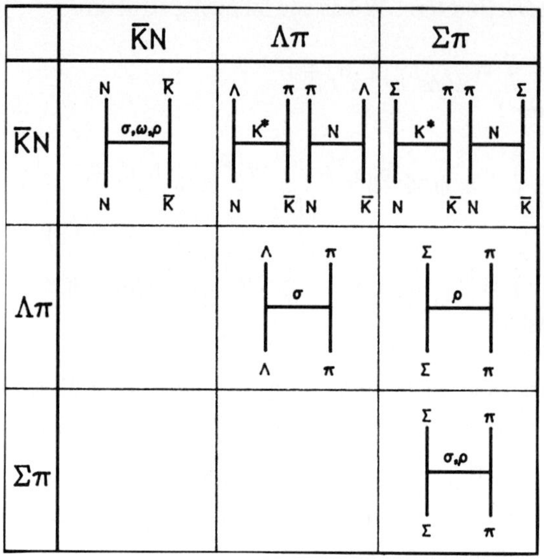

Fig. 2.3: The coupled channel approach to K^-N–scattering.

W that one needs for the calculation of the quasi–potential V(z). We start with the baryon–baryon–meson Lagrangians

$$
\begin{aligned}
L_{BBS} &= g_{BBS}\, \Psi_B(x)\, \Psi_B(x)\, \Phi_S(x) \\
L_{BBP} &= g_{BBP}\, \Psi_B(x)\, i\gamma^5\, \Psi_B(x)\, \Phi_P(x) \\
L_{BBV} &= g_{BBV}\, \Psi_B(x)\, \gamma_\mu \Psi_B(x)\, \Phi_V^\mu(x) \\
&\quad + \frac{f_{BBV}}{4M_N} \bar{\Psi}_B(x)\sigma_{\mu\nu}\Psi_B(x)(\partial^\mu \Psi_V^\nu(x) - \partial^\nu \Phi_V^\mu(x))
\end{aligned}
\tag{2.7}
$$

$$
L_{BDP} = \frac{f_{BDP}}{m_p}(\bar{\Psi}_{D\mu}(x)\Psi_B(x) + \bar{\Psi}_B(x)\Psi_{D\mu}(x))\partial^\mu \Phi_P(x)
$$

$$
L_{BDV} = \frac{f_{BDV}}{m_V} i(\Psi_{D\nu}(x)\gamma^5\gamma_\mu\Psi_B(x) - \bar{\Psi}_B(x)\gamma^5\gamma_\mu\Psi_{D\nu}(x))
(\partial^\mu\Phi_V^\nu(x) - \partial^\nu\Phi_V^\mu(x))
$$

Here $\Psi_B(x)$ ($\Psi_{D\mu}$) is the octet–(decuplet) baryon field operator and Φ_S, Φ_p and Φ_V^μ are the field operators for scalar, pseudoscalar and vector mesons, respectively Note that only the space–spin part is given. The additional SU(3) flavor part, if added, leads to the characteristic coupling constant ratios essentially used in the present paper.

The employed meson–meson–meson couplings are

$$L_{PPS}=g_{PPS}m_P\Phi_P(x)\Phi_P(x)\Phi_S(x)$$
$$L_{PPV}=g_{PPV}\Phi_P(x)\partial_\mu\Phi_P\Phi_V^\mu(x) \qquad (2.8)$$
$$L_{VVP}=\frac{g_{VVP}}{m_V}i\epsilon_{\mu\nu\tau\delta}\partial^\mu\Phi^\nu(x)\partial^\tau\Phi_V^\delta(x)\Phi_P(x)$$

where $\epsilon_{\alpha\beta\gamma\delta}$ is the antisymmetric tensor with $\epsilon^{0123}=1$.

Defining L to be the sum of all these Lagrangians, W given in eq. (2.3) has been determined from $W = -(\int L\, d^3x)$. The resulting vertex functions are modified by introducing form factors F, which describe the extended hadron structure. Their explicit form will be given below. In principle, they are completely determined by the underlying quark–gluon dynamics. However, because of the enormous complexity of QCD in the low energy regime, their derivation is not possible at present. Therefore, they are suitably parametrized, leading to the required suppression of the meson exchange contributions at high momentum transfer. Although in general they depend on all three 4–momenta involved at the vertex, they are usually parametrized in a simple form depending only on the 3–momentum of that particle which is exchanged in the corresponding potentials. Thus, the off–shell dependence appearing in higher iterations is ignored. Furthermore, in the conventional monopole form, the dependence on the zero–th component of q_r is likewise suppressed, *i.e.*

$$F_\alpha(\vec{q}_r^{\,2}) = \left[\frac{\Lambda_\alpha^2 - m_r^2}{\Lambda_\alpha^2 + \vec{q}_r^{\,2}}\right]^{n_\alpha} \qquad (2.9)$$

Such a parametrization is used in the Bonn potential [1] at the NN and NΔ vertices with $n_\alpha = 1$ and $n_\alpha = 2$ for the NΔρ vertex.

For consistency we use the same parametrization, with the same cutoff paramters Λ_α, at corresponding vertices in our model for the KN interaction. Furthermore, the additional form factors which appear at the meson–meson–meson vertices are also parametrized by eq. (2.9), with n_α=1 throughout. The only exception constitutes the form factor which parametrizes the off–shell behaviour of an exchanged hyperon. Here we take a different form,

$$F_\beta(\vec{q}_r^{\,2}) = \left[\frac{\Lambda_\beta^4 + M_r^4}{\Lambda_\beta^4 + (q_r^2)^2}\right] \qquad (2.10)$$

since we want to use the same parametrization, with sufficient convergence and without singularity problems, in corresponding pole graphs appearing in the K^-N–system. Again, Λ_β will be adjusted to the empirical data. Parameters are the coupling constants g, f (eqs. (2.7) and (2.8)) and cutoff masses Λ_α (eq. (2.9)) and Λ_β (eq. (2.19)). Apart from the empirically known masses of the exchanged particles, table 1 shows the parameter values for K^+N scattering. For consistency, coupling constants and cutoff masses belonging to NN and NΔ vertices are taken to be precisely the same as those of the (full) Bonn NN potential, (see [2]). Coupling constants at vertices involving strange baryons ($g_{N\Lambda K}$, $g_{N\Sigma K}$, g_{NY^*K}) have been related by the assumption of SU(6)

Tab. 1: Vertex parameters used in KN scattering.

Vertex	J^P I [a)]	$\frac{g_r}{\sqrt{4\pi}}$ resp. $\frac{f_r}{\sqrt{4\pi}}$	Λ_r [c)] [GeV/c²]
NNπ	0^- 1	3.795	1.3
NNρ [b)]	1^- 1	0.917	1.4
NNω	1^- 0	2.75	1.5
NNσ	0^+ 0	2.385	1.7
NNσ_0	0^+ 0	3.536	2.0
NΔπ	0^- 1	0.473	1.2
NΔρ	1^- 1	4.522	1.4
NΛK	$\frac{1}{2}^+$ 0	–3.944	1.4
NΣK	$\frac{1}{2}^+$ 1	0.759	1.4
NY*K	$\frac{3}{2}^+$ 1	–0.193	2.0
KKρ	1^- 1	0.857	1.55
KKω	1^- 0	0.857	1.6
KKσ	0^+ 0	0.377	1.4
KK*π	0^- 1	0.857	0.8
KK*ρ	1^- 1	0.857	1.1

a) Spin J, parity P, isospin I of the exchanged particle
b) tensor– to–vector coupling ratio is 6.1
c) cutoff mass

symmetry to the empirical NNπ coupling, whereas the meson–meson–meson couplings can be derived from the empirical $\pi\pi\rho$ coupling assuming the same symmetry scheme together with ideal mixing. Since σ–exchange (with a mass of 600 MeV) is meant to be a simple parametrization of correlated 2π–exchange processes, the σ should not be viewed as a real particle. Therefore its coupling strength should not be taken from a symmetry relation, as for the other couplings, but should, in principle, be determined by adjusting it to the result provided by the sum of all correlated 2π processes, as has been done for the NN case. At this stage we have avoided such a complicated calculation; throughout this work we have taken $g_{NN\sigma}$ as well as the remaining cutoff masses to fit the KN data.

The repulsive part of K^+N interaction has been parametrized in several diffent ways. In the first approach we increased the ω–coupling $g_{KK\omega}$ by about 60% over the SU(6) value in order to obtain sufficient short range repulsion. This is in analogy to $g_{NN\omega}$ in the Bonn potential. However, this increased ω–exchange leads to additional repulsion in p– and higher waves, which, on the whole, seems not to be favoured by the empirical data (for details see ref [5]). In the calculations shown here we have adopted a different point of view. If we assume, as we have done for all other couplings, that the real ω–coupling strength is given by SU(6) symmetry, then there is a considerable repulsive part needed which might arise either from additional meson–exchange contributions or possibly explicit quark–gluon effects. From this viewpoint it is quite natural that the additional contribution is much shorter–ranged, which is obviously required by the data. Consequently we keep the SU(6) coupling strength for both $g_{NN\omega}$ and $g_{KK\omega}$, but add instead a very short–ranged, phenomenological, repulsive contribution. For that term, we take the analytical form of σ–exchange, with opposite sign and with an exchanged mass of 1.2 GeV/c^2. The strength of this contribution, denoted by σ_0 in the following, is adjusted to the empirical phase shifts and given, together with other readjusted parameters in table 1. (Of course, only the product of $g_{NN\sigma_0}$ and $g_{KK\sigma_0}$ is fixed by the data; they are arbitrarily taken to be the same).

Our theoretical results are compared with the experimental data in Figs. 2.4–2.6. In Fig. 2.4a we compare the theoretical s and p isoscalar phase shifts with the experimental values, and in Fig. 2.4b the isovector ones. With the exception of p_{03} the agreement is good. The agreement between the theoretical and experimental total cross sections shown in Fig. 2.5 is even better. The deviation beyond $P_{lab} = 850$ MeV is connected with the π threshold that is not yet included in our model. In general the observables are better reproduced by our theory than the phase shifts, which may indicate that there are still problems in these analyses. One example of observables is shown in Fig. 2.6. A complete comparison can be found in ref. [5].

<u>2.5 K^-N Scattering</u>

The basic idea of the investigation of K^-N scattering is to follow as closely as possible the calculations for K^+N scattering. Specifically we will use here the same coupling parameters and form factors with the exception of the phenomenologically introduced repulsive σ_0–exchange. As mentioned before, the ω–meson contribution changes sign for K^-N and becomes attractive. Therefore if we assume SU(6) symmetry the magnitude of the (attractive) ω–meson exchange interaction is fixed. The interesting question now is wether the complete short range repulsion needed in K^+N scattering changes sign or not. Therefore we again introduce a purely phenomenological short range contribution to the K^-N interaction which we describe analytically as exchange of a heavy scalar meson. The mass and form factor of this σ_0–meson are the same as those need in K^+N scattering. The coupling parameter, however, is considered as a free parameter which we adjust to the K^-N scattering data.

We mentioned already in section 2.1 that one has to consider in K^-N scattering pole graphs which are not present in K^+N scattering. Fortunately the pole contributions are far away and do not influence our conclusion about the strength of the ω–meson coupling constant. In Figs. 2.7 and 2.8 we compare our theoretical results with experimental values. The six cross sections shown in Fig. 2.7a and 2.7b are a test for different components of the K^-N interaction.

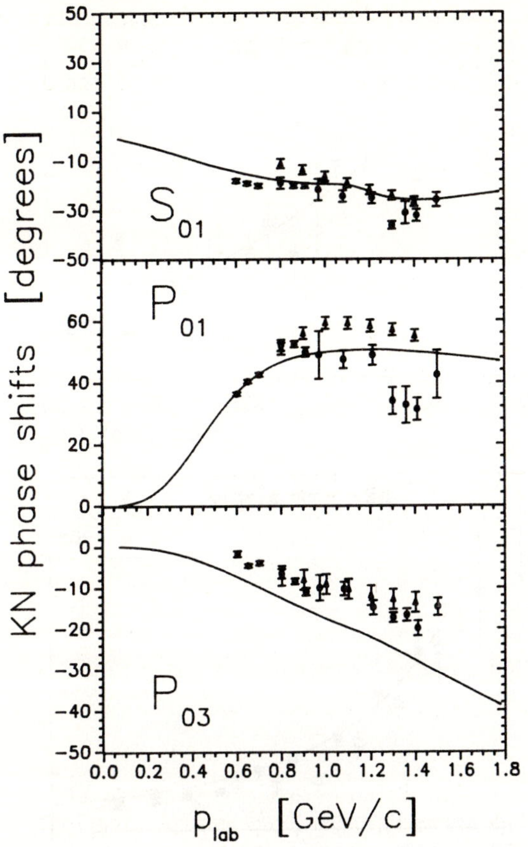

Fig. 2.4a: Isoscalar K⁺N scattering phase shift. The experimental shifts are taken from Hashimoto (circles) [6] and Watts et al. (triangles) [7].

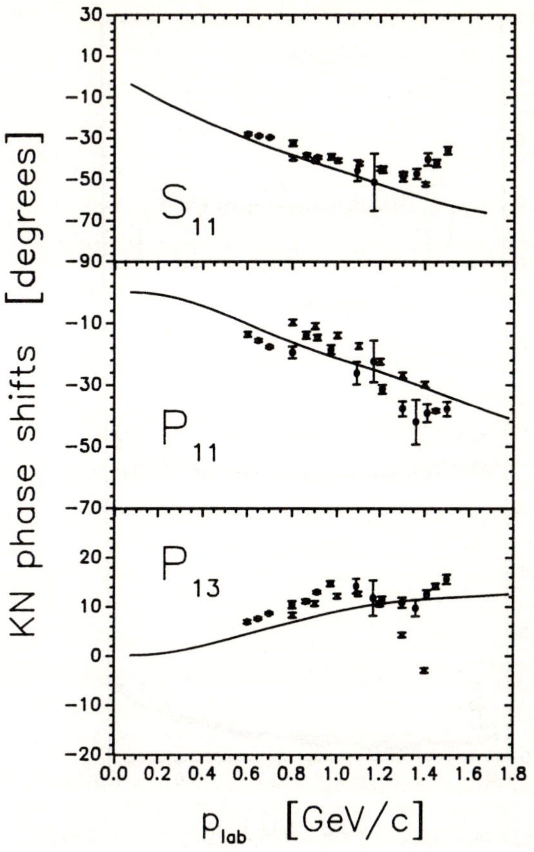

Fig. 2.4b: Same as Fig. 2.4a, for the isovector K⁺N phase shifts.

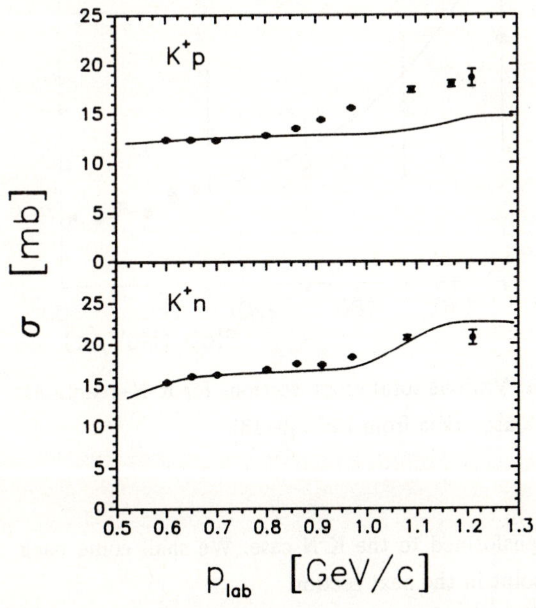

Fig. 2.5: Total cross sections as a function of the kaon lab momentum p_{lab}; for the K^+p and K^+n system. The experimental data are taken from [6].

The "charge–exchange" cross section $pK^- \to nK^0$ shown in the upper part of Fig. 2.7 is sensitive to the isospin–dependent part of the interaction which does not change strangeness. The main contribution to this channel comes from the ρ–meson exchange. The "strangeness–exchange" reactions shown in the middle and lower parts of Fig. 2.7a and 2.7b are directly proportional to the strength of the K–meson exchange (if we neglect box diagrams). Finally, the scattering $pK^- \to pK^-$ shown in the upper part of Fig. 2.7a is mainly a test for the isospin–independent part of the K^-N interaction (because the ρ–contribution has been tested independently). As the attractive σ–meson remains unchanged compared with K^+N scattering, this cross section allows us to investigate the behaviour of the ω–meson and the additional repulsive piece introduced in the K^+N interaction. If we neglect this piece completely and use instead the 60% increase in the ω–coupling (which is in agreement with the K^+N data) and change

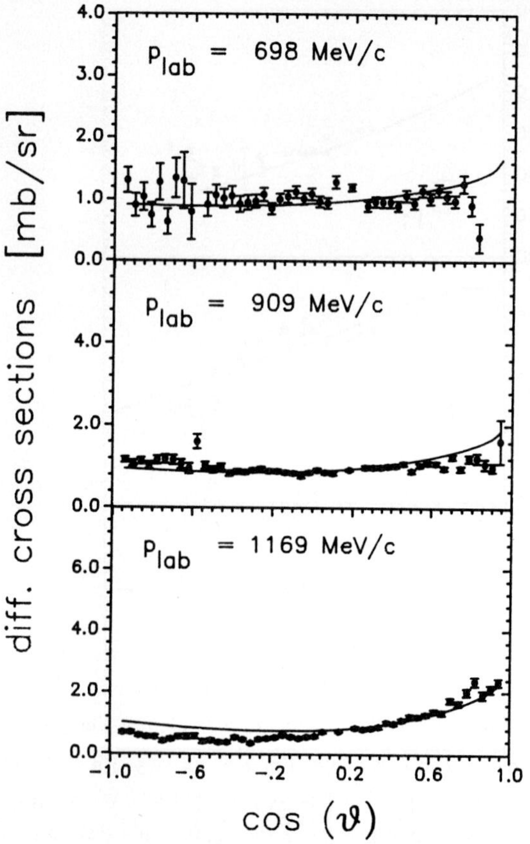

Fig. 2.6: Differential cross sections in the $K^+p \to K^+p$ channel as a function of the scattering angle Θ, for various kaon lab momenta. The experimental data are taken from [8].

the sign in the K^-N case, we obtain a $pK^- \to pK^-$ cross section which is too large by more than a factor of two. This clearly indicates that the increased ω–coupling parametrizes in an effective way a more complicated repulsive contribution to the K^+N interaction. In the calculations which led to the results shown in Fig. 2.7 we chose the SU(6) value for the ω–coupling and adjusted the coupling constant of the (repulsive) σ_0–meson to reproduce the $pK^- \to pK^-$ cross section. The theoretical results depend very sensitively on the magnitude of this parameter, as shown in Fig. 2.8. The experimental data indicate that the repulsive σ_0–contribution has to be a factor of 5 smaller in the K^-N case than in the K^+N case.

This also shows that the additional repulsion needed in the K^+N interaction is the sum of different contributions with opposite G–parity which tend to cancel each other

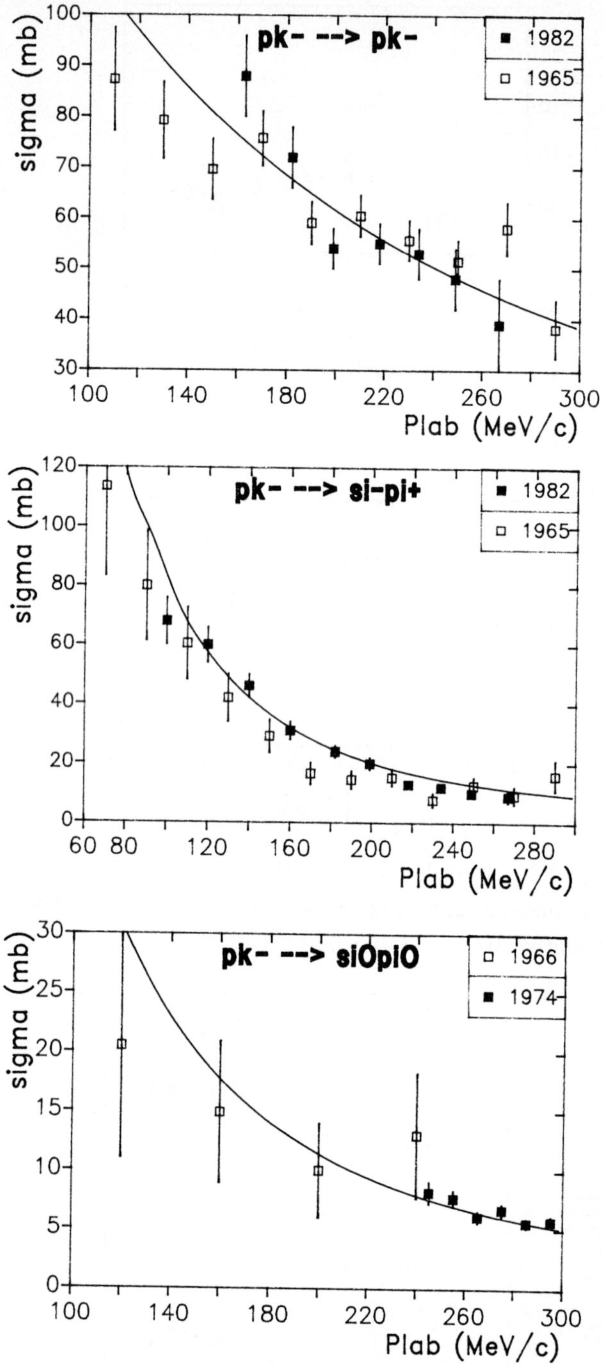

Fig. 2.7a: Various total cross–sections for K^-N–reactions. The data are taken from Refs. [9–13].

when transformed to the K^-N case. We shall come back to this point in the next section.

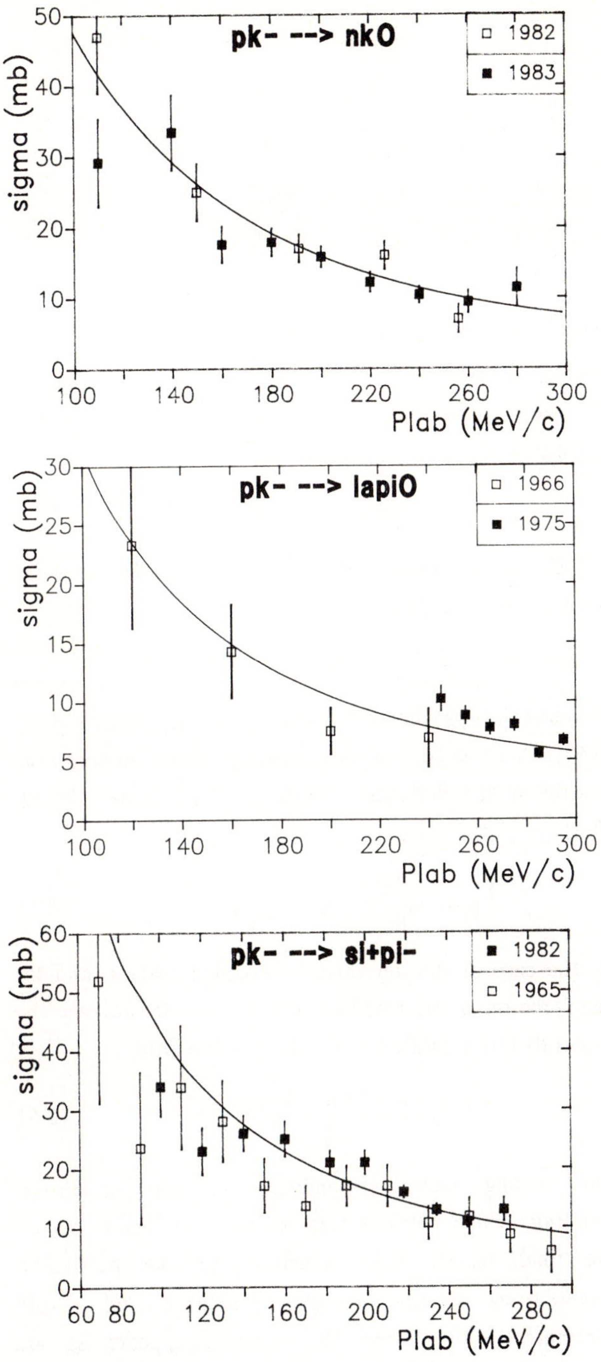

Fig. 2.7b: Same as in Fig. 2.7a.

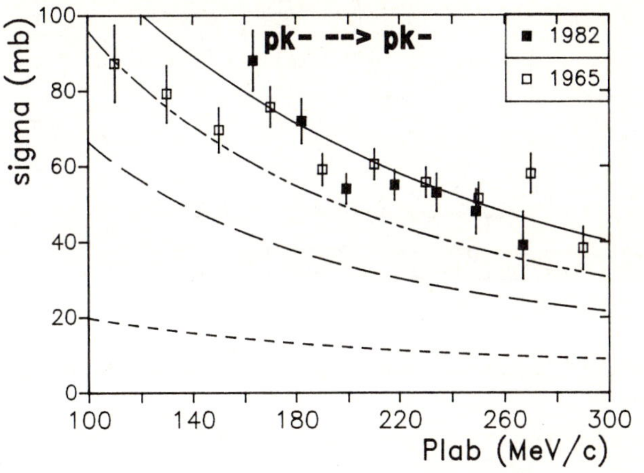

Fig. 2.8: Dependence of the total cross section $pK^- \to pK^-$ on the strength of the repulsive σ_0–meson.

3. Meson–Meson Scattering

3.1 Meson–Meson Interaction

The generalized meson–exchanged model introduced in section 2.1 for kaon–nucleon scattering can be easily extended to meson–meson scattering. We have already defined vertices between mesons in eq. (2.8) which we can now use in the same way for the present case. All the formulas given in section 2.1 are also valid for meson–meson scattering if we replace the corresponding baryon quantum numbers by meson quantum numbers and remember the rules for identical bosons where they have to be applied. The final formulas can be found in ref. [14]. This extended meson–exchange interaction model allows us to:

(i) investigate the strong interaction between two pseudoscalar meson in the frame–work of the meson–exchange model;

(ii) analyse the $\pi\pi$ and $K\pi$ scattering data in a coupled channel approach using this meson–exchange interaction;

(iii) explain the resonance f_0 (975) in the $\pi\pi$ system in a natural way as a quasi–bound state in the coupled $K\overline{K}$ channel;

(iV) estimate the masses of the "genuine" scalar mesons (with and without strangeness) which belong to the scalar meson nonet, and which can be interpreted as one quark–one antiquark systems, to be around 1400 MeV and 1600 MeV, respectively.

We restrict our investigations to the energy range below 1.5 GeV because we neglect possible coupling to $N\overline{N}$ or two vector–meson channels. Since the $K\overline{K}$ channel will turn out to be of extrem importance in the case of $\pi\pi$ scattering, we will treat these channels in a coupled channel framework. The corresponding coupled channels

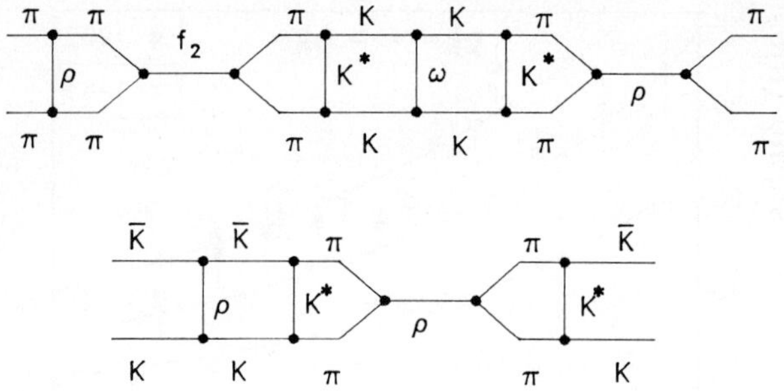

Fig. 3.1: Coupled channel approach for $\pi\pi$–scattering.

are shown in Fig. 3.1. Of course, below the $K\overline{K}$ threshold this coupling occurs only virtually in the $\pi\pi$ system, whereas $K\overline{K}$ scattering, because of real transitions to the $\pi\pi$ system, is inelastic for all energies.

3.2 Coupling parameters

Once the coupling constants and form factors are specified, it only remains to solve the scattering equations (2.5) and (2.6). Our approach is to take coupling constans for the known mesons to be related by SU(3) symmetry. For the 0^+ mesons the coupling constants are adjustable parameters. Thereafter all vertex functions are known and are then used unchanged in the πK scattering system.

For the pseudoscalar–pseudoscalar–vector coupling we determine the coupling constant from the SU(3)–symmetric Lagrangian which gives the following relations between the relevant pseudoscalar–pseudoscalar–vector coupling constants:

$$\begin{aligned} g_{\pi\pi\rho} &= 2G_v \\ g_{KK\rho} &= g_{\overline{K}\overline{K}\rho} = G_v \\ g_{KK\omega_8} &= -g_{\overline{K}\overline{K}\omega_8} = \sqrt{3}G_v \\ g_{K\eta K^*} &= -g_{\overline{K}\eta\overline{K^*}} = -\sqrt{3}G_v \\ g_{\overline{K}\eta\overline{K^*}} &= g_{\overline{K}\eta\overline{K^*}} = -G_v, \end{aligned} \tag{3.1}$$

The corresponding relations for pseudoscalar–pseudoscalar coupling constants are:

$$\begin{aligned} g_{\pi\pi\epsilon_8} &= \frac{2}{\sqrt{3}}G_s^8 \\ g_{KK\epsilon_8} &= g_{\overline{K}\overline{K}\epsilon_8} = \frac{-1}{\sqrt{3}}G_s^8 \\ g_{\eta\eta\epsilon_8} &= \frac{-2}{\sqrt{3}}G_s^8 \\ g_{\pi K\kappa} &= -g_{\pi\overline{K}\overline{\kappa}} = G_s^8 \\ g_{\eta K\kappa} &= g_{\eta\overline{K}\overline{\kappa}} = \frac{-1}{\sqrt{3}}G_s^8 \\ g_{\pi\eta\delta} &= \frac{2}{\sqrt{3}}G_s^8 \\ g_{KK\delta} &= -g_{\overline{K}\overline{K}\delta} = G_s^8, \end{aligned} \tag{3.2}$$

where δ is the I=1, S=0 member of the 0^+ octet, ϵ_8 is the I=0, S=0 member, and κ and $\overline{\kappa}$ are the analogs of K and $\overline{K}$, respectively. For the singlet scalar, the coupling constants are

$$g_{\pi\pi\epsilon_1} = g_{KK\epsilon_1} = g_{\overline{K}\overline{K}\epsilon_1} = G_s^1. \tag{3.3}$$

For the t–channel exchange, we use a form factor of the form given in eq. (2.9).

In addition to the usual pseudoscalar–pseudoscalar–scalar coupling defined in eq. (2.8), there exists in the case of three mesons the possibility of a derivative coupling, which is given in eq. (3.4):

$$L_{pps} = \frac{f_{pps}}{m_p}\partial^\mu\Phi_p(x)\partial_\mu\Phi_p(x)\Phi_s(x) \tag{3.4}$$

In the case of the derivatively–coupled scalars, and for the f_2 meson, a stronger form factor is needed to keep the integrals finite, and we have adopted the form

$$F_1'(\vec{q}^2) = \left[\frac{\Lambda^2 - m^2}{\Lambda^2 + \vec{q}^2}\right]^2. \tag{3.5}$$

The strong energy–dependence of the derivative coupling, which requires this strong form factor, makes the final results quite sensitive to the particular combination of the coupling constants and cutoff parameters (Λ). When the particle appears as an s–channel resonance we take the form given in eq. (2.10).

3.3 Pion–Pion scattering

We restrict our discussion on results of pion–pion scattering to the scalar isoscalar channel $I^G(I^{PC})=0^+(0^{++})$ because only this channel is connected with strangeness. The experimental $\pi\pi$ phase shift in this channel (δ_0^0) shows an 180° jump in a narrow energy range around E ≃ 980 MeV. This resonance ($f_0(975)$) has originally been interpreted as a member of the scalar $q\overline{q}$

nonet. Jaffe [15] has pointed out that this interpretation leads to difficulties and suggested that f_0 (975) and {a_0 (980)} might be 4 quark states (2 quark–2 antiquark), whereas the genuine $q\bar{q}$ scalar mesons should be several hundred MeV higher in energy. We will show that in our model the spectrum below 1 GeV is essentially given as a correlated two–pion and two–kaon system.

In a first approach we solved the T–matrix equation in the $\pi\pi$ channel only. The effects of other mesons are included as box diagrams in the quasipotential V(z). Since the coupling to the $\eta\eta$–channel is very weak, we consider only the coupling to the $K\bar{K}$ pair, which is shown in Fig. 3.2. The effect of this box diagram on the δ_0^0 phase shift is small, as demonstrated in Fig. 3.3. The dashed–dotted line is the result if one considers ρ– and ϵ–exchange (in the t–channel) only. The dashed line is the result where the open $K\bar{K}$ channel is included, but no explicit resonance behavior is included as is seen in the data. In order to explain the experimental phase shift one has to consider explicitly a scalar meson (pole graph) in the s–channel, the coupling constant and mass of which are appropriately chosen to reproduce the experimental data.

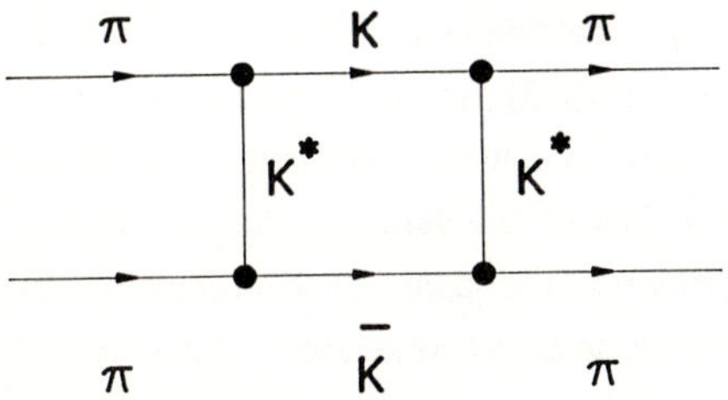

Fig. 3.2: $K\bar{K}$–box–diagram

The t–channel interaction between $K\bar{K}$ pairs for I=0 is rather strong and attractive because all contributions add coherently. Therefore we expect also a larger effect from the $K\bar{K}$ channel if we take this interaction into account. The effects due to the channel coupling on the phase shift and inelasticity are indeed very strong, as demonstrated in Fig. 3.3 and Fig. 3.4. First of all we see in Fig. 3.3 a large jump in the phase shift which comes from a quasi–bound $K\bar{K}$ pair. Because of the attractive interaction in the t–channel, this state is shifted by about 5 MeV below threshold, which gives rise to a strong resonance–like behavior in the phase shift. From Fig. 3.4 we notice that this increase of the phase shift has only little influence on the elasticity η_0^0. It is important to mention that so far only the interaction in the t–channel has been taken into account. We refer to this as Model I. We do not need to consider a genuine scalar resonance around [1 GeV] in order to reproduce the experimental phase shift from the $\pi\pi$ threshold up to 1 GeV. Beyond 1 GeV, however, the theoretical results deviate qualitatively from the more recent experimental phase shift analyses. As we explain the phase jump at 980 MeV as a quasi–bound $K\bar{K}$ pair, it is obvious that

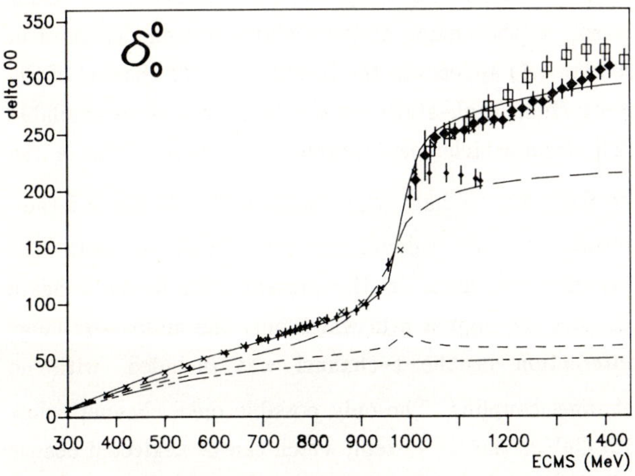

Fig. 3.3: The predictions of four different models for I=0, J=0 $\pi\pi$ phase shifts (see explanation in the text). The experimental data are taken from refs. [16–19].

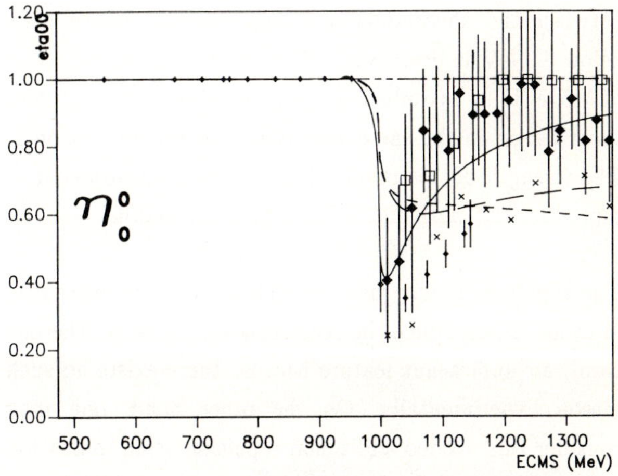

Fig. 3.4: Predictions of the same four different models as in Fig. 3.3 for the elasticity parameter in the I=0, J=0 $\pi\pi$ channel. The data are taken from refs. [16–19].

the further increase of the phase beyond 1 GeV should be connected with the genuine scalar meson which is a member of the scalar SU(3) nonet. (Actually one expects two scalar mesons– one a member of the octet and the other a singlet). Such scalar mesons will give rise to pole terms in the s–channel. The particle data group lists a scalar meson f_0 (1400) (the previous ϵ(1300)) at 1400 MeV with a width of 150–400 MeV that decays mainly into two pions. We consider in the following calculation a scalar meson (which we refer to as ϵ–meson), the bare mass and the coupling consistant of which we take as free parameters that are adjusted to fit the experimental phase shift beyond 1 GeV. Of course one can couple the scalar meson in two ways to the pions. One way is the usual scalar coupling (eq. (2.8)), which is used in section 2. The other possibility is the derivative coupling as defined in eq. (3.4). Both couplings fulfill all symmetry requirements, so it seems a matter of taste which one is used. However the two couplings give rise to quasipotentials with quite different properties. The derivative coupling depends strongly on the pion momentum and this gives rise to a momentum–dependent quasipotential which leaves the phase shift below 1 GeV nearly unaltered. The scalar coupling, on the other hand, is momentum–independent and therfore influences the phase shift over the whole energy range. If we now add to Model I the ϵ–meson, we indeed increase the phase shift above 1 GeV in both cases. For the scalar coupling, however, we obtain also changes in the low energy regime. Now the theoretical scattering length is too large by a factor of 5 and the good agreement between the theoretical and experimental phase shift below 1 GeV is destroyed. We correct this bad behavior of Model II by introducing a repulsive interaction of very short range, in the same way as we did this in section 2. The strength of this repulsive scalar meson is adjusted to reproduce the experimental of scattering length.

The repulsive scalar meson which we have to introduce in order to reproduce the scattering length is, on the one hand, an unpleasant feature because there exists no such meson experimentally. On the other hand, one may consider this needed additional repulsion as an indication of effects beyond the meson–exchange picture. While this may, indeed, be the case, we find that if we replace the scalar coupling between the ϵ–meson and two pions by a derivative coupling, we get essentially the same results without introducing the repulsive scalar meson. The full lines in Figs. 3.3 and 3.4 indicate the theoretical results in this approach. The mass and the width of the ϵ–meson extracted from this analysis using an Argand plot is $m_\kappa \approx$ 1450 MeV, $\Gamma_\kappa \approx$ 300 MeV. The extration of mass wnd width is more difficult in the case of the ϵ, but the mass is approximately 1400 MeV and the width is broad–about 600 MeV. It is certainly very interesting to investigate whether the use of derivative coupling in the KN scattering problem makes the introduction of the repulsive scalar meson unnecessary. Such calculations are in progress.

3.4. Kaon–Pion Scattering

Our model can also be applied in a straightforward way to the $K\pi$ system. Besides the strangeness as a new degree of freedom, kaons have isospin I=1/2 and no definite G–parity. Therefore the isospin of the $K\pi$ system is half–integer and the various channels are characterized by the quantum numbers $I(J^P)$. The $K\pi$ system is of special interest because, at least in principle, the $K\pi$ interaction is fully determined from the previous investigations and the SU(3) symmetry relations. In the calculations we report, the meson–exchange interaction in the t–channel is indeed the same as the one used in the $\pi\pi$–system, whereas the pole contributions in the s–channel have to be adjusted separately. Actually our calculation gives information about the genuine $I(J^P)=1/2(0^+)$ meson with strangeness, usually denoted as κ. Because of the isospin 1/2 of the kaon, the scalar meson κ (the analog of the scalar ϵ–meson discussed in section 4.1) appears in the I=1/2, $J^P = 0^+$ channel of the $K\pi$ system. High–statistics $K\pi$ data are become available [20], from which very reliable $\delta_0^{1/2}$ phase–shifts in the "scalar" channel have been obtained [21]. In Fig. 3.5 the results of our calculation are compared with the experimental data. In the present case Model I again denotes the approach in which only the meson–exchange interaction in the t–channel was included, with no channel coupling. The only possible open channel below 1.5 GeV is the $K\eta$ system, which can be neglected beause of the weak coupling of the η to the other mesons. The theoretical results calculated within Model I are shown as a dashed line. It is clear from that figure that the t–channel interaction is by far the most important part

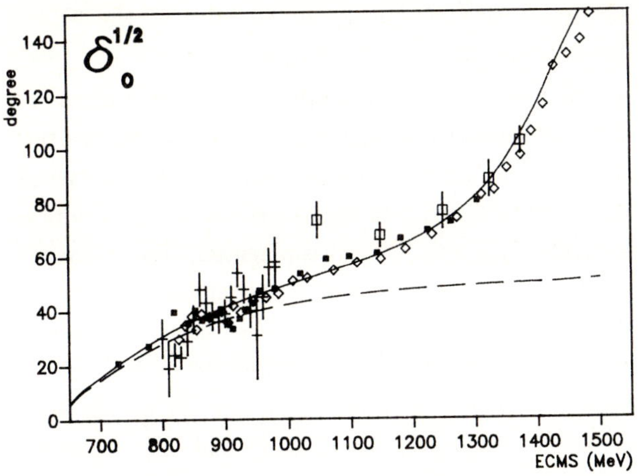

Fig. 3.5: The I=1/2, J=0 Kπ phase shift with (full line) and without (dashed line) the κ–meson contribution. Data are taken from refs.[20, 21].

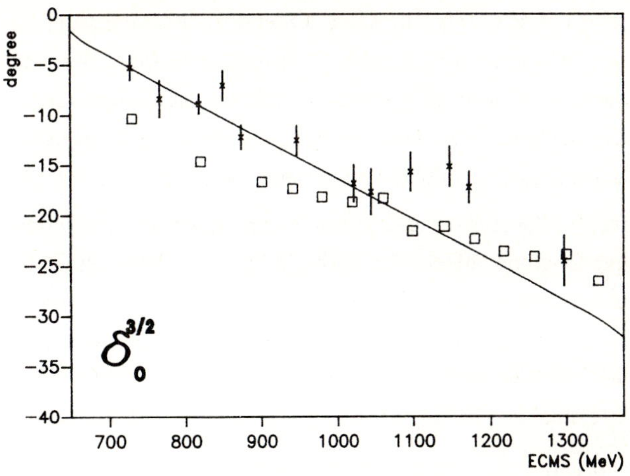

Fig. 3.6: The I=3/2, J=0 Kπ phase shift compared with the analysis of Esterbrooks et al. [21] and Lang [22].

of the Kπ interaction below 1 GeV. We want to point out again that there are no free parameters in Model I; it is exactly the same t–channel interaction as used in $\pi\pi$–scattering.

In Kπ scattering we also have the ambiguity of the coupling of the scalar meson to the two pseudoscalar mesons: one can use scalar or derivative coupling. The results are similar to the ones discussed in section 4.1: if we use scalar coupling we have to introduce an additional short range repulsion whose coupling parameter is adjusted to reproduce the low energy data. Again, the mass of the repulsive scalar meson is not a very sensitive parameter. The $\delta_0^{1/2}$ phase shifts which we get with this model are nearly the same as the ones that we obtain with the gradient coupling, and the corresponding results are shown as full ine in Fig. 3.5. In the actual calculation we used a cut off Λ=3 GeV and a bare mass of m^o= 1600 MeV. The corresponding physical mass of scalar meson K is m_κ= 1430 MeV with a width of Γ= 290 MeV. The scalar channel of the Kπ system is actually simpler than in the $\pi\pi$ system, because here we expect only one genuine ($q\bar{q}$) scalar meson. The good agreement between the experimental and theoretical $\delta_0^{1/2}$ phase shift strongly supports our model. We should, however, also bear in mind that channel coupling may become important around 1.5–1.6 GeV where the ρK* and ωK* thresholds lie. Such investigations are in progress.

A further excellent test for the meson–exchange interaction in the t–channel is the $\delta_0^{3/2}$–phase shift. Like the δ_0^2–phase shift in $\pi\pi$–scattering, all of the effects come from the t–channel interaction. The good agreement between theory and experiment shown in Fig. 3.6 indicate that our model of the interactions works well, especially since all of the relevant parameters have been fixed by the fit in the $\pi\pi$ sector.

4. Summary and Conclusions

In order to establish reliably the validity of the meson exchange picture for low (and possibly medium) energy hadronic systems, corresponding interactions have to be described in a consistent meson theoretic scheme. Only in this way one will have the chance to isolate possible discrepancies and to trace them back to genuine quark–gluon effects.

According to this general program, we have presented meson exchange models for the K^+N–, K^-N–interaction as well as for the $\pi\pi$– and Kπ–interaction. One of the basic questions of these models concerns the interaction at short distances, where we expect deviations from the meson exchange picture due to the underlying quark structure.

A reasonable fit to the K^+–N data required a 60% increase of the KKω–coupling above its symmetry value, which is precisely the same amount as was necessary in the NN system (see [2]). This fact can (and perhaps should) be seen as a strong indication that ω–exchange, as treated so far, is an effective contribution parametrizing, apart from real ω–exchange, either further shorter–ranged mesonic contributions or genuine quark–gluon effects, or both. In fact, the data indicate clearly that this additional contribution is of much shorter range than provided by ω–exchange. Therefore, we alternatively kept the symmetry values for ω–exchange (for $g_{KK\omega}$ as well as $g_{NN\omega}$) but added an short–ranged repulsive contribution (which affects essentially the S–waves only), whose strength was adjusted to the empirical data. In order to investigate this additional repulsive piece we applied the model to K^-N–scattering where the ω–exchange contribution changes its sign and becomes attractive. It turned out that only if we chose the symmetry value for the ω–exchange and reduced the additional repulsive contribution by a factor of 5, could we obtain good agreement with the experimental K^-N–data. This clearly indicates that the "strong" ω–coupling is an effective one which summarizes different contributions which behave differently under G–parity transformation. It is not yet clear, however, whether this behavior can be still explained within the meson–exchange picture or whether it is an effect of the quark–gluon dynamics.

In the second part we report about our theoretical results for $\pi\pi$ and $K\pi$ scattering. In the model we have constructed, meson exchanges, both in the s– and t–channel, are the sole driving factors. All the mesons considered in the model are either well–established particles – π, K, η, ρ, ω, K*, η', Φ, f_2 – or are expected from the quark model – ϵ, κ – and for which there exists considerable experimental evidence. The coupling constants are consistent in almost all cases with SU(3) symmetry, although the use of physical masses and different form factor parameters results in considerable symmetry breaking in the interaction. However the form factor parameters, which we view as an expression of the unknown underlying multi–quark effects, are all in the range expected from hadronic sizes, *i.e.*, 1.5–3 GeV. Application of the model to the $\pi\pi$ and πK systems yields an excellent quantitative fit to all the experimentally measured phase shifts in the energy range up to 1.5 GeV, although discrepancies at the highest energies suggest that it may be necessary to include coupling to channels with thresholds slightly above 1.5 GeV. The model can easily be extended to include such channels without introducing additional parameters. While channels in which resonances are known require an s–channel pole in the quasipotential, the t–channel exchange processes are found to play an important role in the low energy behavior of the phase shifts, and to have substantial influence on the observed widths of the resonances. It is gratifying to note that the partial waves in which resonances are absent and are, therefore, entirely given by t–channel exchange, are also very well described by the model. The data below 1 GeV are dominated by two interacting mesons which correspond in the quark picture as an interaction of two quark–two antiquark pairs. The bare masses of the genuine quark–antiquark states are in all cases beyond 1 GeV. Specifically the members of the scalar nonet where found in our analysis to be between 1400–1600 MeV. It is our hope that the model will find use in explaining processes where many of the same vertices occur, *e.g.* in meson–nucleon interactions, or in interactions requiring a reliable off–shell extrapolation of the meson–meson T–matrix.

References

1) E. Witten, Nucl. Phys. B160 (1979) 57.
2) R. Machleidt, K. Holinde and C. Elster, Physics Reports 149 (1987) 1.
3) R. Büttgen, K. Holinde, B. Holzenkamp and J. Speth, Nucl. Phys. A450 (1986) 403.
4) R. Büttgen, K. Holinde and J. Speth, Phys. Lett. B163 (1985) 305.
5) R. Büttgen, K. Holinde, A. Müller–Groeling, J. Speth and P. Wyborny, submitted to Nucl. Phys.
6) K. Hashimoto, Phys. Rev. C29 (1984) 1377.
7) S. J. Watts et al., Phys. Lett. 95B (1980) 323.
8) B.J. Charles et al., Nucl. Phys. B131 (1977) 7.
9) M. Sakitt et al. Phys. Rev 139 (1965) 719.
10) T.S. Mast et al. Phys. Rev. D11 (1975) 3078.
11) D. Evans et al. J. Phys. G9 (1983) 885.
12) J. Ciborowski et al., J. Phys. G.8 (1982) 13.
13) J.K. Kim, Columbia Univ.Rep., Nevis 149 (1966).

14) D. Lohse, J.W. Durso, K. Holinde and J. Speth, submitted to Nucl. Phys.

15) R. Jaffe, Phys. Rev. D15 (1977) 267

16) W. Ochs, thesis, Universität München, 1973.

17) C.D. Frogatt and J.L. Petersen, Nucl. Phys, B129 (1977) 89.

18) S.D. Protopopescu et al. Phys. Rev. D7 (1973) 1279.

19) M.N. Cason et al., Phys. Rev. D28 (1983) 1583.

20) D. Aston et al., Nucl. Phys. B296 (1988) 293.

21) P. Estabrooks, et al., Nucl. Phys. B133 (1978) 496; M.J. Matison, et al. Phys. Rev. D9 (1974) 1872; A. Firestone et al. Phys. Rev. D5 (1972) 2188.

22) C.B. Lang, Fortschr. der Physik 26 (1978) 509

Z. Phys. C – Particles and Fields 46, S183–189 (1990)

Zeitschrift für Physik C Particles and Fields

Exotic atoms

L.M. Simons

Paul Scherrer Institute, CH-5232 Villigen, Switzerland

Abstract

A review about the present situation in the field of exotic atoms is given. For the different areas in this field, the achievements are described and the necessary future developments are mentioned. Especially the possibility of measuring kaonic hydrogen X–rays at KAON is discussed.

1 Introduction

Exotic atoms are defined to be systems in which a negatively charged particle heavier than the electron is electromagnetically bound to a nucleus. They are formed by decelerating particles like the μ^-,π^-,K^- ,$\bar{p}$ and Σ^-, which live sufficiently long to survive this procedure, in matter. At energies of several tenths of eV, atomic capture will take place by ionization of one of the outer electrons. The path of the particle through the electronic cloud leaves the electronic shell in some state of ionization. This state depends on the atomic number of the atom with which the exotic atom was formed and on the probability that electrons have been refilled from the surroundings. Below the electron shell a quantal cascade dominated by X-ray transitions develops and the atom reaches a state where the electron shell and the finite extension of the nucleus are negligible. In this region, the exotic atom is well described by QED alone. In the case of hadronic atoms the atom will be destroyed in higher orbits as soon as the wavefunction of the hadron overlaps with the nucleus. In muonic atoms the ground state is usually reached and the muon either is captured or it decays by weak interaction.

The observables are conventionally the energies and intensities of the X-rays of the quantal cascade. Differing from the simple idealized picture of a pure hydrogenlike system without contact with the outer world, they show deviations which make the exotic atoms sensitive tools in different respects.

- The observation of X-rays from the early states of the electromagnetic cascade provides information about the capture process and the status of the electron shell(ionization and refilling).The knowledge gained about the processes responsible for the development of the exotic atom is necessary to make precision tools of these systems. Especially in light exotic hydrogen and helium atoms, the influence of the surrounding medium influences the intensities of X-rays strongly. These cascade effects need to be studied in order to fully exploit the physics possibilities of these otherwise extremely simple systems.
- High precision studies in a region where the nucleus still can be considered pointlike permit searches for deviations from Q. E. D.-predictions, or permit determination of particle parameters like mass or magnetic moment.
- In the region where the overlap of the wavefunction of the exotic particle with the nucleus is substantial the interaction with the nucleus leads to new information.
  - In the case of muonic atoms, the different moments (static and dynamic) of the nucleus can be extracted with high accuracy.
  - In hadronic atoms the strong interaction with the nucleus can be measured via line broadening and shift of the last observable transitions.Higher multipole moments of the charge distribution are also measurable with hadronic atoms.

This contribution tries to give a a survey of the knowledge gained for each of the areas listed above together with a list of still open problems and future possibilities. In special illustrative examples, the need for low energy beams of high quality to do physics with exotic atoms is emphasized. A more general introduction to the field of exotic atoms with an extensive list of references can be found in the most recent review article by C. J. Batty[1].

In a necessarily short contribution like this it is obvious that only a rather rough and personal overview with severe omissions can be given. Antiprotonic atoms will not be treated at the length they deserve in spite of the fact that they are at present a source of new information which strongly influences our knowledge of the physics of exotic atoms. Also the examples given reflect very much the personal taste of the author and should be regarded only as proof for the need of an experimental program yet to be fulfilled.

2 Atomic capture and cascade

2.1 Deceleration

Since the bubble chamber experiments of Barkas et al.[2] it is known that the energy loss of oppositely charged particles of the same mass exhibits differences .Later they were measured with counter experiments for both muons[3] and antiprotons/protons [4] with high accuracy as a function of the particle velocity. At velocities lower than $v/c = \alpha = 1/137$ the deceleration time from α to atomic capture has been measured with the muon bottle technique with the result that the deceleration time in hydrogen is $170 \pm 40ns$. In helium, the corresponding time is a factor of three bigger[5].

2.2 Atomic Capture

In contrast to the past, the energy at which capture occurs is rather well known today. In a direct measurement, H. Daniel and coworkers observed muons of energies of 5eV still emerging from a Ag foil[6]. Also, from other more indirect experiments, it follows that the capture occurs at energies of the order of 10 eV by ionising the outer electrons of the atomic shell. Muonic X-rays from states with principal quantum numbers $n \geq 20$ have been observed [7]. The observation of an almost circular cascade in antiprotonic atoms at main quantum numbers ≥ 20 [8] suggests an atomic capture at quantum numbers $n \gg 40$. The influence of the chemical composition on the atomic capture by molecules has been demonstrated in an impressive series of experiments for hydrogen compounds[9][10] in the case of pionic atoms. In special cases, chemical effects could be also established for other compounds and other species of exotic atoms[11][12].

2.3 Capture Probability

Relative capture rates have been measured for mixtures, compounds and alloys in about 500 cases. Compilations of these data[13][14] are available and the measurement of both the energy and intensity of X-rays can in principle be used for a non destructive analysis of the chemical composition of a target[15]. A theoretical description of the capture process and the capture probability was performed by several authors [16],[17],[18]; it is now possible to reproduce the measured periodicity in the relative capture rates rather well [19],[20],[21]. The intensity distribution of measured X-rays is often used to calculate back to unobservable higher states in order to deduce the initial population of principal and angular momentum states. Mostly a statistical distribution over the l- states with only slight deviations fits the data rather well; for the transition element Fe,however, a distribution significantly flatter than statistical could be determined [22].

2.4 Cascade

Most of the information mentioned above was obtained from the intensities of X-ray transitions. One exception is the observation of the emission of Auger electrons from muonic silver [23].Others are the measurement of the circular polarization of muonic X-rays [24][25]and the attempts to determine the anisotropy of X-rays with respect to the beam axis[26]. As a result of these experiments it can be stated the the picture of the cascade process as determined from intensity measurements has been corroborated.

The most important features can be summarized as follows:

The atom is formed at low energies of the incoming particles of the order of tenths of eV. These low energies imply large angle scattering angles and therefore the memory of the beam direction is almost lost. Depending on the mass of the exotic particles, the available number of quantum steps permitted by the gain in binding energy during Auger emission is sufficient to ionize the electron shell in the case of an isolated atom. In the case that electrons can be refilled, an equilibrium ionization state is established depending on the number of electrons in the surrounding material. The complete stripping of muonic and antiprotonic atoms has been shown in a series of experiments with low pressure gases[27][28]. Moreover the data in [28] point to an early development of a circular cascade by which Auger transitions ionize the electrons shell after shell. Hence depending on the principal quantum number of the antiproton, the state of the electron shell is well predicted.

A surprising recent development is the observation of cascade trapping in kaonic liquid helium by the Tokyo group[29]. It shows that unexpected results are possible in seemingly simple systems. This will lead both to a deeper understanding of the cascade process and possibly to practical consequences: namely, the building of monochromatic beams from the decay at rest of the trapped particles.

2.5 Possible Future and Open Problems

The understanding of the atomic capture is only at its beginning due to the lack of systematic experimental research. Energy loss and atomic capture cross sections should be measured as a function of the energy of the incoming particles for elementary targets. Thus, the influence of the state of matter and of the shell structure of the outer electrons could be accessible to a direct experiment. In a second step the influence of chemical binding on the formation of exotic atoms has to be studied. That would finally provide the knowledge to make exotic atoms a tool in material analysis sensitive even to the chemical structure of matter both in the bulk but also at the very surface of matter. The quantities to be measured will be again energies and intensities of X-rays and, with increasing importance,of Auger -electrons. With the beams available now, such a program is hardly feasible.In the future, the development of beams must focus on the highest phase space density at low energies. For most applications an intensity of such beams of $10^{5 \pm 2}$ particles/s is sufficient. The conditions from which a cascade develops will then also be known. Only then can an understanding of the status of exotic atoms at lower states develop, which is of importance for a number of fundamental experiments. These are severely hindered now because of lack of knowledge e.g. of the status of the electron shell in medium or high-Z systems or because the velocity of the system is not well known in exotic hydrogen or helium atoms. As soon as the status of the electron shell is determined in cases where electron refilling can be excluded, the theoretical interpretation of the energy of states of the exotic atoms leads to an increased sensitivity either to Q.E.D. corrections or otherwise to nuclear charge parameters. With highly ionized exotic atoms as a basis, electron– refilling cross sections of interest for plasma physics can be studied by increasing the number of available surrounding electrons, for instance, by increasing the pressure in a gas environment.

3 Electromagnetic Interaction with the nucleus

3.1 Determination of moments of the nucleus

The substantial overlap of the wave functions of the lower states with the nucleus in exotic atoms makes possible the measurement of nuclear charge parameters with high accuracy. Especially muonic atoms served in the past to determine various static and dynamic moments extensively by precision measurements of X–ray transition energies[30]. Hence, for nuclear monopole moments ("radii") a systematic survey was made from lowest Z up to transuranium elements.

The systematics of isotope shifts and to a lesser degree of isotone shifts were worked out in recent years[31]. This work is not only of immediate interest for nuclear physics but provides a necessary calibration for isotope shifts obtained from laser spectroscopy measurements.

The difficulty in combining different methods to obtain information about the charge densities in nuclei lies in the fact, that the quantities measured with one method determine different moments of the charge distribution compared to other methods. Using as many sensitive transitions in muonic atoms as possible, however, the rms radius as well as other momenta can be extracted in a rather model–independent way as was demonstrated in reference [32]. Other methods rely on a combined analysis of muonic,optical and electron scattering data[33].

A recent measurement of the rms radius of muonic carbon measured with a crystal spectrometer [34] and its comparison with electron scattering data very impressively shows the ability of modern muonic X-ray experiments to check the validity of corrections from dispersion effects in the scattering experiment[35]. The analogue of the dispersion effects in muonic atoms ,the nuclear polarization effect ,however, limits the accuracy of the determination of radii in medium and high Z nuclei.

The magnetic moments of nuclei and the distribution of these moments over the nuclear volume were measured in only a few cases in spite of the fact that the deviation of the splitting to be measured from a point nucleus approximation can be a sizeable effect[36]. The quantity to be measured best is the $2s_{1/2} - 2p_{1/2}$–transition, which is observable from about Z=20 on.

Going to higher moments we come to a field where muonic atoms almost exclusively give model–independent information about the spectroscopic quadrupole moment of the nuclear ground state. Accuracies on the percent level can be reached if transitions not affected by nuclear excitation are used; these occur mainly in the 2p–states. Three windows of observation were available. Around Z=13 in the 3d–2p-transitions with a transmission crystal spectrometer, around Z=30 in measuring the $2s - 2p_{3/2}$-transition with solid state detectors and for $Z \geq 40$ the 3d–2p and higher transitions. Again these measurements influence atomic physics experiments by providing the calibration of the electric field gradient at the nucleus which checks the theoretical calculations of the atomic shell configuration [37].

3.2 Muon induced fission

First experiments in muon-induced fission showed the viability of this method to obtain information about the dynamics of the fission process. The muon serves as a trigger for fission during its cascade (prompt fission) or during its stay in the ground state(delayed fission). The presence of the muon, however, does not alter the kinetic energy and mass distribution of the fission products. Therefore the attachment probability to the light fission fragment, which depends on the time duration of the fission process, is a good measure for the nuclear dynamics of this process. First experiments in this direction have been made[38].

3.3 Possible Future and Open Problems

Continuing the measurement of charge radii and especially of isotope shifts with muonic atoms is necessary for the calibration of the laser spectroscopic experiments of long isotope chains. Useful measurements require the precise mesurements of as many sensitive transitions per atom as possible. This checks for inconsistencies stemming from dynamical excitation effects. In addition, nuclear moments can be extracted in a model–independent way. Such a program requires the observation of transitions with low intensities such as those feeding and leaving the 2s–level and crystal spectrometer measurements of transitions where lower resolution measurements would mean an only marginal sensitivity to charge parameters. The use of crystal spectrometers in the reflection mode will lead in addition to a measurement of rms radii for $Z \ll 10$.

An increase in intensity of the secondary beams at a high energy acclerator together with specialized devices such as the cyclotron trap[39] to increase the stop density will be essential to fulfill such a program. In using the transfer process from muonic hydrogen to gas impurities, minute amounts of separated isotope material can be used which will permit working with micrograms of target material in the future.An ambitious development of a phase space compressed secondary muon beam underway at P.S.I. in Switzerland [40] could lead to the use of even less target material.

The experimental accuracy in measuring binding energies cannot be fully exploited because the problem of the nuclear polarization is not yet solved theoretically. Here a new effort is in order.

The measurement of generalized magnetic dipole moments is a field which is still at its beginning and may be stimulated by the advent of similar experiments with one-electron systems at heavy ion accelerators. The determination of quadrupole moments should be extended to even lower Z in order to check theoretical atomic physics calculations in more accessible cases. Precision measurements of the attachment probability after muon induced fission are yet to be done.

4 Strong interaction with the nucleus

The measurement of the strong interaction broadening and shift of the last observable transition has led to an extensive amount of data from pionic and more recently also from antiprotonic atoms. Survey measurements have been performed for kaonic and sigmonic atoms. The pionic data embody about 150 shift and width values, some of which are measured very precisely with a crystal spectrometer.

At first glance the lack of sufficiently precise values for the ground state shift and width in exotic hydrogen and to some extent also in helium atoms is very surprising. In pionic hydrogen and deuterium first measurements with a crystal spectrometer yielded values for the ground state shift only. They are, however, not yet precise enough to allow clear cut comparison with theoretical predictions. In future measurements with a crystal spectrometer of enhanced resolution, this situation could be remedied[41] . In the case of antiprotonic hydrogen where the shift and width of the ground state are measurable with solid–state and gas–filled detectors, both quantities have been measured with reasonable accuracy.The quantities still to be determined are the strong hyperfine splitting of the ground state and a precise measurement of fine- and hyperfine structure of the 2p–states. In kaonic and sigmonic hydrogen, however, reliable data are controversial or still missing. The difficulty in all these experiments is the low yield of the transitions to be observed. The K-transitions and especially the K_α-transition are attenuated strongly. The well known reason for this is the Stark–effect in higher levels resulting from the high electric field that the neutral exotic hydrogen atom experiences in penetrating neighbouring hydrogen molecules. It admixes s-wave amplitudes with higher l–states, which leads to the destruction of the system at an early state of the cascade. Therefore the observation of pionic and especially antiprotonic hydrogen became feasible only with the production of these atoms in low pressure gases. For antiprotons the low energy accelerator ring LEAR at CERN delivered the low–energy high–intensity beams needed to make possible a new series of antiprotonic atom studies and especially the measurement of low Z isotopes.

4.1 Possible Future and Open Problems

Whereas the measurement of strong interaction shifts and widths in higher Z atoms provided sufficient data of high quality, the information about low Z systems is still rather poor. The situation in pionic and also in antiprotonic atoms will be remedied at the existing facilities,but the case of the kaonic atoms must await dedicated secondary beams for instance at the KAON machine.The importance of measurements of strong interaction shifts and widths in kaonic helium and hydrogen has been stressed very recently by C. J. Batty [42].

In the following a possible set-up for an experiment in the gaseous phase of hydrogen is discussed. The same set-up can serve for an experiment in helium as well.

The set–up relies on two facts. First the determination of the intensities in antiprotonic hydrogen for pressures between 16 hPa and 10000 hPa yielded a set of parameters which characterize the cascade process in these atoms in the so–called standard cascade model of Borie and Leon [43]. Assuming this set of values, intensities for pionic hydrogen as a function of pressure could be predicted which agreed well with measurements done at PSI[44]. Hence, it can be concluded that yield values for kaonic hydrogen can also be predicted with sufficient accuracy from the same cascade code. A calculation using the strong interaction parameters from [45] with a 2p-width of .1 meV predicted K_α yields ranging from 4% at 1000 hPa to .7% at 15000 hPa and $7{\cdot}10^{-5}$ for liquid hydrogen. Assuming other strong interaction parameters, as for instance the ones of [46], does not change the predictions by more than 20%. The yield values thus obtained are more than an order of magnitude lower than the values given by C. J. Batty [42]. Taking into account the severe background limitations leads to the consideration of a measurement at equivalent pressures in the range of several 1000 hPa. This can be obtained by working at low temperatures near the evaporating point of liquid hydrogen at pressure values which still allow the low energy K-X-rays to penetrate the target windows.

Secondly, the experience gained from pionic hydrogen X-rays in a recent measurement at PSI can be used to scale the necessary conditions. The necessary stop density at the pressure of 15000 hPa has been reached using the cyclotron trap[39]. For the special beam at PSI, about 10% of the incoming pions could be stopped in .1g of hydrogen. The stop volume was a disk with a radius of 3 cm and a height of 2 cm. The initial beam momentum was 85±5MeV/c. Because such a momentum is too low for kaon beams, the existing cyclotron trap with a maximum accepted momentum of 123 MeV/c cannot be used.In order to achieve a successful injection the beam momentum is limited to 200 MeV/c. With a newly built device, however, beam momenta of 400 MeV/c could be injected. A kaon beam with a resolution of $\Delta p/p = 5\%$ and an emittance of 2000πmm mrad could be stopped in a 30000 hPa hydrogen target of the length of 10cm and a radius of 5cm with a stop efficiency of 5%. It should be noted, that the contaminating pions in the beam will not be stopped in the hydrogen gas at all.

The major difficulty for the measurement is still the low yield of the K_α transition which is about .4% at 30000 hPa. The total yield for the K-transitions at this pressure is 1.3%. That makes the experiment an order of magnitude more difficult compared with the pionic hydrogen measurement with regard to the peak to background ratio. The recent development of charged coupled devices (CCD) for the detection of X-rays will make it possible, however, to enhance this ratio considerably. Even under unfavorable conditions, when such a detector was placed 1000 mm from the stop volume in the pionic hydrogen measurement at PSI, an enhancement of one order of magnitude compared to a specially designed Si(Li)-detector was measured. Under optimal conditions the CCD would be placed as near as possible to the stop volume. The peak will increase quadratically with decreasing distance whereas the background of low energy X-rays comes mainly from shower events which develop according to the geometrical set-up of the experiment and of the detector itself. Because of the pixel structure of the CCD, shower events can be distinguished from genuine X-ray events by their different topology and hence suppressed. The drawback of the CCD detectors is the impossibility to trigger them fast enough. The readout times in the pionic hydrogen experiment were of the order of one minute. Readout times of the order of msec are achieved elsewhere.

Finally it can be stated, that the combination of devices to enhance the stop density such as a cyclotron trap together with state of the art detectors will make the spectroscopy of light kaonic and eventually of also sigmonic atoms for the different hydrogen and helium isotopes accessible.

5 Questions from particle physics

Particle parameters such as the mass or the magnetic moment of negatively charged particle bound electromagnetically to a nucleus have been determined in classical experiments for several particles using exotic atoms. The recent determination of the pion mass[47] made with a crystal spectrometer with an accuracy of $2{\cdot}10^{-6}$ is a good example of the achievable accuracy and the fundamental importance of the measurement; in this case the precision of the muon neutrino mass measurement could be increased.This example also shows the inherent limitation of the method: the missing knowledge of the state of the electron shell will impede measurements with higher resolution. Therefore it illustrates well the statement made above that many of the future experiments should be performed in gas targets where for low Z the electron shell is ionized during the radiative cascade. The same is true for measurements aiming at the precision determination of Q.E.D. corrections. The so-called electron screening correction presently limits the obtainable precision at least in the region of low Z, where measurements with transmission crystal spectrometers have been used [34]. The fundamental experiments to determine Q.E.D. corrections in the hydrogen isotopes with the laser pumping method pioneered by Zavattini [48] in helium still remain to be done. The reason is the low stop rate at the low pressures where the experiment should be performed.

An additional good example of the possibility to do precision experiments is also the study of the muon capture process from the ground state in muonic hydrogen. These experiments in principle allow to measure some of the weak form factors. Measurements in liquid hydrogen [49] and their interpretation [50] give the axial form factor with an accuracy of about 4% and the pseudoscalar coupling constant with a precision of 20%. A radiative muon capture experiment in hydrogen is being performed at TRIUMF and planned for PSI.

In deuterium muon capture experiments determined the capture rate from the doublet state with an accuracy of about 10% [50].

For higher Z nuclei the renormalization of the coupling constants can be tested. The best investigated case up to now is the muon capture in ^{12}C where for the axial (accuracy 8%), pseudoscalar (21%) and magnetic coupling constants (15%) no deviations from the values for free nucleons were found[51]. A series of radiative muon capture experiments as a test for PCAC has been performed at the meson factories with the result of a fair agreement in light and a disagreement in the heavy nuclei [52].

5.1 Possible Future and Open Problems

It is conceivable that the advent of better beams of kaons will yield a new measurement of the mass of the kaon and the sigma. The magnetic moment of the sigma could also be newly determined. With higher stop densities for pion beams, a further refinement of the value of the pion mass should be made. With newly designed muon beams and drastically increased stop densities, the laser spectroscopy in exotic atoms can be done with success and will lead to results of unprecedented precision. The effort put into the development of muon beams with high stop density will immediately pay off in the measurement of the weak coupling constants both in hydrogen and deuterium in low pressure gases. These measurements will find their natural extension to higher Z nuclei as well to a series of measurements for radiative muon capture.

Earlier suggestions for symmetry tests (parity and time reversal invariance)[53][54] in muonic atoms should be followed too. For the sake of illustration a proposal for the measurement of parity violation in muonic atoms is shortly discussed.

The presence of a weak neutral currents in the interaction of the muon with the nucleus will lead to parity impurities in muonic levels. The impurity increases as the spacing between two muonic levels of opposite parity decreases. Thus, early suggestions dealt with the 2s-2p mixing at low Z, where the energy difference between the two levels is only a few eV [55][56][57]. The 2s–1s M1 transition will have then a slight E1 admixture which in turn will produce a pseudoscalar observable such as the circular polarization of X-rays or the forward-backward asymmetry with respect to a spin direction. The magnitude of the effect is at best of the order of several percent. A precise determination with relative accuracies on the percent level would then lead to a high precision check for deviations from the standard model in the muonic sector. Compared to other atomic experiments this would be done in a hydrogen like system which is free from errors in the interpretation.

Soon after the original proposals it became clear, that an observation of the M1 transition was far out of reach [58]. Indeed it has not been observed yet in muonic atoms. The reason is that electrons from the atomic shell open up other decay branches for the 2s-state of higher probability . Even if the electron shell would be ionized as was shown for atoms with $Z \leq 20$ in gas targets, the radiative 2–photon transition 2s–1s would produce background in the region of the M1–transition. Reducing the background requires high resolution dedicated detectors which hinder a high statistics experiment. Furthermore in the experiment one should use atoms with nonzero nuclear spin in order to be able to determine the four model independent parameters completely. The requirement that the effect should not be much lower than the percent level restricts the possibilities to gaseous targets below Z=6. Only boron with Z=5 survives a closer look. It can be produced in the form of diborane gas(B_2H_6). As soon as the muon is in the 2s–state, the atom will be free of electrons. At pressures below 1hPa it remains ionized during the radiative lifetime of the 2s–level of 37 nsec. This time then permits discrimination of prompt competing processes like the 2p–1s transition,which are nearby in energy. Time discrimination requires the information that a muon has stopped, which could be obtained by detecting the L–Xrays. Equally important, however, is the existence of an electronic K–edge of terbium at 51996 eV, about 200 eV below the M1–transition. The observation of delayed muonic X-rays above the K–edge measures a M1–transition with an only 8% contamination of the 2-photon transition. The pseudoscalar observable can then be constructed from the direction of the M1–X–ray together with the direction of the decay electron from the muon decay in the ground state. Working in a magnetic field which also confines the muon beam and allows a sufficient stop rate to be obtained, a μSR time differential study of the parity violation observable can be performed. A Fourier analysis of this will be used to identify the different coupling constants by the rotation frequency of the observable different for each spin component. An experiment of this kind could be begun at the present meson facilities, but the final stage aiming at the utmost precision is conceivable at facilities with still drastically increased stop density.

6 Conclusions

With the advent of new, high intensity low energy kaon beams at the KAON facility the study of kaonic and sigmonic hydrogen will be continued. Systematic measurements at higher Z will then certainly follow. It is very tempting to add a LEAR–type facility at a later stage, because many features of exotic atoms can only be studied with these long–lived particles. The main concern should be however to continue the pion and muon physics at much better conditions. The possibilities in this field reach from so–called bread and butter experiments to dedicated experiments of fundamental interest still to be performed. A program in this field requires low-momentum, high-quality beams yielding high stop densities. The experiments one can contemplate are considerably more complicated than conventional exotic atoms experiments. They certainly require an effort of the whole community of people interested in this field. With new high quality beams it should be possible, however, to interest physicists from other disciplines such as atomic and solid state physics.

References

[1] C.J. Batty, Sov. J. Part. Nucl. 13(1982)71

[2] W.H. Barkas, W. Birnbaum,and F.M. Smith, Phys. Rev. 101(1956)778.

[3] W. Wilhelm, H.Daniel, and F.J. Hartmann, Phys. Lett. 98B(1981)33.

[4] L.H. Andersen et al., Phys.Rev. Lett. 62(1989)1731.

[5] F. Kottmann, Contribution to the II International Symposium on Muon and Pion Interaction with Matter,Dubna 1987.

[6] G. Fottner et al., Z. Phys. A304(1982)333.

[7] F.J. Hartmann et al., Z. Phys. A305(1982)189.

[8] R. Bacher et al., Phys. Rev. A38,(1988)4395.

[9] L.I. Ponomarev, Ann. Rev. Nucl. Sci. 23(1973)395.

[10] D.F. Measday, Invited talk at the workshop on Electromagnetic Cascade and Chemistry of Exotic Atoms, Erice 1989.

[11] H. Schneuwly et al., Phys. Rev. A27(1983)950.

[12] R. Kunselman et al., Phys. Rev. Lett. 36(1976)446.

[13] T. von Egidy, F.J. Hartmann, Phys. Rev. A26(1982)2355.

[14] D. Horvath, F. Entezami, Nucl. Phys. A407(1983)297.

[15] H. Daniel, N.I.M. B3(1984)65.

[16] P. Vogel, P.K. Haff, V. Akylas, A. Winther, Nucl. Phys. A254(1975)445.

[17] M. Leon, R. Seki, Nucl. Phys. A282(1977)445
M. Leon, J.H. Miller, Nucl. Phys. A282(1977)461.

[18] H. Daniel, Phys. Rev. Lett. 35(1975)1649.

[19] T. von Egidy, D.H. Jakubassa-Amundsen, F.J. Hartmann, Phys.Rev. A29(1984)455

[20] H. Schneuwly, V.I. Pokrosky, L.I. Ponomarev, Nucl. Phys. A312(1978)419.

[21] H. Daniel, Z. Phys. A291(1979)29.

[22] F.J. Hartmann et al., Rev. Lett. 37(1976)331.

[23] R. Callies et al., Phys. Lett. 91A(1982)441.

[24] R. Abela et al., Phys.Lett. 71B(1977)290.

[25] T.Yamazaki et al., Phys. Rev. Lett. 39(1977)1462.

[26] R. Abela et al., Hel. Phys. Acta 52(1979)419; G.K. Lum, C.E. Wiegand, G.L. Godfrey, Phys. Rev. Lett. 65B(1976)43.

[27] R. Bacher et al., Phys. Rev. A39(1989)1610.

[28] R. Bacher et al., Phys. Rev. A38(1988)4395.

[29] R.S. Hayano, Invited talk at the workshop on Electromagnetic Cascade and Chemistry of Exotic Atoms, Erice 1989.

[30] C.J. Batty et al., Nucl. Phys. 19(1989)1 .

[31] L.A. Schaller et al., Phys. Rev. C31(1985)1007.

[32] W. Kunold et al., Z. Phys. A312(1983)11 .

[33] D. Rychel, masters thesis, Institut f. Kernphysik, KPH 21/80, Mainz 1980.

[34] W. Ruckstuhl et al., Nucl. Phys. A430(1984)685.

[35] E.A.J.M. Offerman, Dispersion effects in elastic electron scattering from ^{12}C, Thesis,University of Amsterdam (1988).

[36] H.P. Povel, Nucl. Phys. A217(1973)573

[37] E.B. Shera,Proceedings of the 7th International Conference on Atomic Masses and Fundamental Constants, Darmstadt-Seeheim (1984)372, Seeheim.

[38] W. Bertl et al. PSI Annual Report 1988,37.

[39] L.M. Simons, Phys. Scripta T22(1988)90.

[40] D. Taqqu, N.I.M. A247(1986)288 .

[41] SIN proposal R 86-05.1

[42] C. J. Batty, Invited paper at the "12th International Conference on Few Body Problems in Physics", Vancouver, B.C.,Canada, 2-8 July, 1989.

[43] E. Borie, M. Leon, Phys. Rev. A21(1980)1460.

[44] A. J. Rusi EL Hassani et al., Contributed paper at the workshop on Electromagnetic Cascade and Chemistry of Exotic Atoms, Erice 1989.

[45] H.H. Brouwer,J. W. de Maag, L.P. Kok, Z. Phys. A318(1984)199.

[46] A. Deloff, J. Law, Phys. Rev. C20(1979)1597.

[47] B. Jeckelmann et al., Nucl. Phys. A457(1986)709.

[48] G. Carboni et al., Nucl. Phys. A278(1977)381, G. Carboni et al., Phys. Lett. 73B(1978)229.

[49] G. Bardin et al., Nucl. Phys. A352(1981)365.

[50] E. Zavattini, Nucl. Phys. B279(1987)321.

[51] L. P. Roesch et al., Phys. Rev. Lett. 46(1981)1507.

[52] N. C. Mukhopadhyay, Invited talk at the Int. Conf. on Weak and Electromagnetic Interactions in Nuclei, Montreal, 1989 (WEIN-89).

[53] J. Missimer, L.M. Simons, Phys. Rep. 118(4)(1985)179.

[54] J. Deutsch, Proceedings of the workshop on Fundamental Muon Physics:Atoms, Nuclei and Particles LA-10714-C (1986)201.

[55] J. Bernabeu, T.E.O. Ericson, C. Jarlskog, Phys. Lett. 50B(1974)467.

[56] G. Feinberg, M.Y. Chen, Phy. Rev. D10(1974)190.

[57] A.N. Moskalev, JETP Lett. 19(1974)216.

[58] D.P. Grechukhin, A.A. Soldatov, Sov. Phys. JETP 46(1977)15.

Z. Phys. C – Particles and Fields 46, S191–192 (1990)

Zeitschrift für Physik C Particles and Fields

The CF_4/isobutane (80 : 20) gas mixture and high rate proportional chambers

R. S. Henderson*, G. Sheffer, R. Openshaw, W. Faszer, M. Salamon

TRIUMF, 4004 Wesbrook Mall, Vancouver, B.C., Canada V6T 2A3
* also associated with University of Melbourne, Australia

In several laboratories there is considerable interest in developing proportional chambers that are able to operate at high rate and suffer minimal degradation over an extended period of operation. However, the causes of damage and degradation in these devices are still poorly understood. At TRIUMF we have a continuing program to study ageing effects. Using several identical single wire test cells, we have investigated the ageing characteristics of CF_4/Isobutane (80:20), Argon/Ethane (80:20), and Argon/Ethane/CF_4 (48:48:4) [1–3]. With our early results [1] showing a superior ageing performance of the CF_4/Iso mixture, we decided to test a small admixture (4%) of CF_4 to Ar/Et to see if it would inhibit damage. Parameters such as flow rate, gas gain, anode wire current density, and materials in contact with the gas stream have been varied. Some tests have been extended beyond 8 Coulomb/cm of wire. The Ar/Et chambers have shown a high incidence of pulse height degradation, dark currents, cathode foil etching, and deposits on the electrodes. A strong correlation between anode wire current density and rate of damage (%/C/cm) is indicated for Ar/Et chambers. The CF_4/Iso chambers have shown effectively zero pulse height degradation and few other problems to accumulated charges exceeding 5 C/cm. The addition of 4% CF_4 to the Ar/Et mixture has dramatically improved the ageing performance of the Ar/Et/CF_4 cells.

The original motivation for these studies was the need to provide a very high rate MWPC, capable of reliably operating in the secondary pion beams at TRIUMF, where continuous particle rates of $\sim 10^8$/sec over a 2 cm^2 beamspot are common. Calculations based on the useful chamber lifetimes of 0.2 - 0.5 C/cm, reported by some authors [4], indicated chamber survival less than two weeks with the more common gas mixtures. CF_4/Isobutane mixtures had favourable characteristics for high rate applications, including fast electron drift times and high specific ionization enabling thin efficient chambers. Over the past two years we have built five 25 cm^2 MWPC's having small interelectrode distances ($\simeq$0.77 mm) and these have been used in a variety of experiments at TRIUMF. One chamber was operated with maximum rates of 7×10^7 pions/sec and maximum fluxes of $\simeq 3 \times 10^7$ pions/cm^2-sec. After one month of trouble free use the estimated total charge deposited in the central region was $\simeq$0.2 C/cm of wire. Analysis of the anode wires revealed no polymerization, although some microscopic aluminium nodules were found attached to the anode wires adjacent to a slightly etched area of aluminized mylar cathode foil. The short electron drift distances and high drift velocities ($\sim$9 cm/μsec) result in reasonable timing resolution FWHM$\simeq$4 nsec and FWTM$\simeq$9 nsec. A larger active area MWPC (5x18 cm) built at TRIUMF with interelectrode distances of 1.27 mm anode-anode and 3.18 mm wires-cathodes has operated routinely at BNL for long periods. Rates of 15 MHz over $\simeq$20 cm^2 have been usual. We have used a mixture of CF_4/Iso (50:50) in one of the 25 cm^2 MWPC's. Unfortunately, the results with moderate rate beam operation ($\sim$20 MHz) were disastrous; operation was unstable and there were sustained dark currents and gain degradation occurring after less than 0.005 C/cm of wire. This may indicate that the ageing performance of the CF_4/Iso mixture is dependent on the relative concentrations of CF_4 and Isobutane.

The CF_4/Isobutane (80:20) mixture has shown strikingly sucessfull high rate and long lifetime performance in our MWPC's and damage cells. These characteristics make these small interelectrode detectors suitable canditates as beam profile monitiors, position tagging of rare events and front end chambers in spectrometers. Encouraging results in damage cells indicate that Ar/Et/CF_4 (48:48:4) could be a useful and low cost gas mixture in a variety of moderate rate proportional chambers to extend their lifetimes.

References

[1] R. Henderson et al., "A High Rate Proportional Chamber", IEEE Trans. on Nucl. Sci.Vol NS-34(1988)528.

[2] R. Henderson et al., "Wire Chamber Ageing with CF_4/Isobutane and Argon/Ethane Mixtures", IEEE Trans. on Nucl. Sci.Vol NS-34(1988)477.

[3] R. Openshaw et al., "Tests of Wire Chamber Ageing with CF_4/Isobutane (80:20), Argon/Ethane (50:50), and Argon/Ethane/CF_4 (48:48:4)", Presented IEEE symposium, Orlando, Nov 1988.

[4] Proc. of Workshop on Radiation Damage to Wire Chambers, Lawrence Berkeley Labratory 1986.

Z. Phys. C – Particles and Fields 46, S193 (1990)

Zeitschrift für Physik C Particles and Fields

Strangeness production in antiproton annhihilation on nuclei

J. Cugnon, P. Deneye, J. Vandermeulen

Université de Liège, Physique Nucléaire Théorique, Institut de Physique au Sart Tilman, Bâtiment B.5, B-4000 Liège 1, Belgium

Abstract: The strangeness production in antiproton annihilation on nuclei is investigated within a cascade-type model, keeping with a conventional picture of the annihilation on a single nucleon followed by subsequent rescattering proceeding within the hadronic phase. The following hadrons are introduced: N, Λ, Σ, $\overline{\Lambda}$, π, η, K and $\overline{K}$ and, as far as possible, the experimental reaction cross sections are used in our simulation. The numerical results are compared with experimental data up to 4 GeV/c. The $\overline{\Lambda}$ yield is correctly reproduced, while the Λ and K_S yields are overestimated in $\overline{p}$ Ta and $\overline{p}$ Ne cases. On the other hand, the rapidity and transverse momentum distributions are well reproduced. It is shown that the total strange yield is not very much affected by the associated production taking place during the rescattering process. It is also shown that the Λ/K_S ratio is largely due to the strangeness exchange reactions induced by antikaons. In particular, values of the order of 1 to 3 are expected in the energy range investigated here, independently of the detail of the hadron phase dynamics. Finally, it is stressed that rapidity distributions are consistent with the rescattering process. Comparison with other works and implications of our results are examined.

Z. Phys. C – Particles and Fields 46, S195–201 (1990)

Zeitschrift für Physik C Particles and Fields

Summary and concluding remarks

E. Vogt

TRIUMF, 4004 Wesbrook Mall, Vancouver, B.C., Canada V6T 2A3

1. Introduction

As the speaker for the summary spot of this conference I have been allowed some appropriate licence. Although most of the talks and discussion at this conference were devoted to the science of the KAON Factory, it is, after all, the prospect of the imminent emergence of Canada's KAON Factory which has evoked this audience and this meeting. Therefore it seems appropriate not only to provide some remarks on the science which has emerged during the last three days but also to provide a perspective for that science by describing briefly the facility which aroused our interest and especially the circumstances now leading toward decisions for its funding. I want everyone leaving this conference to believe that Canada's KAON is on the way, to understand the basis of that belief and to undertake their personal role in making it all happen. If we leave that way, the joy of physics expressed here will surely multiply in the decade to come.

2. The TRIUMF KAON Facility

Craddock began this conference with a detailed description of the KAON facility planned at TRIUMF. It boosts the present TRIUMF cyclotron energy (~500 MeV) sixty-fold to 30 GeV retaining a beam current of 100 μA. The layout, adjacent to the present TRIUMF, of the two synchrotrons for this purpose and of the various other rings required to appropriately package the beam were all given in Craddock's paper. He also gave an outline of the various kaon, pion, antiproton, muon, proton, neutrino, etc. beams intended for the initial KAON research program. From the European perspective it is a rather complete set of beams, with the possible exception of a low-energy antiproton accumulator not included in the initial program but, of course, allowed for when and if interest demands. The very recent report of the Feshbach Panel in the United States gives a very clear statement of what our field should now expect from a hadron facility and how the Canadian KAON facility nicely answers those needs. The science opportunities of KAON have been addressed in a series of workshops, some of which have had a narrower focus than this one at Bad Honnef. The series is the following:

- Rare Decays and CP Violation (TRIUMF, Vancouver, Canada) Nov 30–Dec 3, 1988
- Spin Physics (TRIUMF) Feb 15–16, 1989
- Hadron Spectroscopy (TRIUMF) Feb 20–21, 1989
- Joint JHP/KAON (KEK, Japan) Apr 3–4, 1989
- Neutrino Physics (Montréal, Canada) May 24, 1989
- Physics at KAON (Bad Honnef, Germany) Jun 7–9, 1989
- Hypernuclear Physics at KAON (KEK, Japan) Jun 17–18, 1989
- Spin and Symmetries (TRIUMF) Jun 30–Jul 2, 1989
- Users Workshop (TRIUMF) Jul 10–11, 1989
- Low Energy Muon Science at Large Accelerators (TRIUMF) Jul 19–21, 1989
- Intense Hadron Sources and Antiproton Physics (Torino, Italy) Oct 23–25, 1989

The particular mix of physics discussed at this Bad Honnef conference should be placed in the broader context of the whole series.

A matter of considerable importance for European scientists planning to work at KAON in Vancouver is the dedication of the laboratory to make life convenient for distant users. For many future European KAON users – especially those, for example, who in the past have worked in low-energy nuclear physics at German universities – it may be a somewhat daunting prospect to work at a laboratory nine time zones away from their home base. It works. World-class physics emerges surprisingly well under these conditions. Our own Vancouver scientists working at European laboratories have discovered this as have many European groups working at the present TRIUMF facility. At such distances the duration of visits to the user facility is usually a week or several weeks, rather than the few days which might pertain to participation at a much nearer user facility, such as CERN, DESY or Jülich. Under these conditions it matters that the laboratory provides appropriate facilities for housing users and their families. TRIUMF has paid extraordinary attention to making users comfortable in Vancouver and it plans to continue to do so with KAON. Further, Vancouver as a major maritime city appears to be particularly congenial to Europeans. Ask those who have been there or come on a trial visit.

3. A Science Perspective For KAON

In order to place the science topics at this international meeting in a proper perspective we discuss very briefly and generally the current issues of subatomic physics. The important issues all arise from the standard model of quarks, leptons and unified forces. The standard model has brought about a startling change in all of subatomic physics. Not only has it provided a basis for the understanding of the wealth of data uncovered in particle physics and nuclear physics, but it has also raised many important new questions for the whole field; some of these are:

- What are the properties of the new gauge bosons, $W^{\pm}$ and Z^0?
- What governs the masses of the quarks and leptons?
- How do systems of quarks behave?
- How does one improve the standard model? Here, for example, one seeks greater unification, that is, to properly bring the strong force into the fold and thus achieve GUT; further, one seeks ways of hiding gauge symmetry which might be better than the Higgs or technicolour methods presently proposed, and one seeks to solve the hierarchy problems which arise when the vastly disparate masses of the different sets of gauge bosons ($W^{\pm}$; X's, etc.) enter perturbatively into matrix elements.
- How does one unify gravity with the other forces?
- Why are there so many particles? Counting leptons, quarks (each in three colours), gauge bosons, Higgs particles, gluons, $\tilde{q}$'s and gravitons, one has still an embarrassing richness of fundamental particles.
- Is there a "theory of everything"?

The new interests of particle physics are to search for improvements in the standard model and for what lies beyond it. The new interests of nuclear physics is to understand how, with QCD, one describes strongly interacting systems and how one uses such systems to study fundamental symmetries.

The physics pursuits of the hadron facilities for the intensity frontier are complementary to, and go hand-in-hand with, the various other components of the whole world network of new major accelerator facilities. This network includes the new hadron-hadron colliders (e.g. the Tevatron, SP$\overline{\text{P}}$S, etc.), the new $e\bar{e}$ machines (TRISTAN, Beijing, Cornell, etc.), the next generation of $e\bar{e}$ colliders (SLC and LEP), the future electron-proton collider (HERA), the proposed new supercolliders (UNK, LHC and SSC), the cw electron accelerator (CEBAF) under construction, the relativistic heavy-ion collider (RHIC) about to be funded, and of course, the intense hadron facilities, KAON and complementary facilities such as JHP. There are also many important new smaller projects. Altogether it is a world system of impressive proportions linking together the industrialized nations of the northern hemisphere. People and ideas flow freely through the links. Almost all the action now takes place at these few large user facilities.

4. The Science of KAON – as Seen at Bad Honnef

Because of the many different kinds of beams provided by the KAON Factory, its science opportunities pertain to many of the current important issues. They range from weak interaction tests of the standard model to the dynamics of complex strongly interacting systems such as hadrons and nuclei. Perhaps its main strength will lie as the world's leading tool to explore the strong interactions (QCD). That was also the bias of this meeting.

The Feshbach Panel in its recent report analysed the science opportunities of the KAON Factory under three main headings: strong interaction physics, antiproton physics and electroweak physics issues. We follow that taxonomy in this summary. All the invited papers at this meeting fall under these headings – if we also add another heading for the description of facilities complementary to the KAON Factory. Huber and Arvieux gave us very interesting accounts of two such facilities.

The strong-interaction physics issues of the KAON Factory can be subdivided among a number of topics as follows (with the names of invited speakers at this meeting given in brackcts):

- HADRON SPECTROSCOPY – including baryon, meson, H-baryons, glueballs, hybrids, "molecules", etc. [Mulders, Bugg, Klempt, Close, Godfrey, Myhrer, Chung, Dunwoodie, Barnes, Speth]
- HYPERNUCLEI – including weak decays, double strangeness, etc. [Paul, Barnes, Chrien]
- K^+-NUCLEUS SCATTERING [Häusser]
- SPIN EFFECTS IN NN SCATTERING [Krisch]
- OTHER [Page]

Two-thirds of the invited papers at this meeting pertained to strong-interaction physics. It is then clear from the list of topics and speakers just given that we have had a very strong emphasis on hadron spectroscopy – almost half of all the papers were devoted to this topic. Among the various KAON workshops this particular focus on hadron spectroscopy constitutes the dominant characteristic of this meeting.

The present state of hadron spectroscopy and the hopes for its development in the KAON era can be placed in the perspective of nuclear spectroscopy, a field which flourished over the past four decades. In both nuclear spectroscopy and hadron spectroscopy one has characterized states by their energy and by their principal quantum numbers such as spin, parity, isospin, etc. In both spectroscopies one has wanted to know more about the nature or structure of the state. In nuclear physics one has described the internal nature of the state primarily in terms of the mean-field orbits of the neutrons and protons. In hadron spectroscopy one would like to know, from experiment and QCD, the quark and gluon content of the states. In either it is the production and decay dynamics which provide the key to the internal nature.

In hadron spectroscopy our knowledge of production and decay dynamics and therefore our understanding of internal structure are presently in a very primitive state. It may therefore be instructive to consider how nuclear physics coped with this problem.

In general a nucleus, such as ^{24}Mg, shown in Fig. 1, has a very large (but denumerable) number of decay channels leading from each state. There is a natural division of the configuration space, as shown in Fig. 1, into an internal region, in which the component states exist, and a channel region. In the interior one has A interacting nucleons. In the channel region one has pairs of noninteracting reaction products – or, alternatively, some simple final-state interactions between the pairs (rather than the fundamental QCD forces between the constituent quarks or nucleons of the pair). What makes this picture work is a simple physical fact which may

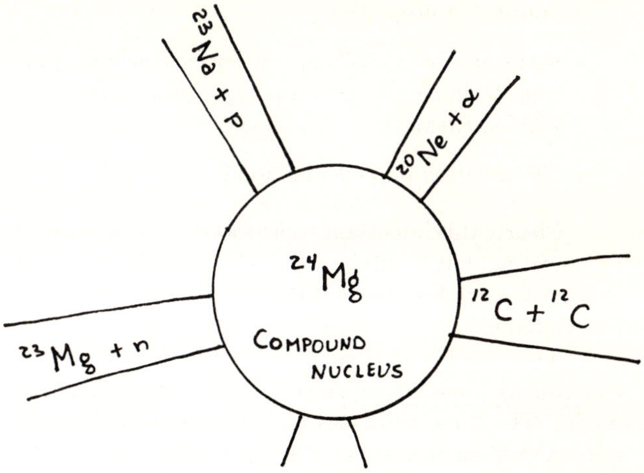

Figure 1: The configuration space of the compound, ^{24}Mg, with some of its decay channels.

not apply nearly as naturally in the case of hadron spectroscopy: if we look at the radial density distribution of a typical nucleus it has a more or less well-defined natural radius – the surface thickness of the nucleus (in which the density falls by e^{-1}) is a small fraction of the nuclear radius. There is then little ambiguity in the division of configuration space into internal and external regions.

The simple topology of the nuclear configuration space then leads to quantitative rules which connect the production and decay rates directly to the desired internal properties. The connection quantities, at the interfaces of the internal region and the channels, are called reduced widths. There are, for example, very powerful sum rules for the reduced widths arising from the very general completeness relations of a known Hilbert space in the internal region. On the basis of such sum rules one knows what it means, for example, to ask whether a state of ^{24}Mg contains, in high percentage, ^{20}Ne plus an alpha particle. It was only when the quantitative description of reaction dynamics was added to the taxonomy of energy levels that nuclear spectroscopy became a significant science.

Whither now hadron spectroscopy? We now have neither the experimental data for the production and decay dynamics nor a proper QCD framework to intelligently enquire about the glueball content of a hadronic state. It was this condition which prompted Dunwoodie to state in his talk: "a future goal for KAON is to provide detailed information on production dynamics in case QCD will some day have something useful to say about this". We should live in hope about the theory and we should then expect that with KAON hadron spectrocopy will also develop into a significant science.

The second broad class of KAON topics in the Feshbach Report pertains to antiproton physics issues. Here there are the following three subtopics with the two (indicated) invited papers at this meeting:

- antiproton properties
- antiproton annihilation – in nuclei and with protons yielding $e\bar{e}$, mesons, hyperons, charm, etc. [Gibbs, Klempt]
- CP violation in $p\bar{p}$ annihilation

Clearly this important topic was only peripheral to this meeting, but it will be the main focus of the second European workshop for KAON to be held in Torino in late October.

The third and final broad topic for KAON is electroweak physics issues involving tests of the standard model. The three subtopics and the corresponding papers at this meeting are as follows:

- neutral kaons and CP violation [Kleinknecht]
- rare kaon decays [Bryman]
- neutrinos [Scheck]

Again these very important issues were not the main focus of this meeting although we had excellent reviews of all three subtopics. The new CP violation experiments may provide confirmation of the standard model but are not the end of the story. KAON with its dedicated neutral kaon beams could be important here. The rare kaon decays will certainly come into their own with KAON. The issues are flavour changing, high masses, etc. With KAON and with a new generation of detectors a number of very important branching ratios should find a significant number of events rather than just upper limits. Scheck gave us a particularly interesting list of current questions for neutrino physics at KAON.

5. Present Funding Status of KAON

KAON is on the way! We at TRIUMF believe we are building the KAON Factory now although a full funding decision by Canada is still almost a year ahead. With that funding decision in mid-1990 KAON should then begin operating five years later, in 1995.

The Canadian KAON Factory is being funded under the HERA model which was pioneered by Germany. To understand the funding status of KAON it will be useful to review how we are employing the HERA Model for the internationalization of KAON.

The user mode is now rampant in subatomic physics. In every country the majority of particle physicists now travel some distance from their home institutions to carry out their experiments. Each country needs a home-base facility if it is to collaborate effectively at user facilities abroad. Such a home-base can be one of the major or less major components of the world network. The Group of Seven (G7) nations (Japan, Canada, U.S.A., Britain, France, Germany, Italy) have their representatives in high-energy physics meet once a year as a working group to co-ordinate activities in this field. One of the earliest decisions of the Working Group was to emphasize the need for home-base facilities. Home-base facilities do serve to bring economic benefits to the host country. Even more, they are vehicles for working effectively in facilities abroad. Thus DESY serves Germany for its world activities in particle physics; the new COSY cooler ring at Jülich will serve German medium-energy scientists as a staging base for work at KAON and elsewhere. For Canada, KAON will be the vehicle which takes Canadian scientists abroad, not only to HERA and COSY, but also to all the pieces of the world network.

There are a variety of models for internationalization of big science projects, many of which have been developed with regard to large accelerator facilities. It is my view that basically three different models now apply. These are:

A. The "CERN MODEL" is one in which a number of countries pool resources, usually through some appropriate formula, and jointly have legal and financial control. This model was pioneered by the European countries for CERN but has since also been used for European Space Agency (ESDA), the European Southern Observatory (ESO), the French-British-German high-flux reactor at Grenoble (ILL), the large European fusion project (JET) and other science projects.

B. The "NATIONAL PLUS MODEL" In this model (whose name we have invented) a single host nation dominates the science but invites other nations to participate. Because the host nation dominates, it usually does not require commitments from foreign partners before making its own funding decision. There are many examples: the SSC project in the United States, the existing TRIUMF project in Canada, UNK in the USSR, KEK in Japan, Gran Sasso in Italy, etc.

C. The "HERA MODEL" applies to projects which are intrinsically international – no nation dominates. It is an alternative to the CERN model. One nation hosts the facility, accepts legal and financial responsibility, but counts on foreign partners for construction contributions. The "HERA MODEL" was pioneered by Germany for its HERA project. It involves rather formal steps which we describe below. Because the host nation does not dominate the science of the facility, it requires a clear statement of intent from its foreign partners before it makes a final funding decision. The "HERA MODEL" probably works only for cases in which the host nation's share of the science lies between 25 and 50%. If more than 50%, it would likely choose the "National Plus" route. If less than 25%, it would likely seek to follow the "CERN MODEL".

The immediate paradox about the HERA model is that it should work at all. Why would any country want to host a facility in which it has a minority interest in the science, and secondly, why would participating countries want to pay for something that they might get free? Increasingly, for science facilities, independent of the model under which they are operated, proposals are accepted on the basis of scientific merit only, and not on whether or not the country pays its dues. These questions have rational answers.

For the host country, there are a number of factors which offset its major financial commitment. It can choose to expand on its existing excellence, as Canada is doing when it bases its KAON factory on TRIUMF. It benefits greatly from the inflow of people and ideas and spin-offs. Every country wishing to participate in the work of the international network of large accelerator facilities, benefits greatly from having a home-base facility through which that work flows. A country which hosts a large facility on the HERA model, even though it has a minority interest in the science, benefits enormously from all of the ideas pouring out from the entire world network. Finally, direct involvement of foreign nations through participation in the funding of the facility is the best manifestation of all of international approval of the facility, and acts therefore as a funding catalyst.

For the participating countries, who have to pay for something that they might think of getting for free, it should be noted first that the contributions are not cash, they are in high technology components, which have impact on high technology at home. Above all, there is the rationalization of internal plans and priorities. Most countries cannot dream now of having a balanced set of large accelerator facilities at home. They must make choices, they must build on excellence, yet most countries have scientists interested in the entire span of ideas. Each country must make critical choices by participating with other countries in funding. A country participating in the HERA Model process has a very cost-effective way of satisfying the needs of its scientists in that field and then getting on with its own home-base facility. It is not altruism on a national scale, but pragmatic judgment which makes the HERA model work so effectively.

There are a number of formal steps associated with the HERA Model. These steps taken in total make it an incredibly effective self-seduction sequence for the governments of countries seeking to establish a major science facility. The five steps are:

(i) The host country has an interesting idea for a big science facility and prepares a proposal. Canada did this for its KAON Facility, in September 1985.

(ii) The host country makes an initial exploration abroad of foreign interest in its facility. Canada did this for KAON in November and December 1987, and found very strong interest abroad.

In its exploration of foreign interest in KAON, a year ago, Canada visited Washington, Tokyo, Bonn, Rome and London. This yielded widespread support for such a Canadian facility. It was viewed as not only excellent physics but also the opportunity for Canada to become a major player in particle physics. The foreign nations encouraged Canada to make a commitment to KAON of sufficient magnitude to initiate the round of consultations now taking place.

(iii) The host country having explored abroad declares it serious intention about the matter. This statement of intention for creating the KAON Factory was the Project Definition Study of July 1988.

In May 1988, the Working Group in High-Energy Physics of the G7 countries again encouraged Canada and defined the step now needed for Canada to be taken seriously. The following extract from the minutes of that Vancouver meeting is self-explanatory: "Canada reported on progress since last year on its proposal to build a KAON Factory at TRIUMF. International collaboration on construction funding has been explored with encouraging results, and a decision by Ottawa appears near on the final Project Definition Studies ($11M) including negotiations with foreign partners. The Working Group reaffirmed last year's conclusion that there is a very good scientific case for a machine of this type for the sound development of high-energy physics. It also concluded that an early decision by Canada to proceed with its KAON Factory would be very welcome and it encouraged Canada to seek interest and engagement from the international community. It was noted that other projects, such as the Japanese Hadron Facility, would explore interesting fields complementary to the KAON Factory."

In July 1988 Canada funded the Project Definition Study and the envisaged consultations are under way. In the first round of visits the Canadian delegation travelled to: Rome (April 17), Bonn (April 19), London (April 21), Paris (April 24), Washington (May 9) and Tokyo (May 17). Separately from the formal visits of the Canadian delegation, TRIUMF scientists are very active in engaging the interests of foreign scientists in the KAON Factory.

(iv) The host country formally consults abroad, intending to achieve something close to letters of intent before it makes its own final commitment. Canada is now in the process of doing this for KAON.

(v) The host country makes its decision and then completes its agreements with its foreign partners.

For Canada's KAON Factory, it is estimated that Canadian scientists will constitute about a third of the total user community of 800 scientists. Canada is clearly in a minority position. The special contributions of foreign partners occur in the construction phase. In the operating phase it is assumed that the host country, Canada, will assume the full normal operating costs – estimated to total about $90M (Canadian) per annum – which apply to accelerator maintenance and electric power and other similar costs. In the operating phase, the foreign partners pay the normal proportion of jointly funded detectors and experimental equipment according to the well-established custom which now prevails for all facilities internationalized under the CERN MODEL, the NATIONAL PLUS MODEL, or the HERA MODEL. The arguments here, then, pertain to the special contributions to accelerator construction relevant to the HERA MODEL.

The total proposed level of foreign contributions to initial construction, in the HERA MODEL, is quite naturally about a third of the total construction cost. The civil components (buildings, tunnels, much of the shielding, etc.) are more easily assigned to the construction firms near the site of the facility, and are therefore domestic. The two-thirds for accelerators and beam lines are the attractive high-technology pieces. The host country clearly wants a major share of this portion to stimulate the domestic economy, and therefore ends up offering about half of it to foreign partners. So it is for Canada's KAON Factory.

The total construction package for KAON is \$571M (1987 Canadian dollars). The foreign participants are anticipated to have two-thirds of the science action, and it is proposed that they contribute accelerator components worth one-third of the total construction costs.

The proportion proposed for each country is based on our estimate of the fraction of the 800 scientific users of the KAON Factory expected to originate from that country. It takes account of the size of each country's medium-energy physics community, of the convenience of the Vancouver location for the community, and of any special interest in the KAON Factory expressed by the scientists of the community.

In its first round of consultations for participation in the KAON Factory, in the spring of 1989, Canada directly approached its partners among the Group of Seven (G7) nations. It is here that the main foreign partners are anticipated, and where the main competitors for our KAON Factory plans existed.

Based on the above general principles, we present the following table for the proposed level of participation in Canada's KAON Factory:

Country	Estimated % of KAON users[a]	Proposed level of participation (Canadian \$)
Canada	30–35	400 M[b]
United States	30–40	90 M
Japan	8–10	50 M
Germany	6–8	30 M
Italy	5–7	30 M
Britain	1–3	[c]
France	1–3	[c]
Other	8–15	

In all of the countries visited by Canada in its first round a very constructive approach to participation in KAON was taken. The situation in various countries at the time of this workshop is as follows:

United States

It is in the United States that recently there has been such spectacular movement toward participation in Canada's KAON Factory. The movement resulted from the Feshbach Report made public about a month ago.

The Feshbach Subcommittee was established by the U.S. Nuclear Sciences Advisory Committee (NSAC) as a direct result of Canada's invitation for U.S. participation in KAON. The charge of this Subcommittee was established by DOE and NSF. It was chaired by Herman Feshbach and also included Peter Barnes, John Domingo, Martin Einhorn, Harold Jackson, Robert Siemann and Michael Zeller. It was, of course, directly concerned with both Canada's KAON and Los Alamos' AHF as well as with the Brookhaven AGS upgrades.

The four elements of the charge to the Feshbach Subcommittee are given in abbreviated form as follows, with the Subcommittee's responses in italics:

• What is the importance of the proposed physics research and its relevance to the U.S. nuclear physics program?

> *The proposed facilities would make a broad range of phenomena of fundamental importance accessible to experimental study. The results would develop significant and informative challenges to the standard model ...*

• What is the adequacy and appropriateness of the proposed KAON Facility to provide the needed experimental capability in this area of hadronic physics?

> *The design of the KAON Facility was judged by the Subcommittee to be conservative. ... No major design problems ... It would certainly provide the needed experimental capability in this area of hadronic physics.*

• What is the impact of the KAON project on the U.S. physics program within the context of the Long Range Plan for Nuclear Science ... ?

> *The KAON Facility would provide a capability which would complement those of CEBAF ... and RHIC. The construction would be completed in a timely fashion ... the total U.S. contribution would be \$75M over a period of five years. The Subcommittee considers such an investment cost effective.*

[a]Total number of users estimated to be about 800.

[b]Includes \$90 M already committed by the Province of B.C.

[c]Amount to be proposed during consultations.

• What about the capabilities of Los Alamos and Brookhaven?

Plans for AHF at Los Alamos are in a preliminary stage requiring considerable R&D before major problems are solved.

As a result of this Report Los Alamos withdrew its AHF from the table. KAON received an enormous boost. As Dr. Treusch mentioned in his opening remarks the situation with regard to the Feshbach Report is very well summarized in an article in the current (May) issue of Physics Today entitled "Subcommittee Encourages U.S. to Join Canadian KAON Factory". As Treusch further remarked one could also consider it to establish the price for international participation as about a third of a million dollars (Canadian) per participating scientist.

The United States is preparing a new NSAC Long Range Plan this summer. We are confident that KAON will have a strong place in that plan and that the United States will then lead the international participation in KAON.

Japan

The Japanese Hadron Project (intense beams of 1 GeV protons with many added arenas for new physics) is entirely complementary to KAON and appears to be far along on its road toward funding. The ties between the Japanese medium-energy community and the corresponding Canadian community are warm. In addition to the joint JHP–KAON workshop held in April, a hypernuclear workshop for KAON will be held at Tsukuba in June. The Ejiri Committee, similar to the Feshbach Subcommittee in the United States, has also reached conclusions about KAON similar to those of the Feshbach Subcommittee.

Germany

Fifteen months ago the German medium-energy physics community was actively pursuing the European Hadron Facility (EHF). The community itself decided, on December 15, 1987, to push participation in Canada's KAON instead. Also, in May 1988 the Specht Committee reported to the German government as follows: "The physics perspective of a hadron facility was viewed as important and interesting. In first priority the Federal Republic shall participate in the planned KAON Factory at TRIUMF in Vancouver, Canada. German experimental groups for KAON shall be supported appropriately under the aegis of Federal Research Funds. Participation in the construction of a European Hadron Facility EHF (with a 100 μA beam intensity) is not recommended. In case KAON is not realized the possibility of a more modest European solution at CERN (with about 10 μA beam intensity) shall be discussed anew."

At this meeting the Bonn government is initiating a poll of Germany nuclear physicists to establish their intent for working at KAON in the middle of the next decade. From the strength of this workshop we are clearly optimistic about Germany involvement.

Italy

Italian scientists have been especially strong proponents of EHF and only very recently have begun consideration of a complementary facility in Italy instead. Because of Italy's very important role not only here but across the board in particle physics, it would be very natural for Italy to play a major role in the KAON Factory despite the distance. The level of participation in construction depends on the level of involvement of the Italian medium-energy physicists which is only now beginning.

Britain

Involvement of Britain in KAON along the lines of its involvement in HERA appears likely.

France

French scientists played only a minor role in EHF, but France has been a leader in particle physics and nuclear physics. It has plans of its own in medium-energy physics. Some scientific interest in KAON is evident and therefore some French involvement is likely.

Other

Many countries other than Canada's G7 partners have expressed interest in KAON and are likely to contribute some components and certainly some scientists.

6. What Next?

Every participant at this workshop should leave it programmed to play their individual role in the creation of KAON. You should now have the clear impression that we are confident that KAON is on the way and that international participation is essential for the creation of KAON. In turn, that international participation will occur only if, in each country, enough scientists clearly express their interest in working at KAON. The vigorous participation in this workshop is one manifestation of such interest. The German scientists are now being given a more explicit opportunity in their response to the questionnaire which Dr. Hartwig is currently circulating. This is a project whose time has come and which will be realised if you now respond positively. We, in Canada, will produce the KAON Facility by 1995 and welcome you to it.

Index of Contributors